MEMS Vibratory Gyroscopes
Structural Approaches to Improve Robustness

MEMS Reference Shelf

Series Editors:

Stephen D. Senturia
Professor of Electrical Engineering, Emeritus
Massachusetts Institute of Technology
Cambridge, Massachusetts

Roger T. Howe
Department of Electrical Engineering
Stanford University
Stanford, California

Antonio J. Ricco
Small Satellite Division
NASA Ames Research Center
Moffett Field, California

MEMS Vibratory Gyroscopes Structural Approaches to Improve Robustness
Cenk Acar and Andrei Shkel
ISBN: 978-0-387-09535-6

BioNanoFluidic MEMS
Peter Hesketh, ed.
ISBN 978-0-387-46281-3

Microfluidic Technologies for Miniaturized Analysis Systems
Edited by Steffen Hardt and Friedhelm Schöenfeld, eds.
ISBN 978-0-387-28597-9

Forthcoming Titles

Self-assembly from Nano to Milli Scales
Karl F. Böhringer
ISBN 978-0-387-30062-7

Photonic Microsystems
Olav Solgaard
ISBN 978-0-387-29022-5

Micro Electro Mechanical Systems: A Design Approach
Kanakasabapathi Subramanian
ISBN 978-0-387-32476-0

Experimental Characterization Techniques for Micro-Nanoscale Devices
Kimberly L. Turner and Peter G. Hartwell
ISBN 978-0-387-30862-3

Microelectroacoustics: Sensing and Actuation
Mark Sheplak and Peter V. Loeppert
ISBN 978-0-387-32471-5

Inertial Microsensors
Andrei M. Shkel
ISBN 978-0-387-35540-5

Cenk Acar and Andrei Shkel

MEMS Vibratory Gyroscopes
Structural Approaches to Improve Robustness

 Springer

Cenk Acar
Systron Donner Automotive
2700 Systron Drive
Concord, CA 94518-1399

Andrei Shkel
University of California, Irvine
Dept. of Mechanical and Aerospace Engineering
4200 Engineering Gateway Building
Irvine, CA 92697-3975

e-978-0-387-09536-3

Printed on acid-free paper.

Printed on acid-free paper.

9 8 7 6 5 4 3 2 1

springer.com

To my beloved wife Şebnem Acar, and my dear parents.

Preface

Merging electrical and mechanical systems at a micro scale, Microelectromechanical Systems (MEMS) technology has revolutionized inertial sensors. Since the first demonstration of a micromachined gyroscope by the Draper Laboratory in 1991, various micromachined gyroscope designs fabricated in surface micromachining, bulk micromachining, hybrid surface-bulk micromachining technologies or alternative fabrication techniques have been reported. Inspired by the promising success of micromachined accelerometers in the same era, extensive research efforts towards commercial micromachined gyroscopes led to several innovative gyroscope topologies, fabrication and integration approaches, and detection techniques. Consequently, vibratory micromachined gyroscopes that utilize vibrating elements to induce and detect Coriolis force have been effectively implemented and demonstrated in various micromachining-based batch fabrication processes. However, achieving robustness against fabrication variations and environmental fluctuations still remains as one of the greatest challenges in commercialization and high-volume production of micromachined vibratory rate gyroscopes.

The limitations of the photolithography-based micromachining technologies define the upper-bound on the performance and robustness of micromachined gyroscopes. Conventional gyroscope designs based on matching or near-matching the drive and sense mode resonant frequencies are quite sensitive to variations in oscillatory system parameters. Thus, producing stable and reliable vibratory micromachined gyroscopes have proven to be extremely challenging, primarily due to the high sensitivity of the dynamical system response to fabrication and environmental variations.

In the first part of this book, we review the Coriolis effect and angular rate sensors, and fundamental operational principles of micromachined vibratory gyroscopes. We review basic mechanical and electrical design and implementation practices, system-level architectures, and common fabrication methods utilized for MEMS gyroscopes and inertial sensors in general. We also discuss electrical and mechanical parasitic effects such as structural imperfections, and analyze their impact on the sensing element dynamics.

In the second part, we review recent results of the study on design concepts that explore the possibility of shifting the complexity from the control electronics to the structural design of the gyroscope dynamical system. The fundamental approach is to develop structural designs and dynamical systems for micromachined gyroscopes that provide inherent robustness against structural and environmental parameter variations. In this context, we primarily focus on obtaining a gain and phase stable region in the drive and sense-mode frequency responses in order to achieve overall system robustness. Operating in the stable drive and sense frequency regions provides improved bias stability, temperature stability, and immunity to environmental and fabrication variations. Toward this goal, two major design concepts are investigated: expanding the dynamic system design space by increasing the degree-of-freedom of the drive and sense mode oscillatory system, and utilizing an array of drive-mode oscillators with incrementally spaced resonant frequencies.

This book provides a solid foundation in the fundamental theory, design and implementation of micromachined vibratory rate gyroscopes, and introduces a new paradigm in MEMS gyroscope sensing element design, where disturbance-rejection capability is achieved by the mechanical system instead of active control and compensation strategies. The micromachined gyroscopes of this class are expected to lead to reliable, robust and high performance angular-rate sensors with low production costs and high yields, fitting into or enabling many applications in the aerospace/defense, automotive and consumer electronics markets.

June 2008   *Cenk Acar, Andrei Shkel*

Contents

Part I
Fundamentals of Micromachined Vibratory Gyroscopes

Chapter 1
Introduction

In this chapter, we present a brief overview of the Coriolis effect and angular rate sensors, micromachining and the MEMS technology, implementation of vibratory gyroscopes at the micro-scale, and a chronological survey of the prior work on micromachined gyroscopes.

1.1 The Coriolis Effect

The Coriolis effect, which defies common sense and intuition, has been observed but not fully understood for centuries. Found on many archaeological sites, the ancient toy spinning top (Figure 1.1) is an excellent example that the Coriolis effect was part of the daily life over three thousand years before Gaspard Gustave Coriolis first derived the mathematical expression of the Coriolis force in his paper *"Mémoire sur les équations du mouvement relatif des systémes de corps"* [1] investigating moving particles in rotating systems in 1835.

Fig. 1.1 A wooden decorated spinning top from the 14th century BC found in the tomb of Tutankhamun, currently at the Egyptian Museum. One of the most beloved toys of Egyptian children in ancient times, the spinning top relies on the Coriolis effect to spin upright and slowly starts precessing as it loses angular momentum [40].

The Coriolis effect arises from the fictitious Coriolis force, which appears to act on an object only when the motion is observed in a rotating non-inertial reference frame. The Foucault pendulum (Figure 1.2) demonstrates this phenomenon very well: When a swinging pendulum attached to a rotating platform such as earth is observed by a stationary observer in space, the pendulum oscillates along a constant straight line. However, an observer on earth observes that the line of oscillation precesses. In the dynamics with respect to the rotating frame, the precession of the pendulum can only be explained by including the Coriolis force in the equations of motion.

Fig. 1.2 The Foucault pendulum, invented by Jean Bernard Léon Foucault in 1851 as an experiment to demonstrate the rotation of the earth. The swinging direction of the pendulum rotates with time at a rate proportional to the sine of the latitude due to earth's rotation [41].

1.2 Gyroscopes

In simplest terms, gyroscope is the sensor that measures the rate of rotation of an object. The name "gyroscope" originated from Léon Foucault, combining the Greek word "skopeein" meaning to see and the Greek word "gyros" meaning rotation, during his experiments to measure the rotation of the Earth.

The earliest gyroscopes, such as the Sperry gyroscope, and many modern gyroscopes utilize a rotating momentum wheel attached to a gimbal structure. However, rotating wheel gyroscopes came with many disadvantages, primarily concerning bearing friction and wear. Vibrating gyroscopes, such as the Hemispherical Resonator Gyroscope (HRG) and Tuning-Fork Gyroscopes presented an effective solution to the bearing problems by eliminating rotating parts.

Alternative high-performance technologies such as the Fiber-Optic Gyroscope (FOG) and Ring Laser Gyroscope (RLG) based on the Sagnac effect have also been developed. By eliminating virtually all mechanical limitations such as vibration and shock sensitivity and friction, these optical gyroscopes found many high-end applications despite their high costs.

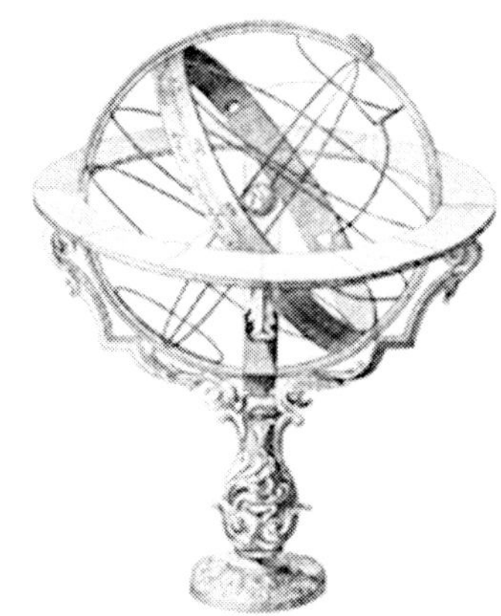

Fig. 1.3 One of the first examples of the gyrocompass, developed in the early 1800s. The gyrocompass gained popularity, especially in steel ships, since steel blocked the ability of magnetic compasses to find magnetic north.

1.3 The MEMS Technology

As the name implies, Microelectromechanical Systems (MEMS) is the technology that combines electrical and mechanical systems at a micro scale. Practically, any device fabricated using photo-lithography based techniques with micrometer (1μm $= 10^{-6}$m) scale features that utilizes both electrical and mechanical functions could be considered MEMS.

Evolved from the semiconductor fabrication technologies, the most striking feature of the MEMS technology is that it allows building moving micro-structures on a substrate. With this capability, extremely complex mechanical and electrical systems can be created. Masses, flexures, actuators, detectors, levers, linkages, gears, dampers, and many other functional building blocks can be combined to build complete sophisticated systems on a chip. Inertial sensors such as accelerometers and gyroscopes utilize this capability to its fullest.

Photolithography based pattern transfer methods and successive patterning of thin structural layers adapted from standard IC fabrication processes are the enabling technologies behind micromachining. By dramatically miniaturizing and batch processing complete electro-mechanical systems, substantial reductions in device size, weight and cost are achieved.

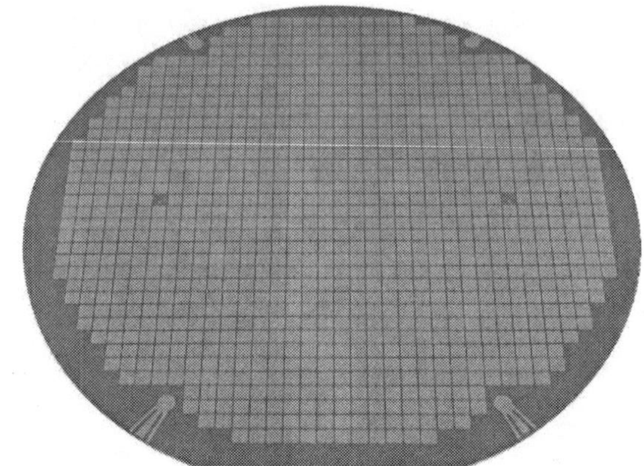

Fig. 1.4 A 150mm wafer from a gyroscope prototyping run. In a typical production process, it is common to have well over 2000 devices on a 150mm wafer.

1.4 Micromachined Vibratory Rate Gyroscopes

Even though an extensive variety of micromachined gyroscope designs and operation principles exist, majority of the reported micromachined gyroscopes use vibrating mechanical elements to sense angular rate. The concept of utilizing vibrating elements to induce and detect Coriolis force presents many advantages by involving no rotating parts that require bearings and eliminating friction and wear. That is the primary reason why vibratory gyroscopes have been successfully miniaturized by the use of micromachining processes, and have become an attractive alternative to their macro-scale counterparts.

The fundamental operation principle of micromachined vibratory gyroscopes relies on the sinusoidal Coriolis force induced due to the combination of vibration of a proof-mass and an orthogonal angular-rate input. The proof mass is generally suspended above the substrate by a suspension system consisting of flexible beams. The overall dynamical system is typically a two degrees-of-freedom (2-DOF) mass-spring-damper system, where the rotation-induced Coriolis force causes

Fig. 1.5 Singulated micromachined gyroscope dice designed and fabricated at UCI Microsystems Laboratory. Courtesy of Alexander A. Trusov.

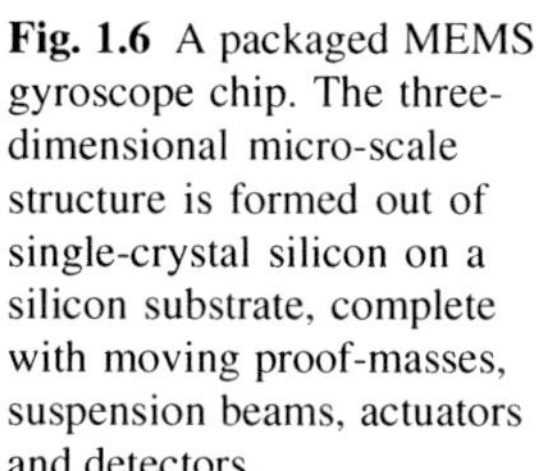

Fig. 1.6 A packaged MEMS gyroscope chip. The three-dimensional micro-scale structure is formed out of single-crystal silicon on a silicon substrate, complete with moving proof-masses, suspension beams, actuators and detectors.

energy transfer to the sense-mode proportional to the angular rate input. In most of the reported micromachined vibratory rate gyroscopes, the proof mass is driven into resonance in the drive direction by an external sinusoidal electrostatic or electro-magnetic force. When the gyroscope is subjected to an angular rotation, a sinusoidal Coriolis force at the driving frequency is induced in the direction orthogonal to both the drive-mode oscillation and the angular rate axis.

Ideally, it is desired to utilize resonance in both the drive and the sense modes in order to attain the maximum possible response gain and sensitivity. This is typically achieved by designing and if needed tuning the drive and sense resonant frequencies to match. Alternatively, the sense-mode is designed to be slightly shifted from the drive-mode to improve robustness and thermal stability, while intentionally sacrific-ing gain and sensitivity.

Even though increasing the spacing between the drive and sense frequencies re-duces the impact of variations in oscillatory system parameters that shift the natural

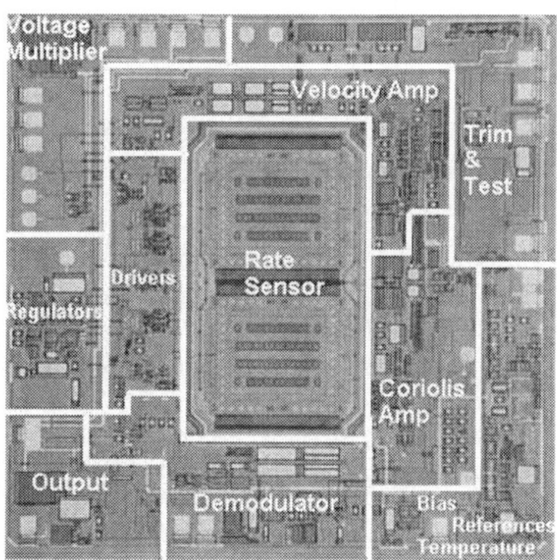

Fig. 1.7 The iMEMS ADXRS angular rate sensor by Analog Devices is an excellent example of a micromachined vibratory gyroscope, which integrates the angular rate sensing element and signal processing electronics on the same die. Courtesy of Analog Devices.

frequencies and damping values, the resulting errors still require compensation by advanced control and signal processing architectures.

1.5 Applications of MEMS Gyroscopes

As their performance keeps constantly improving in time, micromachined gyroscopes are becoming a viable alternative to expensive and bulky conventional inertial sensors. High-performance angular rate sensors such as precision fiber-optic gyroscopes, ring laser gyroscopes, and conventional rotating wheel gyroscopes are usually too expensive and too large for use in most emerging applications. With micromachining processes that allow batch production of micro-electro-mechanical systems on a chip similar to integrated circuits, unit costs unimaginable in any other technology are achieved. Moreover, advances in the fabrication techniques that allow electronics to be integrated on the same silicon chip together with the mechanical sensor elements provide an unmatched integration capability. Consequently, miniaturization of vibratory gyroscopes with innovative micro-fabrication processes and gyroscope designs is already becoming an attractive solution to current inertial sensing market needs, and even opening new market opportunities.

With their dramatically reduced cost, size, and weight, MEMS gyroscopes potentially have a wide application spectrum in the aerospace industry, military, automotive and consumer electronics markets. The automotive industry applications are diverse, including advanced automotive safety systems such as electronic stability control (ESC), high performance navigation and guidance systems, ride stabilization, roll-over detection and prevention, and next generation airbag and brake systems. A wide range of consumer electronics applications with very high volumes include image stabilization in digital cameras and camcorders, virtual reality products, inertial pointing devices, and computer gaming industry. Miniaturization of gyroscopes also enable higher-end applications including micro-satellites, micro-robotics, and even implantable devices to cure vestibular disorders.

1.6 Gyroscope Performance Specifications

The specifications and test procedures for rate gyroscopes are outlined in the *IEEE Standard Specification Format Guide and Test Procedure for Coriolis Vibratory Gyros* [2]. The following is a summary of important specifications and definitions from *IEEE Standard for Inertial Sensor Terminology* [3].

Scale factor:
The ratio of a change in output to a change in the input intended to be measured, typically specified in mV/°/sec, and evaluated as the slope of the least squares straight line fit to input-output data. Scale factor error specifications include:

Linearity error: The deviation of the output from a least-squares linear fit of the input-output data. It is generally expressed as a percentage of full scale, or percent of output.

Nonlinearity: The systematic deviation from the straight line that defines the nominal input-output relationship.

Scale factor temperature and acceleration sensitivity: The change in scale factor resulting from a change in steady state operating temperature and a constant acceleration.

Asymmetry error: The difference between the scale factor measured with positive input and that measured with negative input, specified as a fraction of the scale factor measured over the input range.

Scale factor stability: The variation in scale factor over a specified time of continuous operation. Ambient temperature, power supply and additional factors pertinent to the particular application should be specified.

Bias (zero rate output):

The average over a specified time of gyro output measured at specified operating conditions that has no correlation with input rotation. Bias is typically expressed in °/sec or °/hr. The zero-rate output drift rate specifications include:

Random drift rate: The random time-varying component of drift rate. Random drift rate is usually defined in terms of the Allan variance components:

a) Angle Random Walk: The angular error buildup with time that is due to white noise in angular rate, typically expressed in $°/\sqrt{hr}$ or $°/s/\sqrt{hr}$.

b) Bias Instability: The random variation in bias as computed over specified finite sample time and averaging time intervals, characterized by a $1/f$ power spectral density, typically expressed in °/hr.

c) Rate Random Walk: The drift rate error buildup with time that is due to white noise in angular acceleration, typically expressed in $°/hr/\sqrt{hr}$.

Environmentally sensitive drift rate: Components of drift rate dependent on environmental parameters, including acceleration sensitivity, temperature sensitivity, temperature gradient sensitivity, temperature hysteresis and vibration sensitivity.

Operating range (input rate limits):

Range of positive and negative angular rates that can be detected without saturation.

Resolution:

The largest value of the minimum change in input, for inputs greater than the noise level, that produces a change in output equal to some specified percentage (at least 50%) of the change in output expected using the nominal scale factor.

Bandwidth:

The range of frequency of the angular rate input that the gyroscope can detect. Typically specified as the cutoff frequency coinciding to the -3dB point. Alternatively, the frequency response or transfer function could be specified.

Turn-on time:
The time from the initial application of power until a sensor produces a specified useful output, though not necessarily at the accuracy of full specification performance.

Linear and angular vibration sensitivity:
The ratio of the change in output due to linear and angular vibration about a sensor axis to the amplitude of the angular vibration causing it.

Shock resistance:
Maximum shock that the operating or non-operating device can endure without failure, and conform to all performance requirements after exposure. Pulse duration and shape have to be specified. Full recovery time after exposure can also be specified.

Reliability requirements such as operating life, operating temperature range, thermal shock, thermal cycling, humidity, electrostatic discharge (ESD) immunity, and electromagnetic emissions and susceptibilities are also typically specified in many applications.

1.7 A Survey of Prior Work on MEMS Gyroscopes

Since the first demonstration of a micromachined gyroscope by the Draper Laboratory in 1991 [6], various micromachined gyroscope designs fabricated in a variety of processes including surface, bulk and hybrid surface-bulk micromachining technologies or alternative fabrication techniques have been reported in the literature. The development of miniaturized piezoelectric gyroscopes, for example the quartz tuning-fork by Systron Donner [7] and the fused-quartz HRG by Delco [8], date back to the early 1980's. Incompatibility of quartz devices with IC fabrication technologies and the know-how generated from micromachined accelerometers in the same era led to several successful academic and commercial silicon-based microgyroscopes over the following decades.

1.7.0.1 Important Development Milestones

The evolution of the design and performance of silicon micromachined gyroscopes is better understood by investigating the important development milestones in chronological order:

- Draper Laboratory reported the first micromachined gyroscope in 1991, utilizing a double-gimbal single crystal silicon structure suspended by torsional flexures; and demonstrated $4°/s/\sqrt{Hz}$ resolution at 60Hz bandwidth [6].
- In 1993, Draper Laboratory reported their next generation silicon-on-glass tuning fork gyroscope with $1°/s/\sqrt{Hz}$ resolution. The glass substrate aimed to minimize stray capacitance. The tuning fork proof masses were driven out of-phase

Fig. 1.8 The scanning electron micrograph image of the first working prototype tuning fork gyroscope from the Draper Laboratory. The device utilizes single-crystal silicon as the structural material, fabricated with a dissolved wafer process [9].

electrostatically with comb-drives, and the sense resoponse in the out-of-plane rocking mode was detected [9].

- University of Michigan developed a vibrating ring gyroscope with $0.5°/s/\sqrt{Hz}$ resolution in 1994, fabricated by metal electroforming [10]. The in-plane elliptically shaped primary mode of the ring was electrostatically excited, and the transfer of energy to the secondary flexural mode due to the Coriolis force was detected.

- British Aerospace Systems reported a single crystal silicon ring gyroscope in 1994. The sensor structure was formed on glass substrate by deep dry etching of a 100μm silicon wafer. Silicon Sensing Systems and Sumitomo Precision Products have commercialized this sensor with a resolution of $0.5°/s/\sqrt{Hz}$ over a 100Hz bandwidth [11].

- Murata developed a lateral axis (x or y) surface-micromachined polysilicon gyroscope in 1995. The sensing electrodes underneath the perforated polysilicon resonator of the gyroscope were formed by diffusing phosphorus into the substrate. A resolution of $2°/s/\sqrt{Hz}$ was reported [12].

- Berkeley Sensor and Actuator Center (BASC) utilized the integrated surface micromachining process iMEMS by Analog Devices Inc. to develop an integrated z-axis gyroscope in 1996 [13], and an x-y dual axis gyroscope in 1997 [14]. The z-axis gyroscope with a resolution of $1°/s/\sqrt{Hz}$ employed a single proof-mass driven into resonance in-plane, and sensitive to Coriolis motion in the in-plane orthogonal direction. Drive and sense modes were electrostatically tuned to match, and the quadrature error due to structural imperfections were compensated electrostatically. The x-y dual axis gyroscope with a 2μm thick polysilicon rotor disc utilized torsional drive-mode excitation and two orthogonal torsional sense modes to achieve a resolution of $0.24°/s/\sqrt{Hz}$.

- In 1997, Robert Bosch Gmbh. reported z-axis micromachined tuning-fork gyroscope design that utilizes electromagnetic drive and capacitive sensing for automotive applications, with a resolution of $0.4°/s/\sqrt{Hz}$ [15]. Through the use of a permanent magnet inside the sensor package, drive-mode amplitudes in the order of 50μm were achieved.

- Jet Propulsion Laboratory (JPL) developed a bulk micromachined clover-leaf shaped gyroscope in 1997 together with UCLA. The device had a metal post epoxied inside a hole on the silicon resonator to increase the rotational inertia of the sensing element. A resolution of $70°/hr/\sqrt{Hz}$ was demonstrated [16].

- Delphi reported a vibratory ring gyroscope with an electroplated metal ring structure in 1997. The ring was built on top of CMOS chips, and suspended by semicircular rings. The measured noise floor was $0.1°/s/\sqrt{Hz}$ with 25 Hz bandwidth [17].

- In 1997, Samsung presented a 7.5μm thick low-pressure chemical vapor deposited polysilicon gyroscope with 0.3μm polysilicon lower sensing electrodes [18], similar to Murata's sensor. The device exhibited $0.1°/s/\sqrt{Hz}$ resolution with vacuum-packaging. An in-plane device with four fish-hook spring suspension was also demonstrated with the same resolution [19].

- Daimler Benz reported an SOI-based bulk-micromachined tuning-fork gyroscope with piezoelectric drive and piezoresistive detection in 1997. Piezoelectric aluminum nitride was deposited on one of the tines as the actuator layer, and the rotation induced shear stress in the step of the tuning fork was piezoresistively detected [20].

- Allied Signal developed bulk-micromachined single crystal silicon sensors in 1998, and demonstrated a resolution of $18°/hr/\sqrt{Hz}$ at 100Hz bandwidth [21].

- Draper Laboratories reported a 10μm thick surface-micromachined polysilicon gyroscope in 1998. The resolution was improved to $10°/hr/\sqrt{Hz}$ at 60Hz bandwidth in 1993, with temperature compensation and better control techniques [22].

- In 1999, Murata developed a DRIE-based 50μm thick bulk micromachined single crystal silicon gyroscope with independent beams for drive and detection modes, which aimed to minimize undesired coupling between the drive and sense modes. A resolution of $0.07°/s/\sqrt{Hz}$ was demonstrated at 10Hz bandwidth [23].

- Robert Bosch Gmbh. developed a surface micromachined gyroscope with thick polysilicon structural layer in 1999. The device with 12μm thick polysilicon layer demonstrated a $0.4°/s/\sqrt{Hz}$ resolution at 100Hz bandwidth [24].

- Samsung demonstrated a wafer-level vacuum packaged 40μm thick bulk micromachined single crystal silicon sensor with mode decoupling in 2000, and reported a resolution of $0.013°/s/\sqrt{Hz}$ [25].

- Seoul National University reported a hybrid surface-bulk micromachining process in 2000. The device with 40μm thick single crystal silicon demonstrated a resolution of $9°/hr/\sqrt{Hz}$ at 100Hz bandwidth [26].

- In 2000, a z-axis vibratory gyroscope with digital output was developed at BSAC, utilizing the CMOS-compatible IMEMS process by Sandia National Laboratories. Parallel-plate electrostatic actuation provided low actuation voltages with limited drive-mode amplitude. $3°/s/\sqrt{Hz}$ resolution was demonstrated at atmospheric pressure [27].

- Carnegie-Mellon University demonstrated both lateral-axis [28] and z-axis [29] integrated gyroscopes with noise floor of about $0.5°/s/\sqrt{Hz}$ using a maskless post-CMOS micromachining process in 2001. The lateral-axis gyroscope with 5 μm thick structure was fabricated by a thin-film CMOS process, starting with Agilent 0.5μm three-metal CMOS. Excessive curling was observed due to the residual stress and thermal expansion coefficient mismatch in the structure, and limited the device size. The 8μm thick z-axis integrated gyroscope was fabricated starting with UMC 0.18μm six copper layer CMOS.

- HSG-IMIT reported in 2002 a gyroscope with excellent structural decoupling of drive and sense modes, fabricated in the standard Bosch fabrication process featuring 10μm thick polysilicon structural layer. A resolution of $25°/hr/\sqrt{Hz}$ with 100Hz bandwidth was reported [30].

- Analog Devices Inc. developed a dual-resonator z-axis gyroscope in 2002, fabricated in the iMEMS process by ADI with a 4μm thick polysilicon structural layer. The device utilized two identical proof masses driven into resonance in opposite directions to reject external linear accelerations, and the differential output of the two Coriolis signals was detected. On-chip control and detection electronics provided self oscillation, phase control, demodulation and temperature compensation. This first commercial integrated micromachined gyroscope had a measured noise floor of $0.05°/s/\sqrt{Hz}$ at 100Hz bandwidth [31].

- An integrated micromachined gyroscope with resonant sensing was reported in 2002 by BSAC. Fabricated in the IMEMS process by Sandia National Laboratories, the device utilized frequency shift of double-ended tuning forks (DETF) due to the generated Coriolis force. A resolution of $0.3°/s/\sqrt{Hz}$ was demonstrated with the on-chip integrated electronics [32].

- In 2002, University of Michigan reported their 150μm thick bulk micromachined single crystal silicon vibrating ring gyroscope, with $10.4°/hr/\sqrt{Hz}$ resolution [33].

- In 2003, Carnegie-Mellon University demonstrated a DRIE CMOS-MEMS lateral axis gyroscope with a measured noise floor of $0.02°/s/\sqrt{Hz}$ at 5 Hz, fabricated by post-CMOS micromachining that uses interconnect metal layers to mask the structural etch steps. The device employs a combination of 1.8μm thin-film structures for springs with out-of-plane compliance and 60μm bulk silicon structures defined by DRIE for the proof mass and springs with out-of-plane stiffness, with on-chip CMOS circuitry. Complete etch removal of selective silicon regions provides electrical isolation of bulk silicon to obtain individually controllable comb fingers. Excessive curling is eliminated in the device, which was problematic in prior thin-film CMOS-MEMS gyroscopes [34].

- In 2004, Honeywell presented the experimental results on commercial development of MEMS vibratory gyroscopes [35], the adaptation of the tuning fork architecture originally developed by Draper's Laboratory. The demonstrated performance of the gyro was $1440°/s$ operation range, less than $30°/hr$ bias in-run stability, and $0.05°/\sqrt{hr}$ angle random walk.

- In 2005, a bulk micromachined gyroscope with bandwidth of 58 Hz and $0.3°/hr$ bias stability tested in 10 mTorr pressure was presented by Seoul National University [36], however not enough details on design and testing conditions were given to independently verify the performance characteristics reported. There were no subsequent publications on the design supporting the data.
- In 2006, Microsystems Laboratory at UCIrvine introduced a design architecture of vibratory gyroscope with 1-DOF drive-mode and 2-DOF sense-mode [63]. The architecture provided a gain and phase stable operation region in the sense-mode frequency response to achieve inherent robustness at the sensing element level. The gyroscope exhibited a measured noise floor of $0.64°/s/\sqrt{Hz}$ at 50 Hz in atmospheric pressure with external discrete electronics.
- In 2007, Georgia Institute of Technology demonstrated a vibratory silicon gyroscope in a tuning fork arrangement to achieve $0.2°/hr$ bias drift with automatic mode-matching and sense-mode Quality factor of 36,000. The sense mode is automatically tuned down by the ASIC until the zero-rate output is maximized [37]. On the same device, $5.4°/hr$ bias drift and 1.5 Hz bandwidth for 2 Hz mode-mismatch and Quality factor of 10,000 at fixed temperature, and $0.96°/hr$ bias drift and 0.4bHz bandwidth for 0 Hz mode-mismatch and Quality factor of 40,000 were previously reported [38].
- In 2008, Microsystems Laboratory at UCIrvine improved the design architecture of structurally robust MEMS gyroscopes [151] and demonstrated high operational frequency devices (over 2.5kHz) and bandwidth over 250 Hz, with the uncompensated temperature coefficients of bias and scale factor of $313°/hr/°C$ and 351 ppm/$°C$, respectively. With off-chip detection electronics, the measured resolution was $0.09 °/s/\sqrt{Hz}$ and the bias drift was $0.08 °/s$.

1.8 The Robustness Challenge

The tolerancing capabilities of the current photolithography processes and microfabrication techniques are inadequate compared to the requirements for production of high-performance inertial sensors. The resulting inherent imperfections in the mechanical structure significantly limits the performance, stability, and robustness of MEMS gyroscopes [45, 61]. Thus, fabrication and commercialization of high-performance and reliable MEMS gyroscopes that require picometer-scale displacement measurements of a vibratory mass have proven to be extremely challenging [4, 43].

In micromachined vibratory rate gyroscopes, the mode-matching requirement renders the system response very sensitive to variations in system parameters due to fabrication imperfections and fluctuations in operating conditions. Inevitable fabrication imperfections affect both the geometry and the material properties of MEMS devices [61], and shift the drive and sense-mode resonant frequencies. The dynamical system characteristics are observed to deviate drastically from the designed values and also from device to device, due to slight variations in photolithography steps, etching processes, deposition conditions or residual stresses. Process control becomes extremely critical to minimize die-to-die, wafer-to-wafer, and lot-to-lot variations.

Fluctuations in the temperature of the structure also perturb the dynamical system parameters due to the temperature dependence of Young's Modulus and thermally induced localized stresses. Temperature also drastically affects the damping and the Q factor in the drive and sense modes.

Extensive research has focused on design of symmetric suspensions and resonator systems that provide mode-matching and minimize temperature dependence [91, 92]. Various symmetric gyroscope designs based on enhancing performance by mode-matching have been reported. However, especially for lightly-damped devices, the requirement for mode-matching is well beyond fabrication tolerances; and none of the symmetric designs can provide the required degree of mode-matching without active tuning and closed-loop feedback control [46, 47]. Also the gain is affected significantly by fluctuations in damping conditions, which makes the device very vulnerable to any possible vacuum leak in the hermetic package seal or outgassing within the cavity.

Fabrication imperfections also introduce anisoelasticities due to extremely small imbalances in the gyroscope suspension. This results in mechanical interference between the modes and undesired mode coupling often much larger than the Coriolis motion. In order to suppress coupled oscillation and drift, various devices have been reported employing independent suspension beams for the drive and sense modes [85, 87–89, 91, 99].

Consequently, the mechanical sensing elements of micromachined gyroscopes are required to provide excellent performance, stability, and robustness to meet demanding specifications. Fabrication imperfections and variations, and fluctuations in the ambient temperature or pressure during the operation time of these devices introduce significant errors, which typically require electronic compensation. Closed-loop force-feedback implementations in the sense-mode are known to alleviate the sensitivity to frequency and damping variations, and increase the sensor bandwidth. However, a closed-loop sense-mode requires additional feedback electrodes, and increases the cost and complexity of both the MEMS device and the electronics. Thus, it is desirable to achieve inherent robustness at the sensing element to minimize compensation requirements.

1.9 Inherently Robust Systems

In recent years, a number of gyroscope designs with multiple proof-masses and different operation principles have been proposed to enhance performance and robustness of MEMS gyroscopes [49, 50, 54, 85, 87, 88, 99]. Most of these designs rely on constraining the oscillation degree-of-freedom of the driven mass to lie only in the drive direction. In these designs, either a part of the Coriolis Force induced on the driven mass is transferred to the sensing mass while suppressing the motion of the sensing mass in drive direction [49, 54, 85, 99]; or the drive direction oscillation of the driven mass is transferred to the sensing mass while the driven mass is not allowed to oscillate in the sense direction [50, 87, 88]. These designs are still virtually two degrees-of-freedom systems, however, they offer various advantages from the drive and sense mode decoupling and mode-matching points of view.

Multiple degrees-of-freedom resonators providing larger drive-direction amplitudes for improving the performance of vibratory MEMS devices have also been recently reported [52, 53, 56, 89, 93]. Two degrees-of-freedom oscillators utilizing mechanical amplification of motion for large oscillation amplitudes have been proposed, however, no results on integration of this oscillator system into MEMS gyroscopes have been indicated [52, 53, 93].

1.10 Overview

This book is organized in two parts. The first part reviews the fundamental operational principles of micromachined vibratory gyroscopes, mechanical sensing element design and practical implementation aspects, electrical design and system-level architectures for actuation and detection, basics of microfabrication methods used for MEMS gyroscopes, and test and characterization techniques. The second part reviews new dynamical systems and structural designs for micromachined gyroscopes, that will provide inherent robustness against structural and environmental parameter variations, and require less demanding active compensation schemes. The basic approach is to achieve a frequency response with an operating frequency region where the response gain and phase are stable, in contrast to a resonant conventional system.

Chapter 2
Fundamentals of Micromachined Gyroscopes

In this chapter, we review the fundamental operational principles of micromachined vibratory rate gyroscopes. First, the dynamics of linear and torsional vibratory gyroscope sensing elements are developed. Then, oscillation patterns and the characteristics of the gyroscope response to the rotation-induced Coriolis force are analyzed, considering basic phase relations and oscillation patterns.

2.1 Dynamics of Vibratory Rate Gyroscopes

The basic architecture of a vibratory gyroscope is comprised of a drive-mode oscillator that generates and maintains a constant linear or angular momentum, coupled to a sense-mode Coriolis accelerometer that measures the sinusoidal Coriolis force induced due to the combination of the drive vibration and an angular rate input.

The vast majority of reported micromachined rate gyroscopes utilizes a vibratory proof mass suspended by flexible beams above a substrate. The primary objective of the dynamical system is to form a vibratory drive oscillator, coupled to an orthogonal sense accelerometer by the Coriolis force.

Both the drive-mode oscillator and the sense-mode accelerometer can be implemented as either linear or torsional resonators. In the case of a linear vibratory gyroscope, a Coriolis force is induced due to linear drive oscillations, while in a torsional vibratory gyroscope, a Coriolis torque is induced due to rotary drive oscillations. The dynamics and operational principles of linear and torsional gyroscopes are outlined below.

2.1.1 Linear Gyroscope Dynamics

The most basic implementation for a micromachined vibratory rate gyroscope is a single proof mass suspended above the substrate. The proof mass is supported by anchored flexures, which serve as the flexible suspension between the proof mass and the substrate, making the mass free to oscillate in two orthogonal directions - the drive and the sense directions (Figure 2.1).

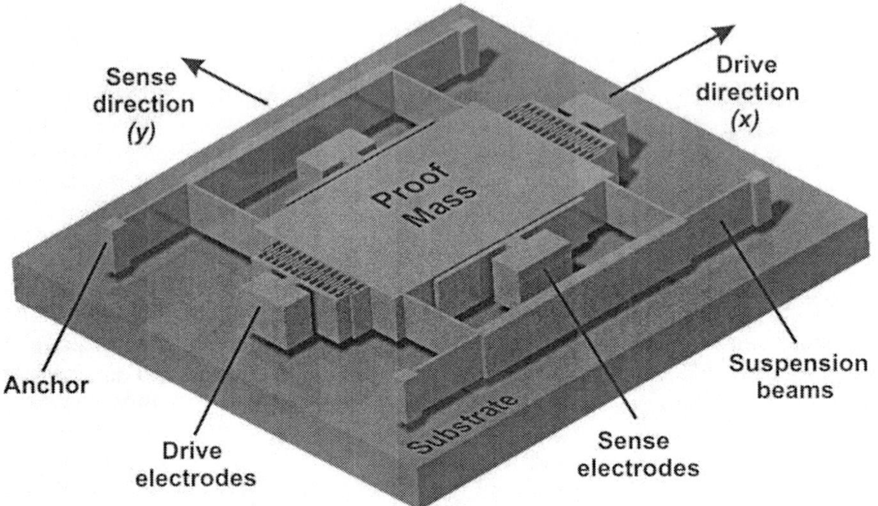

Fig. 2.1 A generic MEMS implementation of a linear vibratory rate gyroscope. A proof-mass is suspended above a substrate using a suspension system comprised of flexible beams, anchored to the substrate. One set of electrodes is needed to excite the drive-mode oscillator, and another set of electrodes detects the sense-mode response.

The drive-mode oscillator is comprised of the proof-mass, the suspension system that allows the proof-mass to oscillate in the drive direction, and the drive-mode actuation and feedback electrodes. The proof-mass is driven into resonance in the drive direction by an external sinusoidal force at the drive-mode resonant frequency.

The sense-mode accelerometer is formed by the proof-mass, the suspension system that allows the proof-mass to oscillate in the sense direction, and the sense-mode detection electrodes. When the gyroscope is subjected to an angular rotation, a sinusoidal Coriolis force at the frequency of drive-mode oscillation is induced in the sense direction. The Coriolis force excites the sense-mode accelerometer, causing the proof-mass to respond in the sense direction. This sinusoidal Coriolis response is picked up by the detection electrodes.

For a generic z-Axis gyroscope, the proof mass is required to be free to oscillate in two orthogonal directions: the drive direction (x-Axis) to form the vibratory oscillator, and the sense direction (y-Axis) to form the Coriolis accelerometer. The overall dynamical system becomes simply a two degrees-of-freedom (2-DOF) mass-spring-damper system (Fig. 2.2).

The dynamics and principle of operation can be best understood by considering the rotation-induced Coriolis force acting on a body that is observed in a rotating reference frame. One of the most intuitive methods to obtain the equations of motion is taking the second time derivative of the position vector to express the acceleration of a body moving with the rotating reference frame.

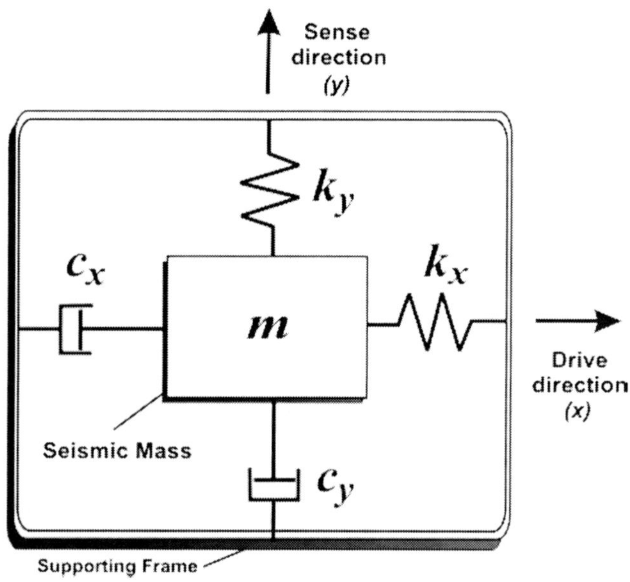

Fig. 2.2 A vibratory rate gyroscope is comprised of a proof mass which is free to oscillate in two principle orthogonal directions: drive and sense.

The accelerations experienced by a moving body in a rotating reference frame can be conveniently derived starting with the following definitions:

A : Inertial (stationary) frame
B : Non-inertial (rotating) reference frame
$\mathbf{r_A}$: Position vector relative to inertial frame A
$\mathbf{r_B}$: Position vector relative to rotating frame B
θ : Orientation vector of rotating frame B relative to inertial frame A
Ω : Angular velocity vector of rotating frame B, $\Omega = \dot{\theta}$
$\mathbf{R}$: Position vector of rotating frame B

The time derivative of a vector $\mathbf{r}$, which is defined in the two reference frames A and B as $\mathbf{r_A}$ and $\mathbf{r_B}$, respectively, is given as

$$\dot{\mathbf{r}}_\mathbf{A}(t) = \dot{\mathbf{r}}_\mathbf{B}(t) + \dot{\theta} \times \mathbf{r_B}(t) \qquad (2.1)$$

Taking the second time derivative of the position vector $\mathbf{r}$, the acceleration of a body moving with the rotating reference frame can be calculated as

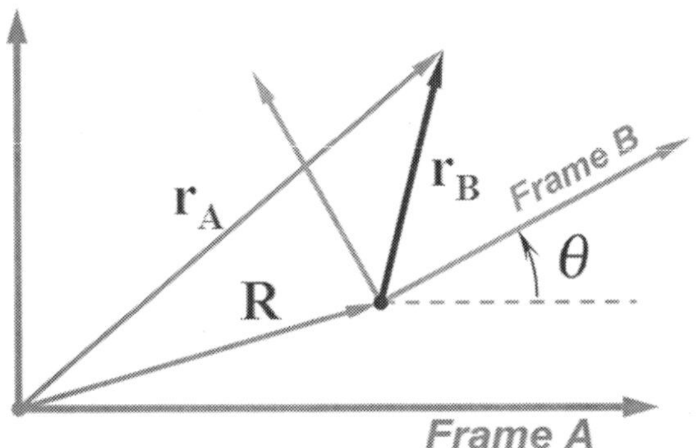

Fig. 2.3 Representation of the position vector relative to the inertial frame A and the rotating reference frame B.

$$\mathbf{r_A}(t) = \mathbf{R}(t) + \mathbf{r_B}(t) \tag{2.2}$$

$$\dot{\mathbf{r}}_\mathbf{A}(t) = \dot{\mathbf{R}}(t) + \dot{\mathbf{r}}_\mathbf{B}(t) + \dot{\theta} \times \mathbf{r_B}(t) \tag{2.3}$$

$$\ddot{\mathbf{r}}_\mathbf{A}(t) = \ddot{\mathbf{R}}(t) + \ddot{\mathbf{r}}_\mathbf{B}(t) + \dot{\theta} \times \dot{\mathbf{r}}_\mathbf{B}(t) + \dot{\theta} \times (\dot{\theta} \times \mathbf{r_B}(t)) + \ddot{\theta} \times \mathbf{r_B}(t) + \dot{\theta} \times \dot{\mathbf{r}}_\mathbf{B}(t) \tag{2.4}$$

With the definition of $\mathbf{v_B}$ and $\mathbf{a_B}$ as the velocity and acceleration vectors with respect to the rotating reference frame B, $\mathbf{a_A}$ as the acceleration vector with respect to the inertial frame A, $\mathbf{A}$ as the linear acceleration of the reference frame B, and Ω as the angular velocity vector of the reference frame B; the expression for acceleration reduces to

$$\mathbf{a_A} = \mathbf{A} + \mathbf{a_B} + \dot{\Omega} \times \mathbf{r_B} + \Omega \times (\Omega \times \mathbf{r_B}) + 2\Omega \times \mathbf{v_B} \tag{2.5}$$

In this equation, $(\mathbf{A} + \mathbf{a_B} + \dot{\Omega} \times \mathbf{r_B})$ is the local acceleration, and $\Omega \times (\Omega \times \mathbf{r_B})$ is the centripetal acceleration. The last term $2\Omega \times \mathbf{v_B}$ is the Coriolis acceleration, which is the primary mechanism that scales and converts the rotation rate of the rotating reference frame B into a fictitious inertial force when observed in the rotating frame.

When applied to the position vector of a vibratory gyroscope proof-mass, this analysis yields the dynamics of the gyroscope attached to a rotating object. The equation of motion of the proof-mass can be derived by expressing the acceleration vector of the proof mass with respect to the inertial frame A by taking the second time derivative of the position vector (Figure 2.4),

$$\mathbf{F_{ext}} = m \left[\mathbf{A} + \mathbf{a_B} + \dot{\Omega} \times \mathbf{r_B} + \Omega \times (\Omega \times \mathbf{r_B}) + 2\Omega \times \mathbf{v_B} \right] \tag{2.6}$$

where $\mathbf{A}$ is the linear acceleration and Ω is the angular velocity of the rotating gyroscope frame, $\mathbf{v_B}$ and $\mathbf{a_B}$ are the velocity and acceleration vectors of the proof mass with respect to the reference frame, and $\mathbf{F_{ext}}$ is the total external force applied on the proof mass.

In a z-Axis gyroscope, the two principle oscillation directions are the drive direction along the x-axis and the sense direction along the y-axis. Decomposing the motion into the two principle oscillation directions and assuming that the linear accelerations are negligible, the two equations of motion along the drive and sense axes can be expressed as

$$m\ddot{x} + c_x\dot{x} + (k_x - m(\Omega_y^2 + \Omega_z^2))x + m(\Omega_x\Omega_y - \dot{\Omega}_z)y = \tau_x + 2m\Omega_z\dot{y}$$
$$m\ddot{y} + c_y\dot{y} + (k_y - m(\Omega_x^2 + \Omega_z^2))y + m(\Omega_x\Omega_y + \dot{\Omega}_z)x = \tau_y - 2m\Omega_z\dot{x} \qquad (2.7)$$

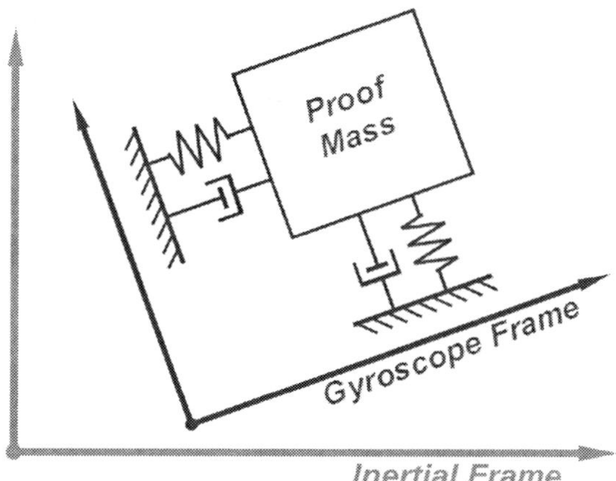

Fig. 2.4 Schematic illustration of the gyroscope frame rotating with respect to the inertial frame.

For a constant angular rate input $\dot{\Omega}_z = 0$, and for angular rates at much lower frequencies than the operating frequency of the gyroscope, the terms Ω_x^2, Ω_y^2, and $\Omega_x\Omega_y$ become negligible. Since the sense-mode response is usually orders of magnitude smaller in amplitude than the drive motion, the Coriolis term $2m\Omega_z\dot{y}$ is also negligible. Thus, the practical simplified 2-DOF equations of motion of a vibratory rate gyroscope become:

$$m\ddot{x} + c_x\dot{x} + k_x x = \tau_x$$
$$m\ddot{y} + c_y\dot{y} + k_y y = \tau_y - 2m\Omega_z\dot{x} \qquad (2.8)$$

where τ_x is the external force in the drive direction, which is usually a sinusoidal drive excitation force, and τ_y is the total external force in the sense direction, comprised of parasitic and external inertial forces. The term $2m\Omega_z\dot{x}$ in the sense-mode equation is the rotation-induced Coriolis force, which causes the sense-mode response proportional to the angular rate.

A very common method to amplify the mechanical response of the sense-mode accelerometer to the Coriolis force is to design the resonant frequency of the sense-mode accelerometer close to the frequency of the Coriolis force. If the Coriolis force frequency, and thus the drive-mode resonant frequency, is matched with the sense-mode resonant frequency, the Coriolis force excites the system into resonance in the sense direction. This allows to amplify the resulting oscillation amplitude in the sense direction by the sense-mode Q factor, which could mean orders of magnitude improvement in sensitivity as explained in the following sections.

2.1.2 Torsional Gyroscope Dynamics

Even though conservation of both linear and angular momentum are required to express the complete dynamics of a gyroscope proof mass, linear gyroscope systems can be modeled based on conservation of linear momentum only, assuming negligible angular deflections. Similarly, the dynamics of a torsional gyroscope can be analyzed based on conservation of angular momentum with the assumption that the linear deflections are negligible.

The angular momentum of a mass with an inertia tensor $\mathbf{I}$ and an angular velocity vector ω is $\mathbf{H} = \mathbf{I}\omega$. Expressing the angular momentum $\mathbf{H}$ in an inertial frame, angular momentum balance under the presence of an external moment $\mathbf{M}$ is

$$\frac{d\mathbf{H}}{dt} = \mathbf{M} \tag{2.9}$$

When the angular momentum is expressed in a non-inertial coordinate frame which rotates with the same angular velocity ω as the mass, the inertia tensor $\mathbf{I}$ becomes constant and diagonal, and the angular momentum balance becomes

$$\mathbf{I}\dot{\omega} + \omega \times (\mathbf{I}\omega) = \mathbf{M} \tag{2.10}$$

If we consider a simple case of a z-axis torsional gyroscope with a gimbal, the dynamics of each rotary proof-mass in the torsional gyroscope system is best understood by attaching non-inertial coordinate frames to the center-of-mass of each proof-mass and the substrate (Figure 2.5). The angular momentum equation for each mass will be expressed in the coordinate frame associated with that mass. This allows the inertia matrix of each mass to be expressed in a diagonal and time-invariant form. The absolute angular velocity of each mass in the coordinate frame of that mass will be obtained using the appropriate transformations. By conservation of angular momentum, the equations of motion of the masses are

$$\mathbf{I}_s\dot{\omega}_s^s + \omega_s^s \times (\mathbf{I}_s\omega_s^s) = \tau_{se} + \tau_{sd} \tag{2.11}$$
$$\mathbf{I}_d\dot{\omega}_d^d + \omega_d^d \times (\mathbf{I}_d\omega_d^d) = \tau_{de} + \tau_{dd} + M_d \tag{2.12}$$

where $\mathbf{I}_s$ and $\mathbf{I}_d$ denote the diagonal and time-invariant inertia matrices of the sensing mass and the drive gimbal, respectively, with respect to the associated body

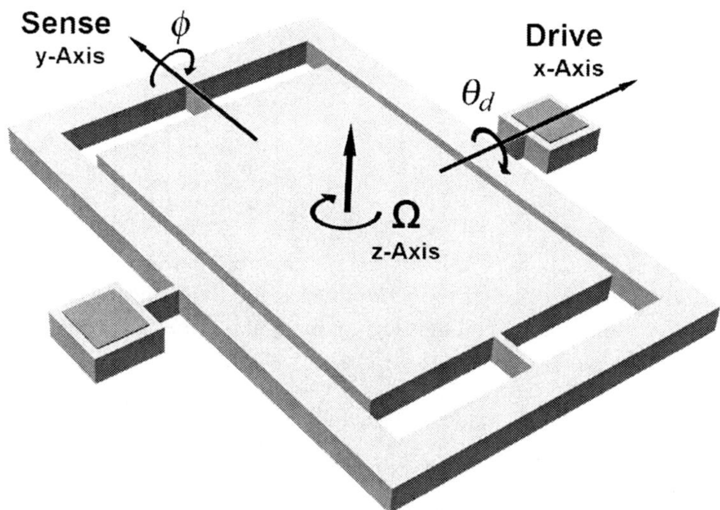

Fig. 2.5 Torsional z-Axis gyroscope with drive gimbal structure. The drive-mode deflection angle of the gimbal is θ_d, and the sense-mode deflection angle of the sensing mass is ϕ.

attached frames. Similarly, ω_s^s and ω_d^d are the absolute angular velocity of the sensing mass and the gimbal, respectively, expressed in the associated body frames. The external torques τ_{se} and τ_{de} are the elastic torques, and τ_{sd} and τ_{dd} are the damping torques acting on the associated mass, whereas M_d is the driving electrostatic torque applied to the drive gimbal.

If we denote the drive direction deflection angle of the drive gimbal by θ_d, the sense direction deflection angle of the sensing mass by ϕ (with respect to the substrate), and the absolute angular velocity of the substrate about the z-axis by Ω_z as in Figure 2.5, the homogeneous rotation matrices from the substrate to drive gimbal ($R_{sub \to d}$), and from drive gimbal to the sensing mass ($R_{d \to s}$), respectively, become

$$R_{sub \to d} = \begin{bmatrix} 1 & 0 & 0 \\ 0 & cos\theta_d & -sin\theta_d \\ 0 & sin\theta_d & cos\theta_d \end{bmatrix} \tag{2.13}$$

$$R_{d \to s} = \begin{bmatrix} cos\phi & 0 & sin\phi \\ 0 & 1 & 0 \\ -sin\phi & 0 & cos\phi \end{bmatrix} \tag{2.14}$$

Using the obtained transformations, the total absolute angular velocity vectors of the drive gimbal and the sensing mass can be expressed in the non-inertial sensing mass coordinate frame as

$$\omega_d^d = \begin{bmatrix} \dot{\theta}_d \\ 0 \\ 0 \end{bmatrix} + R_{sub \to d} \begin{bmatrix} 0 \\ 0 \\ \Omega_z \end{bmatrix} \tag{2.15}$$

$$\omega_s^s = \begin{bmatrix} 0 \\ \dot{\phi} \\ 0 \end{bmatrix} + R_{d \to s} \begin{bmatrix} \dot{\theta}_d \\ 0 \\ 0 \end{bmatrix} + R_{d \to s} R_{sub \to d} \begin{bmatrix} 0 \\ 0 \\ \Omega_z \end{bmatrix} \tag{2.16}$$

Substitution of the angular velocity vectors into the derived angular momentum equations with small-angle approximation yields the dynamics of the drive gimbal about the drive axis (x-axis) and the sensing mass about the sense axis (y-axis) as

$$I_y^s \ddot{\phi} + D_y^s \dot{\phi} + [K_y^s + (\Omega_z^2 - \dot{\theta}_d^2)(I_z^s - I_x^s)]\phi =$$
$$(I_z^s + I_y^s - I_x^s)\dot{\theta}_d \Omega_z + I_y^s \theta_d \dot{\Omega}_z + (I_z^s - I_x^s)\phi^2 \theta_d \Omega_z \tag{2.17}$$
$$(I_x^d + I_x^s)\ddot{\theta}_d + (D_x^d + D_x^s)\dot{\theta}_d + [K_x^d + (I_y^d - I_z^d + I_y^s - I_z^s)\Omega_z^2]\theta_d =$$
$$-(I_z^s + I_x^s - I_y^s)\dot{\phi}\Omega_z - I_x^s \phi \dot{\Omega}_z + M_d \tag{2.18}$$

where I_x^s, I_y^s, and I_z^s denote the moments of inertia of the sensing plate; I_x^d, I_y^d, and I_z^d are the moments of inertia of the drive gimbal; D_x^s and D_x^d are the drive-direction damping ratios, and D_y^s is the sense-direction damping ratio of the sensing mass; K_y^s is the torsional stiffness of the suspension beam connecting the sensing plate to the drive gimbal, and K_x^d is the torsional stiffness of the suspension beam connecting the drive gimbal to the substrate.

With the assumptions that the angular rate input is constant, i.e. $\dot{\Omega}_z = 0$, and the oscillation angles are small, the rotational equations of motion can be further simplified, yielding

$$I_y^s \ddot{\phi} + D_y^s \dot{\phi} + K_y^s \phi = (I_z^s + I_y^s - I_x^s)\dot{\theta}_d \Omega_z \tag{2.19}$$
$$(I_x^d + I_x^s)\ddot{\theta}_d + (D_x^d + D_x^s)\dot{\theta}_d + K_x^d \theta_d = M_d \tag{2.20}$$

The term $(I_z^s + I_y^s - I_x^s)\dot{\theta}_d \Omega_z$ is the Coriolis torque that excites the sensing mass about the sense axis, with ϕ being the detected deflection angle about the sense axis for angular rate measurement.

The dynamics and fundamental operation of torsional gyroscopes are similar to that of linear gyroscopes. Without any loss of generality, we will be primarily illustrating the basic operational principles on linear gyroscopes in the following sections. All obtained results are directly applicable to torsional gyroscopes as well.

2.2 Resonance Characteristics

Vast majority of micromachined vibratory gyroscopes employ a combination of proof-masses and flexures to form 1 degree-of-freedom (1-DOF) resonators in both the drive and sense directions. Thus, understanding the dynamics and response characteristics of a generic 1-DOF resonator is critical in the design of both the drive and the sense-mode oscillators of a vibratory gyroscope.

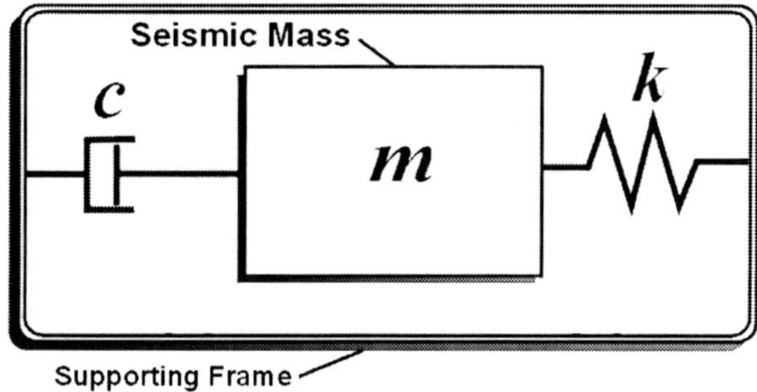

Fig. 2.6 The lumped mass-spring-damper model of a typical 1-DOF resonator.

Let us start by investigating the dynamics of a typical 1-DOF resonator as in Figure 2.6. The equation of motion of the resonator with a proof-mass m, a combined stiffness of k, and a damping factor of c is

$$m\ddot{x} + c\dot{x} + kx = F(t) \tag{2.21}$$

With the definition of the undamped natural frequency ω_n and the damping factor ξ which represents the ratio of damping to critical damping ($2\sqrt{km}$), the equation of motion becomes

$$\ddot{x} + 2\xi\omega_n\dot{x} + \omega_n^2 x = \frac{F(t)}{m} \tag{2.22}$$

$$\omega_n = \sqrt{\frac{k}{m}} \tag{2.23}$$

$$\xi = \frac{c}{c_c} = \frac{c}{2\sqrt{km}} = \frac{c}{2m\omega_n} \tag{2.24}$$

In the Laplace domain, the transfer function of the 1-DOF resonator is simply

$$\frac{X(s)}{F(s)} = \frac{\frac{1}{m}}{s^2 + 2\xi\omega_n s + \omega_n^2} \tag{2.25}$$

When the resonator is excited with a harmonic force $F = F_0 \sin \omega t$ at the frequency ω, the steady-state component of the response is also harmonic, of the form

$$x = x_0 \sin(\omega t + \phi) \tag{2.26}$$

$$x_0 = \frac{\frac{F_0}{k}}{\sqrt{\left[1 - \left(\frac{\omega}{\omega_n}\right)^2\right]^2 + \left[2\xi \frac{\omega}{\omega_n}\right]^2}} \tag{2.27}$$

$$\phi = -\tan^{-1} \frac{2\xi \frac{\omega}{\omega_n}}{1 - \left(\frac{\omega}{\omega_n}\right)^2} \tag{2.28}$$

In the presence of considerable damping, the amplitude expression is maximized at the frequency

$$\omega_r = \omega_n \sqrt{1 - 2\xi^2} \tag{2.29}$$

For lightly damped systems, i.e. $\xi \ll 1$, the amplitude is maximized at the natural frequency ω_n, and the amplitude at resonance becomes

$$|x_0|_{res} = \frac{F_0}{2k\xi} = \frac{F_0}{c\omega_n} \tag{2.30}$$

The *Quality factor* of the system is defined as maximum ratio of the amplitude to the static deflection, which is F_0/k. Taking the ratio of the amplitude at resonance to the static deflection, the Q factor of a lightly damped system reduces to

$$Q = \frac{1}{2\xi} = \frac{m\omega_n}{c} \tag{2.31}$$

It should be noticed that the Quality factor is one of the most important parameters of a resonator, since it directly scales the amplitude at resonance. For example, for a resonator with a known Q factor, the oscillation amplitude at resonance can be found as

$$|x_0|_{res} = Q \frac{F_0}{k} \tag{2.32}$$

At the resonant frequency, the phase is $-90°$ shifted from the forcing function phase. At frequencies lower than the resonant frequency, the phase approaches $0°$, meaning that the position follows the forcing function closely. At frequencies higher than the resonant frequency, the phase approaches $-180°$. The transition from $0°$ to $-180°$ around the resonant frequency becomes more abrupt for higher Q values.

The *bandwidth* or the *half-power bandwidth* of the system is defined as the difference between the frequencies where the power is half of the resonance power. Since the power is proportional to the square of the oscillation amplitude, the half-power frequencies are solved by equating the amplitude expression to $1/\sqrt{2}$ times the resonance amplitude. For small values of damping, the bandwidth is approximated as

$$BW \approx \frac{\omega_n}{Q} \tag{2.33}$$

This analysis forms the background for the following discussions on the dynamics and response of the drive and sense oscillators in vibratory gyroscopes.

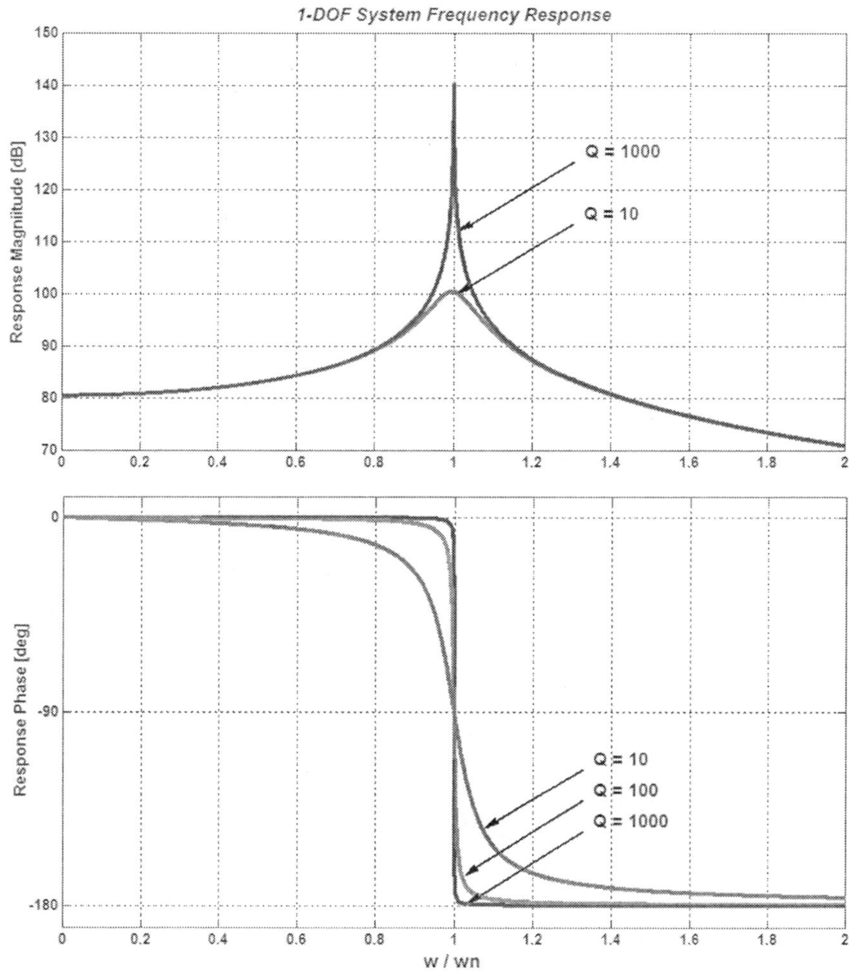

Fig. 2.7 The Bode plot of a typical 1-DOF resonator.

2.3 Drive-Mode Operation

Since the Coriolis effect is based on conservation of momentum, every gyroscopic system requires a mechanical subsystem that generates momentum. In vibratory gyroscopes, the drive-mode oscillator, which is comprised of a proof-mass driven into a harmonic oscillation, is the source of momentum. The drive-mode oscillator is most commonly a 1 degree-of-freedom (1-DOF) resonator, which can be modeled as a mass-spring-damper system consisting of the drive proof-mass m_d, the drive-mode suspension system providing the drive stiffness k_d, and the drive damping c_d consisting of viscous and thermoelastic damping. With a sinusoidal drive-mode excitation force, the drive equation of motion along the x-axis becomes

$$m_d\ddot{x} + c_d\dot{x} + k_d x = F_d \sin \omega t \tag{2.34}$$

With the definition of the drive-mode resonant frequency ω_d and the drive-mode Quality factor Q_d the amplitude and phase of the drive-mode steady-state response $x = x_0 \sin(\omega t + \phi_d)$ becomes:

$$x_0 = \cfrac{F_d}{k_d \sqrt{\left[1 - \left(\frac{\omega}{\omega_d}\right)^2\right]^2 + \left[\frac{1}{Q_d}\frac{\omega}{\omega_d}\right]^2}} \tag{2.35}$$

$$\phi_d = -\tan^{-1} \cfrac{\frac{1}{Q_d}\frac{\omega}{\omega_d}}{1 - \left(\frac{\omega}{\omega_d}\right)^2} \tag{2.36}$$

where

$$\omega_d = \sqrt{\frac{k_d}{m_d}} \tag{2.37}$$

$$Q_d = \frac{m_d \omega_d}{c_d} \tag{2.38}$$

The scale factor of the gyroscope is directly proportional to the drive-mode oscillation amplitude. The phase and the frequency of the drive oscillation directly determines the phase and the frequency of the Coriolis force, and subsequently the sense-mode response. Thus, it is extremely critical to maintain a drive-mode oscillation with stable amplitude, phase and frequency.

For these reasons, almost all reported gyroscopes operate exactly at the drive-mode resonant frequency in practical implementations. At resonance, the drive-mode phase becomes $-90°$, and the amplitude simply reduces to

$$x_{0res} = Q_d \frac{F_d}{m_d \omega_d^2} \tag{2.39}$$

Self resonance by the use of an amplitude regulated positive feedback loop (Figure 2.8) is a common and convenient method to achieve a stable drive-mode amplitude and phase. The positive feedback loop destabilizes the resonator, and locks the operational frequency to the drive-mode resonant frequency. This allows to set the oscillation phase exactly 90° from the excitation signal. An Automatic Gain Control (AGC) loop detects the oscillation amplitude, compares it with a reference amplitude signal, and adjusts the gain of the positive feedback to match the reference amplitude. Operating at resonance in the drive mode also allows to minimize the excitation voltages during steady-state operation.

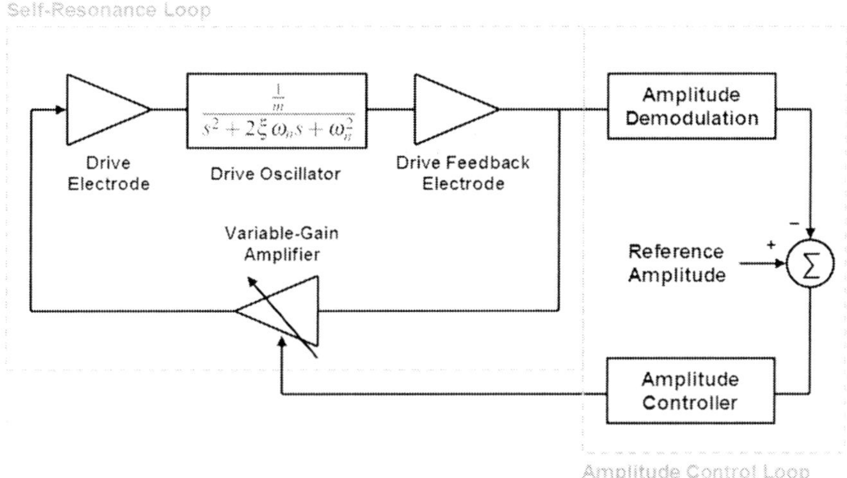

Fig. 2.8 A typical implementation of an Automatic Gain Control (AGC) loop, which drives the drive-mode oscillator into self-resonance and regulates the oscillation amplitude.

2.4 The Coriolis Response

The Coriolis response in the sense direction is best understood starting with the assumption that the drive-mode is operated at drive resonant frequency ω_d, and the drive motion is amplitude regulated to be of the form $x = x_0 \sin(\omega_d t + \phi_d)$ with a constant amplitude x_0. The Coriolis force that excites the sense-mode oscillator is

$$F_C = -2m_C\Omega_z\dot{x} = -2m_C\Omega_z x_0 \omega_d \cos(\omega_d t + \phi_d) \qquad (2.40)$$

where m_C is the portion of the driven proof mass that contributes to the Coriolis force. In a simple single-mass design, m_C is usually equal to m_d. Here it should be emphasized again that the Coriolis force amplitude, which sets the scale factor of the gyroscope, is directly proportional to the drive-mode oscillation amplitude. Thus, the drive amplitude has to be well regulated to achieve a stable scale factor.

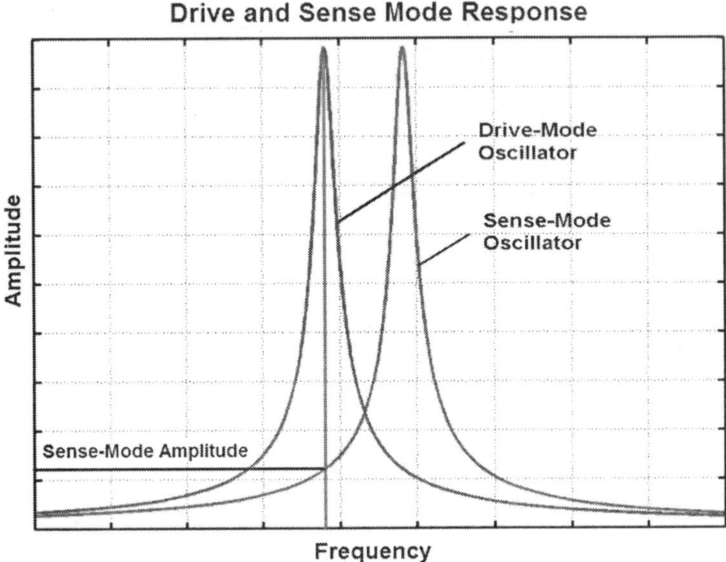

Fig. 2.9 The complete gyroscope system can be viewed as a combination of a 1-DOF drive-mode oscillator and a 1-DOF sense-mode oscillator.

Similar to the drive-mode oscillator, the sense-mode oscillator is also often a 1-DOF resonator. The primary reason for utilizing a 1-DOF resonator as the sense-mode Coriolis accelerometer is to take advantage of resonance to amplify the mechanical response to the Coriolis force. The equation of motion of the 1-DOF sense mode oscillator is

$$m_s \ddot{y} + c_s \dot{y} + k_s y = -2m_C x_0 \omega_d \Omega_z \cos(\omega_d t + \phi_d) \tag{2.41}$$

where m_s is the portion of the proof-mass that responds to the Coriolis force. Again, in a simple single-mass design, m_C, m_d and m_s are equal.

The amplitude and phase of the steady-state sense-mode Coriolis response in a linear system, defining the sense-mode resonant frequency ω_s and the sense-mode Quality factor Q_s, become:

$$y_0 = \Omega_z \frac{m_C \omega_d}{m_s \omega_s^2} \frac{2x_0}{\sqrt{\left[1 - \left(\frac{\omega_d}{\omega_s}\right)^2\right]^2 + \left[\frac{1}{Q_s}\frac{\omega_d}{\omega_s}\right]^2}} \tag{2.42}$$

$$\phi_s = -\tan^{-1}\frac{\frac{1}{Q_s}\frac{\omega_d}{\omega_s}}{1 - \left(\frac{\omega_d}{\omega_s}\right)^2} + \phi_d \tag{2.43}$$

where

$$\omega_s = \sqrt{\frac{k_s}{m_s}} \tag{2.44}$$

$$Q_s = \frac{m_s \omega_s}{c_s} \tag{2.45}$$

To achieve the maximum possible gain in the sense-mode, it is generally desirable to operate at or near the peak of the sense-mode response curve. This is typically achieved by matching drive and sense resonant frequencies. When operating at sense-mode resonance, i.e. $\omega_d = \omega_s$, the sense-mode phase becomes $-90°$ from the drive velocity, and the amplitude reduces to

$$y_{0res} = \Omega_z \frac{2Q_s x_0 m_C}{m_s \omega_s} \tag{2.46}$$

Investigating the resonant amplitude of the sense-mode Coriolis response, the sensitivity of the gyroscope to the angular rate input Ω_z can be improved by

- Increasing the drive-mode oscillation amplitude x_0
- Increasing Q_s by decreasing damping, usually by vacuum packaging
- Maximizing the mass m_C that generates the Coriolis force, while minimizing the total mass m_s excited by the Coriolis force.

In torsional systems, the fundamental resonant operation principle is the same as linear systems. Similarly, if we assume an amplitude regulated torsional drive motion of the form

$$\theta_d = \theta_d^\circ \sin(\omega_d t + \angle\theta_d) \tag{2.47}$$

the steady-state sense-mode Coriolis response, with ϕ being the detected deflection angle about the sense axis for angular rate measurement, can be expressed as

$$\phi = \phi^\circ \sin(\omega_d t + \angle\phi) \tag{2.48}$$

where

$$\phi^\circ = \Omega_z \frac{(I_z^s + I_y^s - I_x^s)\omega_d}{I_y^s \omega_s^2} \frac{\theta_d^\circ}{\sqrt{\left[1 - \left(\frac{\omega_d}{\omega_s}\right)^2\right]^2 + \left[\frac{1}{Q_s}\frac{\omega_d}{\omega_s}\right]^2}}$$

$$\angle\phi = -\tan^{-1}\frac{\frac{1}{Q_s}\frac{\omega_d}{\omega_s}}{1 - \left(\frac{\omega_d}{\omega_s}\right)^2} + \angle\theta_d \tag{2.49}$$

The sense-mode resonant frequency ω_s and the sense-mode Quality factor Q_s are defined as

$$\omega_s = \sqrt{\frac{K_y^s}{I_y^s}} \tag{2.50}$$

$$Q_s = \frac{I_y^s \omega_s}{D_y^s} \tag{2.51}$$

2.4.1 Mode-Matching and Δf

Even though matching the drive and sense-mode resonant frequencies greatly enhances the sense-mode mechanical response to angular rate input, it comes with many disadvantages. Operating close to the sense resonant peak also makes the system very sensitive to variations in system parameters that cause a shift in the resonant frequencies or damping.

To illustrate this effect, let us consider a sense-mode system with a resonant frequency of $\omega_s = 10\text{kHz}$ and a Q factor of $Q_s = 1000$ (Figure 2.10). When the operating frequency matches the sense-mode resonant frequency ω_s the amplification factor is 1000, equal to the Q factor. If there is only 5Hz relative shift between the operating frequency and the sense-mode resonant frequency, the gain drops by 29.3%. For a 10Hz relative shift, the gain drop is 55%.

Under higher quality factor conditions the gain is higher, however, the bandwidth becomes even narrower. For example, let us take the same sense-mode system with a resonant frequency of $\omega_s = 10\text{kHz}$, and increase the Q factor to $Q_s = 10,000$ (Figure 2.11). The amplification factor at resonance directly increases to 10,000. However, the half-power bandwidth becomes:

$$BW = \frac{\omega_s}{Q_s} = \frac{10,000\text{Hz}}{10,000} = 1\text{Hz} \tag{2.52}$$

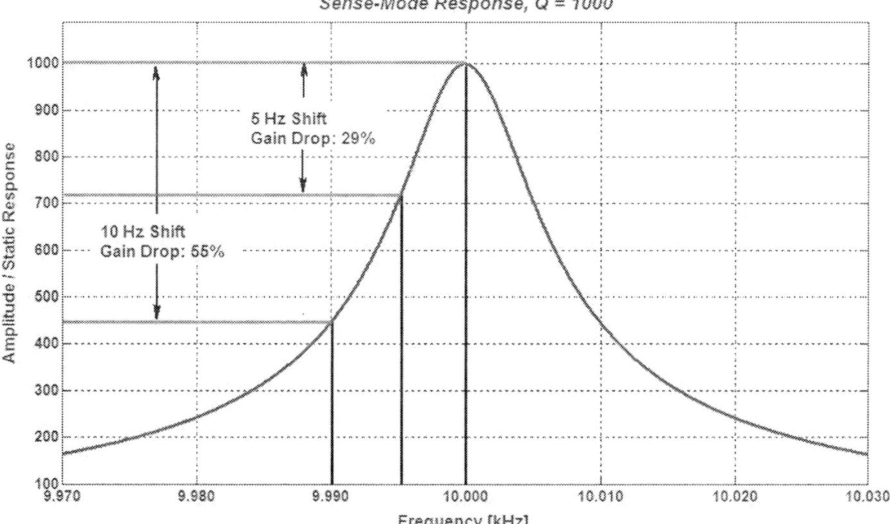

Fig. 2.10 The sense-mode amplification of a sense-mode system with a resonant frequency of $\omega_s = 10\text{kHz}$ and a Q factor of $Q_s = 1000$. For a 5Hz relative shift between the operating frequency and the sense-mode resonant frequency, the gain drop is 29%.

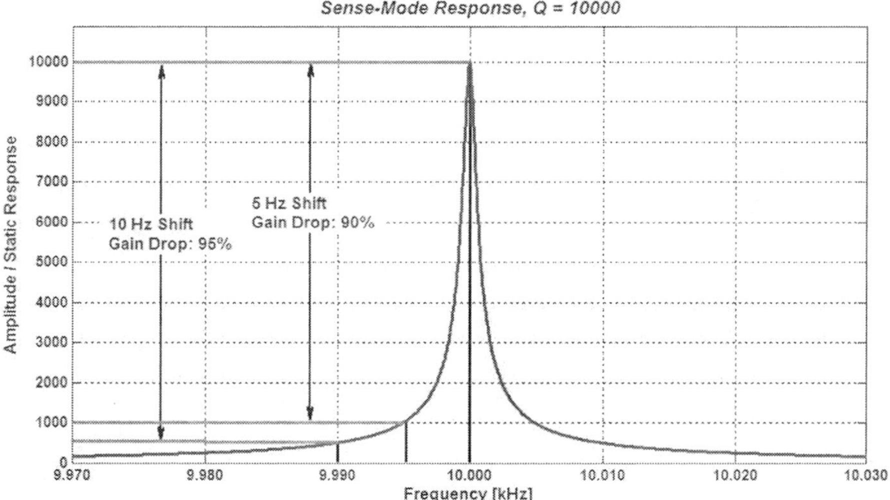

Fig. 2.11 High quality factor devices provide higher gains, but become more sensitive to frequency variations. The sense-mode amplification of a system with $\omega_s = 10\text{kHz}$ and $Q_s = 10,000$ drops by over 90% for a 5Hz relative shift between the operating frequency and the sense-mode resonant frequency.

This means that for only 0.5Hz mismatch, the gain drops by $1/\sqrt{2}$, i.e. 29.3%. If the mismatch is 5Hz, which is only 0.05% of the 10kHz resonant frequency, the gain drops by 90%. Thus, especially in vacuum-packaged devices, the relative position of the sense-mode resonant frequency with respect to the operating frequency has to be controlled with extreme precision.

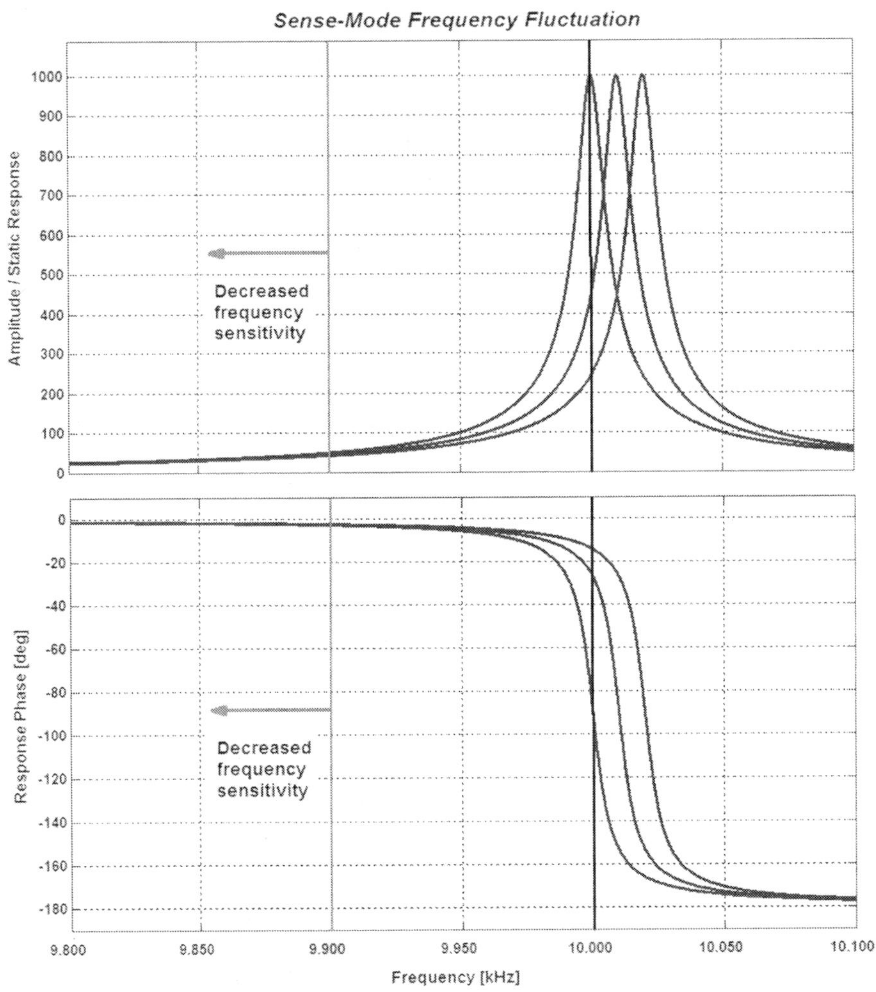

Fig. 2.12 To reduce sensitivity of response gain and phase to frequency fluctuations, the operation frequency is usually set away from the sense-mode frequency.

Fabrication imperfections are inevitable, and affect material properties and geometry of MEMS structures. In surface micromachined gyroscopes, the thickness of the suspension elements is determined by deposition process, and the width is affected by etching process. In addition, Young's Modulus of the structure is affected by deposition conditions [62]. In bulk-micromachined devices, the width and cross-section of the suspension beams often vary due to lateral over-etching. Variations in these parameters cause the resonant frequencies to vary drastically from device to device. Furthermore, fluctuations in the ambient temperature and stresses result in frequency fluctuations during operation.

Given the structural and environmental effects that result in quite large variations in the resonant frequencies, it is extremely difficult to control the drive and sense frequencies with the precision required for mode-matched devices. Thus, it is a common practice to operate away from the resonant frequency of the sense-mode, where the frequency variations have reduced effect on the output gain and phase (Figure 2.12). This is achieved by setting the sense-mode frequency ω_s spaced by a certain percentage away from the drive-mode frequency ω_d. This frequency separation is commonly referred as

$$\Delta f = f_s - f_d = \frac{1}{2\pi}(\omega_s - \omega_d) \tag{2.53}$$

In addition to frequency variations, the gain is affected significantly by fluctuations in damping (Figure 2.13). This is a very important factor, since it means that the Q factor of the sense-mode response also has to be controlled tightly. Thus, maintaining a constant pressure around the device using a leak-free hermetic packaging is crucial. However, the damping also changes with temperature. To achieve a good temperature bias stability, operating away from the resonance peak becomes a practical solution.

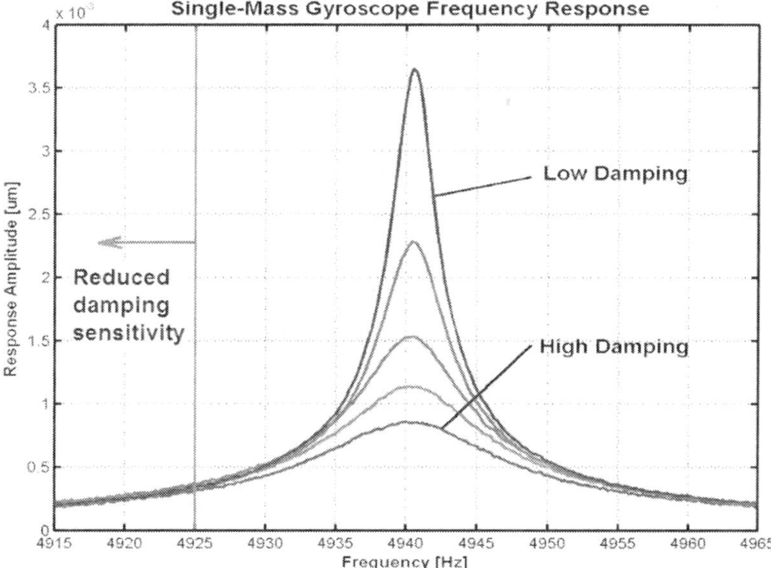

Fig. 2.13 When operating close to the resonance peak, the sense-mode gain is very sensitive to damping.

2.4.2 Phase Relations and Proof-Mass Trajectory

As explained in previous sections, the drive-mode oscillator is usually operated at resonance, and the drive-mode position phase is $-90°$ relative to the drive AC signal. However, the sense-mode position phase depends heavily on the drive and sense frequency separation Δf, and damping. In the steady state, the drive-mode position is of the form

$$x(t) = x_0 \sin(\omega_d t + \phi_d) \tag{2.54}$$

The drive-mode velocity can also be expressed as a sine signal, by taking the time derivative of the position and inverting the sign of cosine, resulting in a $90°$ phase shift from the drive position

$$\dot{x}(t) = x_0 \omega_d \cos(\omega_d t + \phi_d) = x_0 \omega_d \sin(\omega_d t + \phi_d + 90°) \tag{2.55}$$

The Coriolis force generated by the drive-mode oscillator will also be a sine signal at the same frequency and phase. The Coriolis force in the positive direction can be obtained by adding a $-180°$ phase shift

$$F_C(t) = -2m_C \Omega_z x_0 \omega_d \sin(\omega_d t + \phi_d + 90°)$$
$$F_C(t) = 2m_C \Omega_z x_0 \omega_d \sin(\omega_d t + \phi_d - 90°) \tag{2.56}$$

Thus, the sense-mode position will then be a sine signal of the same form as the drive mode position

$$y(t) = y_0 \sin(\omega_d t + \phi_s) \tag{2.57}$$

with the phase

$$\phi_s = \phi_d - 90° - \tan^{-1} \frac{\frac{1}{Q_s} \frac{\omega_d}{\omega_d + \Delta f}}{1 - \left(\frac{\omega_d}{\omega_d + \Delta f}\right)^2} \tag{2.58}$$

This phase relation between the drive and sense modes is very important, since it determines the Coriolis demodulation phase as explained in the following chapters. It should be noticed that it is independent of ϕ_d and solely dependent on Q_s, ω_d and Δf. Now let us investigate how the phase relation is affected by Δf.

When the drive and sense modes are matched, i.e. $\Delta f = 0$, the sense-mode phase becomes (Figure 2.14)

$$\phi_s = \phi_d - 180° \tag{2.59}$$

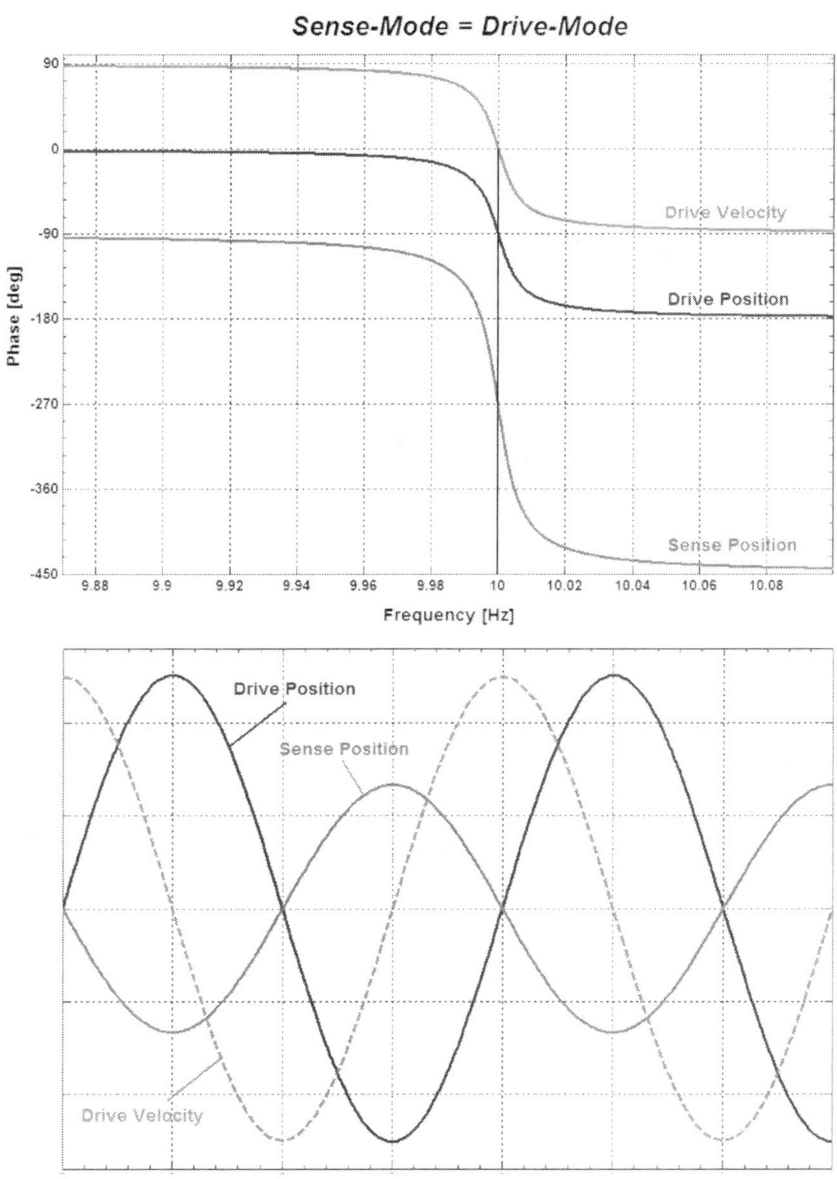

Fig. 2.14 The drive position, drive velocity and sense position phase relations when $\Delta f = 0$.

When the sense-mode is higher than drive, i.e. $\Delta f > 0$, the sense-mode phase converges to the Coriolis force phase, which is $-90°$ shifted from the drive position (Figure 2.15).

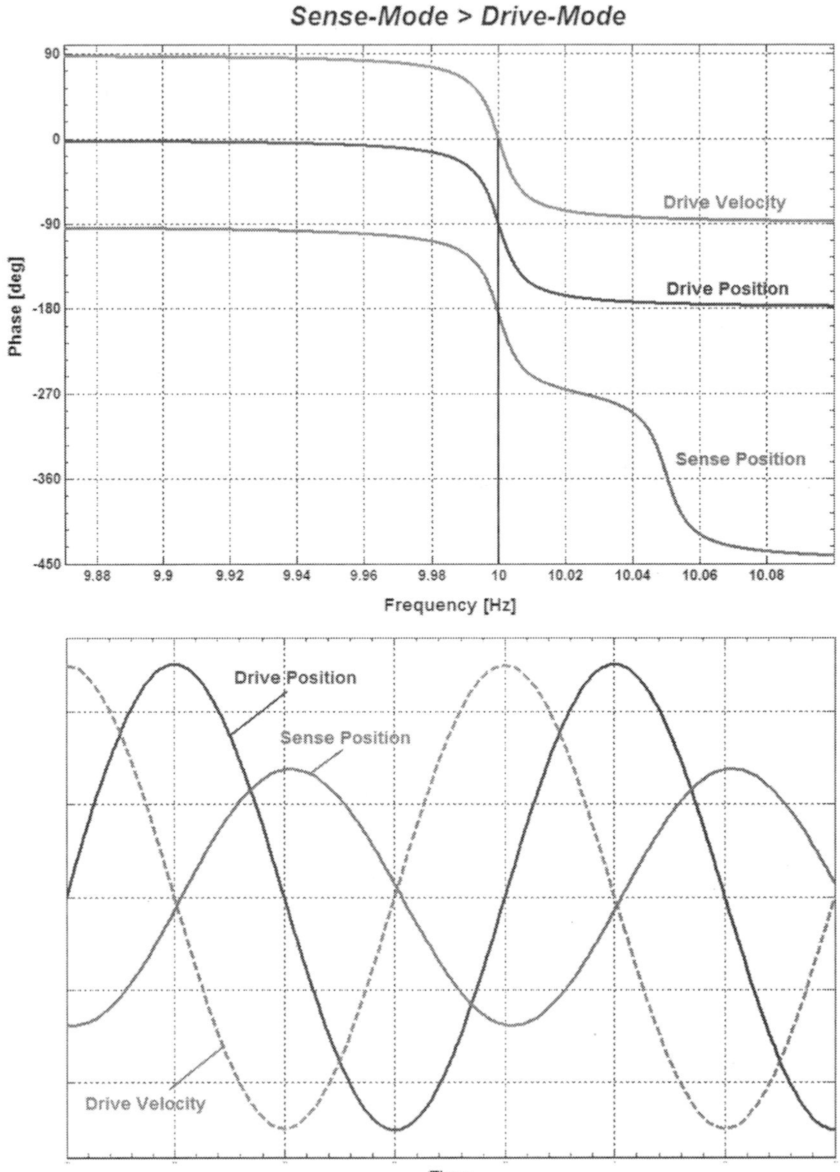

Fig. 2.15 The drive position, drive velocity and sense position phase relations when $\Delta f > 0$.

When the sense-mode is lower than drive, i.e. $\Delta f < 0$, the sense-mode phase converges to $-180°$ from the Coriolis force phase, which is $-270°$ from drive position (Figure 2.16).

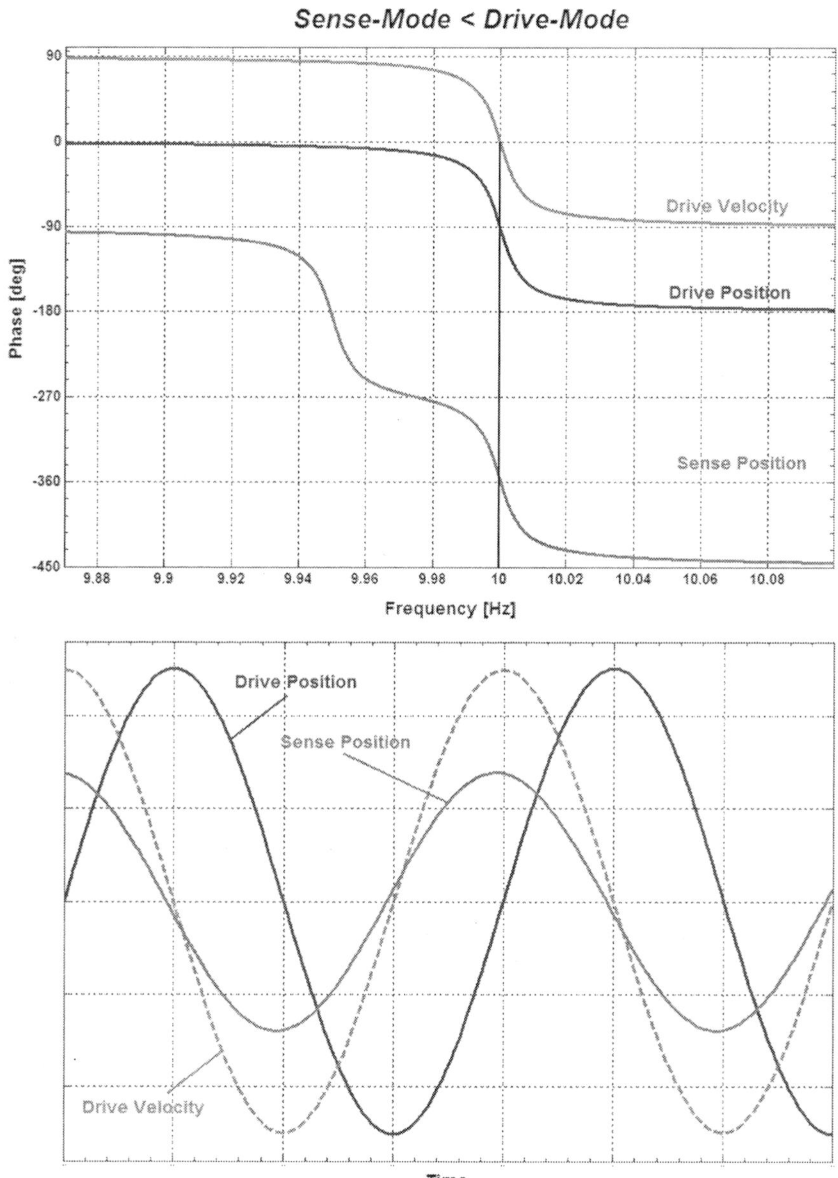

Fig. 2.16 The drive position, drive velocity and sense position phase relations when $\Delta f < 0$.

It should be noticed that the phase transition when the modes are mismatched is heavily dependent on the sense-mode damping. If the sense quality factor Q_s is high, the phase transition happens abruptly. If Q_s is low, the phase transition happens over a wider frequency range (Fig. 2.17). Thus, the sense-mode phase becomes more and more dependent on damping and the frequency separation Δf.

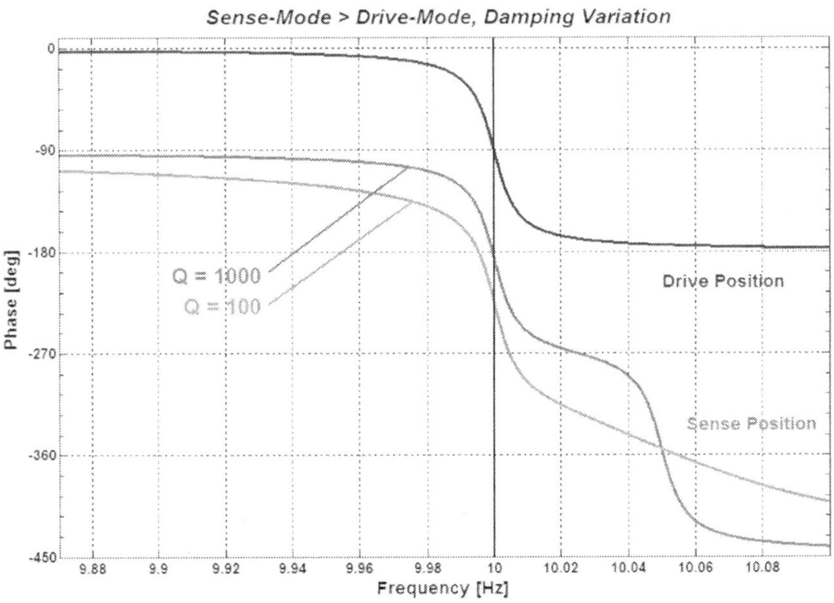

Fig. 2.17 For lower Q factor values, the sense-mode phase becomes more sensitive to damping and frequency variations.

Usually the rate output of the gyroscope is obtained by synchronous demodulation of the Coriolis response signal at the operating frequency, using the drive signal as a reference. Thus, the stability of the phase difference becomes extremely critical for the stability of the gyroscope output signal.

The oscillation pattern (or the trajectory) of the proof-mass also depends on the phase difference between the drive and sense direction positions.

If the phase difference $\phi_s - \phi_d = -180°$ as in the mode-matched case, then the drive and sense positions reach their extremes at the same time, and the steady-state trajectory is a straight line (Fig. 2.18). The slope of the straight line is determined by the Coriolis force, and thus the input rotation. Since y_0 is proportional to the excitation Coriolis force F_C, which is also proportional to the drive amplitude x_0, the slope of the straight line of oscillation does not depend on drive direction oscillation amplitude.

If the phase difference $\phi_s - \phi_d \neq -180°$, then the trajectory turns into an ellipse, which means y does not have its maximum at the time x is at its extreme points (Fig. 2.19). Thus, the oscillation pattern deviates from a straight line.

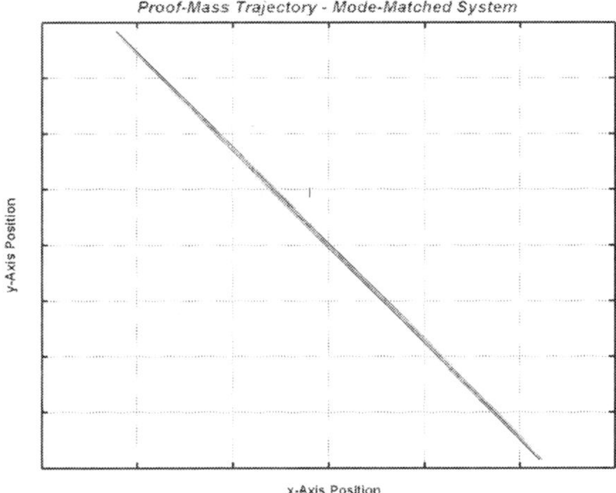

Fig. 2.18 The oscillation pattern when $\omega_d = \omega_s$.

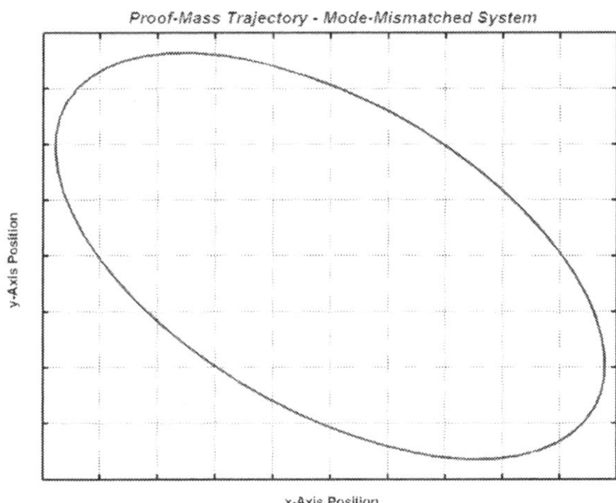

Fig. 2.19 The oscillation pattern when $\omega_d \neq \omega_s$.

Since the phase difference $\phi_s - \phi_d$ deviates from $-180°$ as the mode-mismatch Δf increases, the ellipticity of the oscillation pattern increases with increasing Δf (Fig. 2.20).

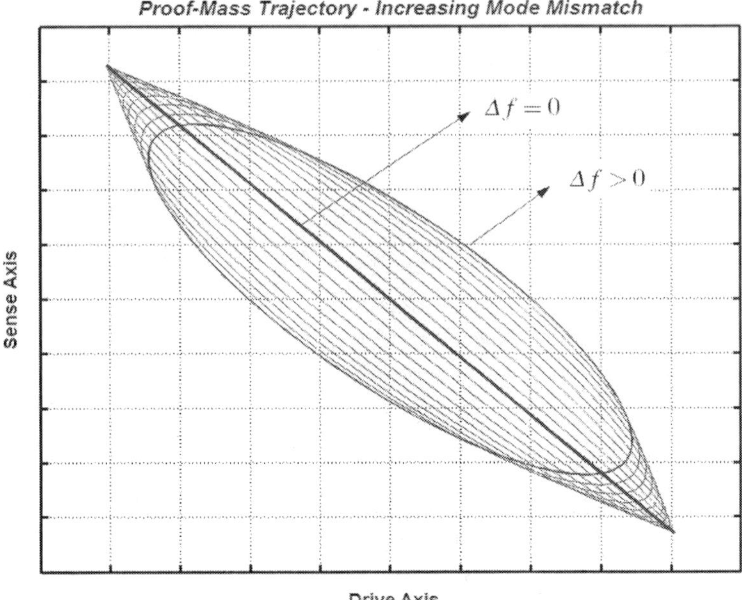

Fig. 2.20 The ellipticity of the straight line of oscillation grows as the driving frequency ω_d deviates from the sense direction resonant frequency ω_s.

2.5 Summary

In this chapter, the dynamics of a generic vibratory rate gyroscope was developed, and the principle of operation of a MEMS implementation was discussed. Explaining the basics of the drive and sense-mode oscillators, it was illustrated that the performance of the gyroscope is very sensitive to variations in system parameters that shift the drive or sense resonant frequencies. Under high quality factor conditions the gain is high, however, the bandwidth is extremely narrow, which also results in abrupt phase changes. In addition, the gain and the phases are affected significantly by fluctuations in damping conditions. To achieve a stable response, the drive and sense mode resonant frequencies are usually separated by design.

Chapter 3
Fabrication Technologies

In this chapter, we give a brief summary of fundamental micro-fabrication processes and methods commonly used in inertial sensor production. We then present an overview of representative process flows applicable to inertial sensors. We finally discuss electronics integration and packaging challenges.

3.1 Microfabrication Techniques

Inertial sensors require moving parts to detect inertial phenomena. Micromachining technologies have revolutionized inertial sensing by allowing to fabricate moving mechanical systems at the micro scale.

Integrated circuit (IC) microfabrication has been the primary enabling technology for MEMS. Originated from semiconductor fabrication techniques, micromachining technologies have made it possible to merge micro-scale mechanical and electrical components. Over the last decades, many techniques and process flows were explored to generate free-standing movable mechanisms with micro-scale features. The most significant capabilities common to all micromachining technologies are dramatically miniaturizing complex electro-mechanical systems and batch processing hundreds to thousands of devices on a single wafer.

The essence of all micromachining techniques is successive patterning of thin structural layers on a substrate. Photolithography based pattern transfer methods adapted from standard IC fabrication processes are the fundamental technologies behind micromachining. The following is a brief overview of basic micromachining techniques, used as common building blocks in every microfabrication process.

3.1.1 Photolithography

Identical to almost every CMOS integrated circuit fabrication process, micromachining processes rely exclusively on photolithography to transfer the desired patterns into the structural material layers. The fundamental principle behind photolithography, also known as lithography, is defining patterns that will be deposited, removed or doped on the wafer, by precisely patterning a light sensitive film. Prior to processing, a photolithography mask that carries the wafer-level layout of a layer is generated. Then the image on the mask is projected onto a photosensitive material deposited on the wafer, commonly known as *photoresist*.

A widely used method to deposit the photoresist on wafers is *spinning* or *spin coating*. In the spinning process, a sufficient amount of photoresist is dispensed at the center of the wafer, and the wafer is spun with a controlled speed, acceleration and duration. The thickness of the resulting photoresist layer primarily depends on the spinning speed and the photoresist viscosity. Photoresist suppliers usually provide a spin rate versus thickness chart to determine the spinning parameters for a desired thickness. Prior to photoresist coating, the wafer is cleaned using common techniques such as RCA cleaning (developed at RCA laboratories) and piranha cleaning, and dehydrated by baking to remove organic compounds, contaminants and moisture. To achieve better photoresist adhesion, primers or adhesion promoters such as hexamethyldisilazane (HMDS) can also be used. Following the photoresist spin-coat, a soft bake is performed to evaporate solvents and improve the adhesion and stress qualities of the resist film.

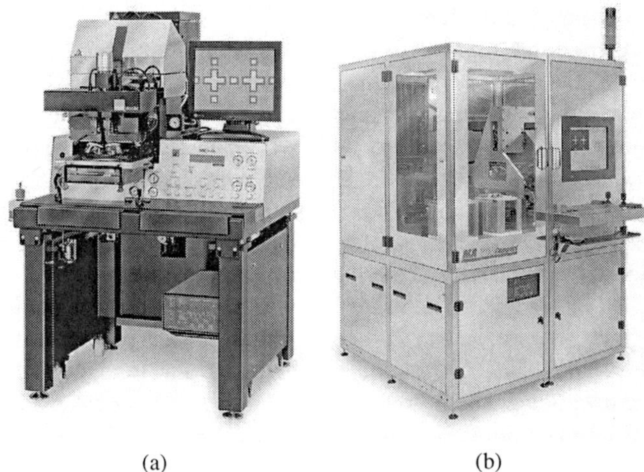

(a)   (b)

Fig. 3.1 (a) MA6 by Suss MicroTec is a common manual alignment and exposure tool for research environment. (b) MA200 by Suss MicroTec is a production cassette-to-cassette alignment and exposure tool with high throughput [128]. Courtesy of Suss MicroTec.

When the photoresist coating on the wafer is complete, the wafer is exposed to UV light through the photolithography mask. The photoresist areas that are beneath the transparent areas of the mask are exposed to light, altering the polymer structure. The areas beneath the dark areas of the mask are not exposed. For positive photoresists, the exposed areas become soluble in developer. Negative photoresists are initially soluble, and the exposed areas become insoluble.

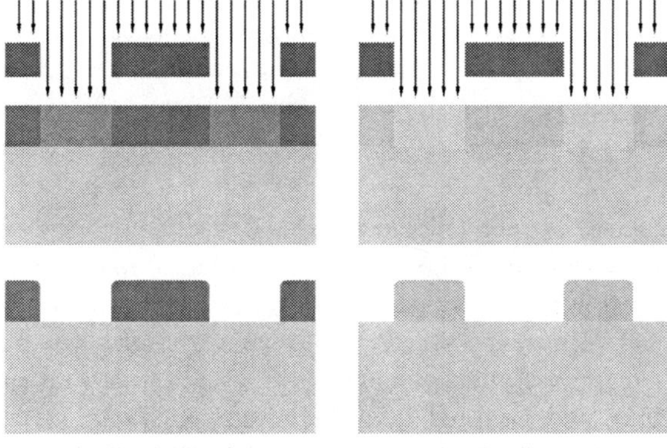

Fig. 3.2 Photolithographical patterning of positive and negative photoresists by exposure and developing.

Most fabrication processes involve multiple patterned layers. Thus, the mask has to be aligned precisely with respect to existing patterns on the wafer during exposure. Modern aligners as in Figure 3.1 feature top and bottom side alignment which allows to align and expose both sides of the wafer, and even infrared alignment to track patterns within wafer stacks. The aligner also sets the spacing between the mask and the photoresist surface. Contact printing, where the mask is in full contact with the resist layer, provides the best level of resolution. To reduce mask wear and photoresist defect problems, proximity printing is used, which places the mask 20μm or more above the photoresist. The resolution is lower due to diffraction, but the mask life is improved. Projection printing utilizes objective lenses between the mask and the wafer to combine advantages of non-contact printing with higher resolution.

The exposure step is followed by development, where the wafer is immersed in a developer solution. Exposed areas of positive photoresists are dissolved by the developer, and the remaining photoresist pattern is identical to the mask pattern. For negative photoresists, the unexposed areas are dissolved, resulting in the negative image of the mask (Figure 3.2).

3.1.2 Deposition

Micromachined devices employ patterned thin films as structural materials. Most process flows start with a blank wafer, and use successive deposition and patterning of multiple structural layers. These layers can be intended for moving structures, interconnect or electrode areas, or dielectric layers for electrical isolation. Depending on the material, layer thickness, or conformal coverage requirements, different deposition techniques such as Chemical Vapor Deposition (CVD), Physical Vapor Deposition (PVD), or electroplating may be used.

3.1.2.1 Silicon

Due to its excellent electrical and mechanical properties, silicon is the primary structural material in both the semiconductor industry and MEMS. Especially in micromachined gyroscopes which require constant oscillation, silicon is the ideal material. Silicon is perfectly brittle and experiences no plastic deformation. This provides exceptional mechanical stability, with no hysteresis and negligible energy dissipation, and minimal fatigue and creep. Silicon is used in MEMS devices in either single crystal form as a complete bonded layer, or in polycrystalline form deposited on a substrate.

Polysilicon, or polycrystalline silicon is one of the most common structural materials used in MEMS, especially in surface micromachining. It can be used either as the moving structural layers, or conductive interconnect layers in inertial sensors. Most modern microfabrication processes utilize Low-Pressure Chemical Vapor Deposition (LPCVD) for polysilicon deposition, which provides improved film uniformity across the wafer. LPCVD process thermally decomposes silane (SiH_4) under vacuum and high temperature to form polysilicon. Film thickness is usually limited to 2μm. To deposit *in situ* doped polysilicon, dopants such as phosphine or arsine can be added in the LPCVD process. When polysilicon is intended to be used as a released structural layer, annealing usually follows LPCVD to reduce residual stresses.

Epitaxial growth of polysilicon is commonly used to achieve thicker polysilicon layers than LPCVD, or better control of doping concentrations. Epitaxy can be used to thicken existing thin polysilicon layers, to as much as over 10μm. When the epitaxially grown film is the same material as the starting layer material, the process is called homoepitaxy. Epitaxy allows to grow crystalline structures as well. In homoepitaxy, the deposited film follows the identical lattice structure and orientation identical to the seed crystal. Thus, it also becomes a powerful tool to thicken single-crystal silicon layers.

3.1.2.2 Silicon Nitride and Silicon Oxide

LPCVD is commonly used to deposit the most common dielectric materials in MEMS: Silicon Nitride and Silicon Oxide. Both materials are known to provide excellent electrical isolation. They can also be used as intermediate masking layers, called hard-masks, which can withstand harsh etching conditions that deteriorate photoresist. Silicon Nitride can also be deposited using Plasma Enhanced Chemical Vapor Deposition (PECVD) at substantially lower temperatures, providing better compatibility with CMOS processes.

Another popular Silicon Oxide deposition technique is thermal growth, which is based on oxidizing an existing silicon layer surface at high temperatures in an oxygen environment. During thermal oxide growth, 0.45 μm silicon is consumed for 1 μm grown oxide. This method is convenient especially for growing conformal electrical isolation layers on blank or patterned silicon substrates.

3.1.2.3 Metals

Various metals such as Gold, Aluminum, Nickel, Silver, Chromium, or Titanium are often used in microfabrication processes. In inertial sensors, metals are primarily used for interconnection lines or electrodes. Some fabrication technologies also utilize metals as structural layers, even though metals lack some of the superior mechanical properties of silicon.

Many processes are available for metal deposition, or metallization, such as sputtering, evaporation, e-beam evaporation, CVD and electroplating. For thin films, sputtering is the most common technique. Sputtering is a Physical Vapor Deposition (PVD) process, in which a target or the source material is bombarded by high-energy ions, and the atoms sputtered from the target condense on the substrate, forming a thin metal film.

In sputtering, the material changes phase to vapor by a mechanical process. This allows to sputter virtually any material, including very high melting point materials, and even non-conductive materials. Sputtering also provides excellent adhesion and uniformity. Since directionality of the sputtering process is low, it is not ideal for shadow-masking and lift-off processes. In evaporation, the source material is transformed into the gaseous phase thermally or by an electron beam in high vacuum, and then condense on the substrate to form a thin uniform coating.

For thicker metal deposition, electroplating or electrodeposition is the most common process. Usually a thin continuous seed layer of chromium and gold or titanium is deposited on the whole wafer. The wafer and the source metal are placed in an electrolyte bath that allows flow of current, and a voltage is applied. The metal ions are dissolved from the source metal, transferred through the electrolyte, and deposited on the wafer. When electrical contact to the deposition areas on the wafer is not possible, electroless deposition is used, which utilizes a more specific electrolyte solution and no external current source. High aspect ratio structures can be obtained by electroplating metals into a patterned mold, usually made of photoresist or other sacrificial materials that are selectively dissolved after the electroplating process. The Lithographie Galvanoformung Abformung (LIGA) process is an excellent example, which is one of the pioneering high aspect ratio fabrication methods that provided external lateral precision.

3.1.3 Etching

Etching is the fundamental subtractive technique to transfer a photoresist or mask pattern into structural layers. In almost all fabrication technologies, a photoresist or hard mask layer is patterned by photolithography, and etching transfers this pattern into the actual structural materials and defines the geometry of the device by selective material removal.

Traditionally, etching has been divided into two primary categories: wet etching and dry etching. As the name implies, wet etching uses a liquid chemical solution. In dry etching, either a vapor phase etchant or reactive ions are used. In MEMS, many factors such as the desired sidewall and bottom surface profiles, isotropy, or stiction issues determine the required etching method.

3.1.3.1 Wet Etching

Wet etching has been heavily used in MEMS for decades due to its simplicity and low cost. In wet etching, after a masking layer is coated on the wafer and patterned, the wafer is immersed into an etchant solution specific to the material to be etched. The masking material and etchant have to be chosen to provide good selectivity, and to be compatible with other materials on the wafer. The exposed areas that are not covered by masking are attacked and etched by the solution. The etching process occurs first by the diffusion of the etchant onto the exposed material, then the chemical reaction between the etchant and the material, and finally the diffusion of the byproducts from the wafer into the solution.

Wet etching of polysilicon, silicon oxide, silicon nitride and metals are isotropic, meaning that the etch rate is the same in all directions. However, wet etching of crystalline structures such as single crystal silicon or quartz can be either isotropic or anisotropic. In anisotropic etching, the etch rate depends on the crystallographic orientation.

For single crystal silicon, Tetramethylammonium hydroxide (TMAH) and potassium hydroxide (KOH) are the most common anisotropic etchants. With KOH, the $<111>$ planes etch up to 400 times slower than the $<100>$ planes. This effect is usually exploited to achieve three dimensional structures. The most prominent example is anisotropic etching of (100) silicon wafers. The etch profile is bounded by the $<111>$ planes, resulting in a pyramid shape cross-section with 54.7° sidewalls. Continued etching results in a V-groove where two $<111>$ planes intersect at the bottom. Arbitrary mask patterns ultimately etch into a rectangular top-view, under-

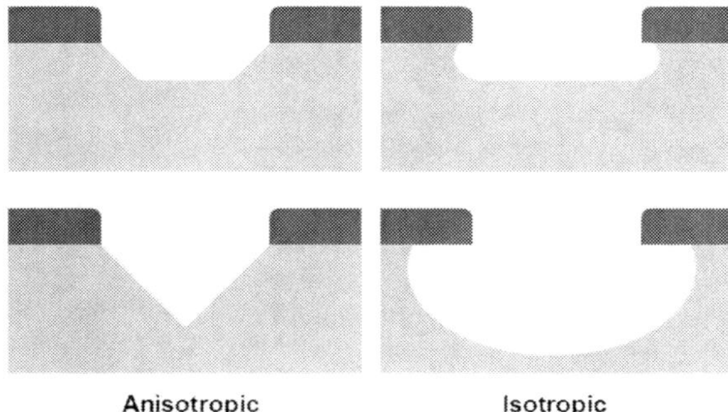

Fig. 3.3 Anisotropic and isotropic etching of (100) Silicon. The anisotropic etch yields 54.7° sidewalls bounded by the <111> planes, while isotropic etch rate is identical in all directions.

cutting the mask until bounded by the intersecting <111> planes. Timed etch of wide openings provide a perfectly flat bottom, allowing to form accurate shallow cavities.

Wet etching also provides high selectivity for doped silicon regions. Since the etch rate of heavily p-type doped silicon is much lower than n-type and lightly doped p-type, boron doping has been widely used as an etch stop, or to define suspended structures.

3.1.3.2 Dry Etching

Several dry etching tools and methods are available for isotropic and anisotropic etching. Dry etching methods utilize physical etch, chemical etch or a combination of both. Regardless of the process, the primary advantages of dry etching are flexibility in etch profile parameters, and avoiding the surface tension effects of liquid etchants that often cause stiction.

Reactive Ion Etching (RIE)
RIE is among the most common and versatile dry etching methods. It combines isotropic chemical etch with directional physical etch. The process is performed in a reactor in which an RF power source generates a plasma that converts a gas mixture into ions. The chemical etch occurs when the ions reacts with the wafer surface. Usually the ions are also accelerated towards the wafer with high energy to break atoms out of the surface material in a directional manner. This is the physical etch of RIE, and it determines the degree of etch anisotropy. The sidewall profile is adjusted by tuning the parameters of the chemical and physical etch.

Deep Reactive Ion Etching (DRIE)
DRIE is the process that revolutionized the fabrication of micromachined inertial devices. Allowing to etch deep and high aspect ratio trenches with vertical sidewalls, DRIE enabled complex and advanced inertial sensor designs. Especially for capacitive micromachined gyroscopes, DRIE became the ideal process by providing

optimal suspension beam geometries that suppress undesired modes and electrodes with large overlap areas and small gaps.

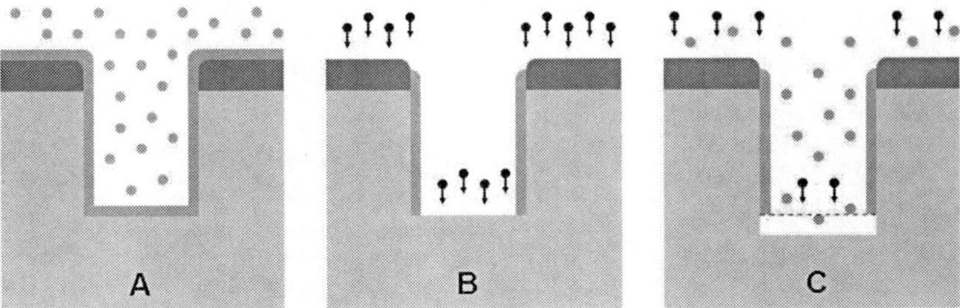

Fig. 3.4 The typical DRIE cycles. A- The C_4F_8 plasma deposits polymer passivation layer. B- The accelerated ions directionally remove passivation at the base. C- SF_6 plasma etches the exposed silicon areas.

The DRIE technology originated from the Bosch process developed in the late 1980's. It is essentially based on repetitive cycles of time-multiplexed etching in an inductively-coupled plasma (ICP) chamber. In one cycle, the DRIE tool alternates between a plasma etch step and a passivation step that protects the sidewalls [134].

The modern DRIE tools utilize three distinct phases: In the passivation cycle, C_4F_8 plasma in the chamber deposits a protective polymer conformally on all surfaces. In the etch cycle, first the accelerated ions directionally remove the passivation layer from the base of the trenches, but not the sidewalls. Then SF_6 plasma in the chamber attacks the exposed silicon areas at the base of the trenches. The passivation and etch cycles are repeated until the desired etch depth is achieved.

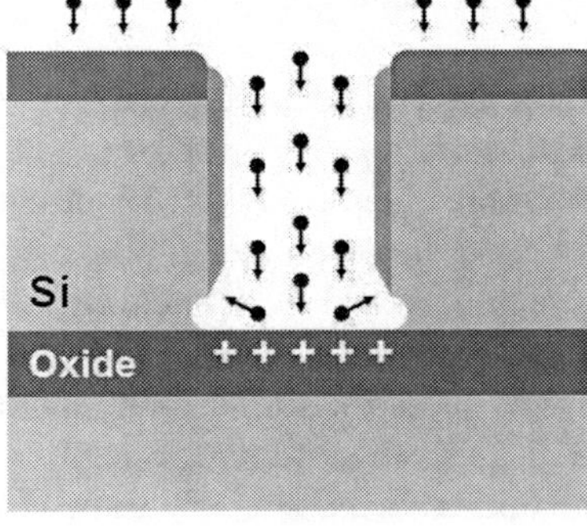

Fig. 3.5 Formation of the notching effect in DRIE. The incident ions are deflected due to the surface charge accumulated on the oxide layer.

In recent years, DRIE tools evolved to minimize many process dependent defects that impact performance of inertial sensors. One of the important issues is the notching effect observed on SOI wafers. Notching occurs due to the surface charge accumulated on the oxide interface. The charge deflects incident ions towards the corners, causing sidewall passivation breakdown and thus lateral etching of silicon at the bottom of the trenches. An effective method to minimize notching is to reduce charging of oxide by using a time modulated platen RF to allow charge dissipation.

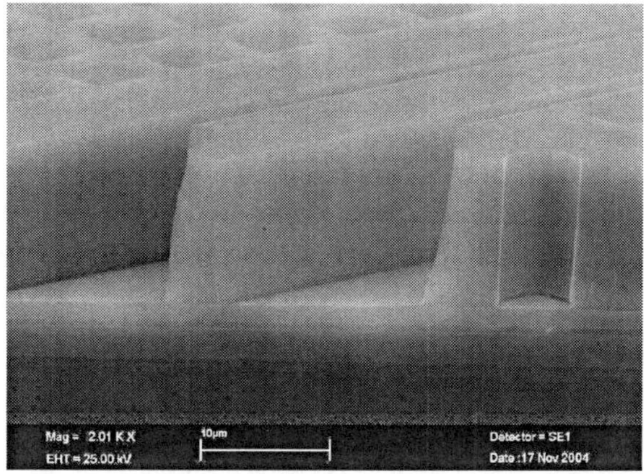

Fig. 3.6 SEM of an SOI device etched with notching-compensated DRIE process. Courtesy of STS.

3.1.4 Wafer Bonding

Wafer bonding is among the core technologies that enabled bulk micromachining processes and wafer-level packaging. Wafer bonding allows to incorporate thick and even single crystal structural layers into the process flow, not possible with any deposition process. Especially for capacitive inertial sensors, this capability dramatically enhances the overall device performance.

Several wafer bonding techniques have been developed over the years that allow bonding of various wafer types. The following is a summary of important bonding techniques widely used in inertial sensors.

3.1.4.1 Anodic Bonding

One of the oldest and most common bonding methods, anodic bonding allows to bond silicon and glass. It is based on simultaneously utilizing temperature and electrostatic attraction to realize the bond. Usually very large voltages in the order of kV's are applied, resulting in large electrostatic forces between the wafers. This allows relatively low bonding temperatures, usually from 200°C to 500°C. Glass wafers such as Pyrex, which offer thermal expansion coefficients close to silicon are commonly used in this process. Anodic bonding is known to provide good hermeticity, making it suitable for wafer-level packaging.

3.1.4.2 Silicon Fusion Bonding

Fusion bonding is a high-temperature method used to bond two silicon wafers. It is also called direct silicon bonding, since no intermediate bonding layers are required. The silicon wafers are first pre-bonded at room temperature empoying van der Waals forces. Thus, excellent surface quality on both wafers and surface hydration is required for maximizing molecular attraction to achieve a good pre-bond. The wafers are then annealed at high temperatures, usually over 1000°C to form the chemical bond. Fusion bonding also allows bonding of oxidized silicon wafers, which could be utilized as electrical isolation. For low temperature silicon fusion binding, plasma activation systems could be utilized to realize plasma-enhanced fusion bonding, which allows lower stresses and CMOS integration capability.

3.1.4.3 Eutectic Bonding

Eutectic bonding typically utilizes intermediate bonding layers deposited on the two wafers, so it is a quite versatile bonding process. Similar to a soldering process, a eutectic combination of two materials is employed, which has lower melting temperature than both materials. Several eutectic combinations are used, but the most common in MEMS processes are gold/tin and silicon/gold. In the bonding process, the two materials are brought to contact, and the temperature is raised slightly over the eutectic temperature. By diffusion, eutectic composition forms and grows, while the eutectic compound liquidifies. The temperature is then reduced slowly, solidifying the bond. Since the eutectic compound is reflowed, a greatly continuous seal is achieved, providing excellent hermeticity.

3.2 Bulk Micromachining Processes

The basic process building blocks presented above can be combined in many ways to attain moving micro-electro-mechanical systems. Depending on the structural layer forming technique, micromachining processes are usually divided into two major categories: Surface micromachining and Bulk micromachining. As the micromachining process demands evolve, modern process flows adapt characteristics of both categories, and the distinction becomes more vague.

Bulk micromachining traditionally implies the use of subtractive processes to pattern thick structural layers. In most bulk micromachining processes, two or more wafers are bonded, and the moving structures are made out of the whole thickness of a silicon wafer.

For many decades, bulk micromachining relied on wet etching, which pioneered the fabrication of inertial sensors. The major limitation of wet anisotropic etching is that the resulting geometry primarily depends on the internal crystalline structure, and fabricating complex free-form geometry is not possible. Development of the DRIE process in recent years provided design flexibility unimaginable with wet etching, and dramatically accelerated the development of bulk micromachined inertial sensors.

Since bulk micromachining provides thick structural layers, it offers many advantages for inertial micromachined devices. Larger device thickness increases the mass and overlap area of capacitive electrodes, directly improving gyroscope performance. Thicker suspension beams provide higher out-of-plane stiffness, which reduces shock and vibration susceptibility, and minimizes the risk of stiction to the substrate. It also allows the use of single crystal silicon as the device material, which provides excellent mechanical stability.

Bulk micromachining can be implemented in many different fabrication technologies. Below, we present simple representative process flows that illustrate the fundamentals of inertial sensor implementation in bulk micromachining.

3.2.1 SOI-Based Bulk Micromachining

Silicon-on-Insulator (SOI) wafers are excellent starting materials for bulk micromachining. The silicon device layer comes bonded on an insulator layer. Simply by patterning the device layer and the oxide layer underneath, electrically isolated and mechanically anchored free-standing structures can be formed. SOI wafers with a wide range of device, oxide and handle layer thicknesses and conductivities are readily available that could fit into any specific device design.

The simplest form of an SOI-based bulk micromachining process is a one-mask process. The process relies on deep-reactive ion etching (DRIE) through the whole thickness of the device layer of an SOI wafer, and front-side release of the structures by etching the Oxide layer in HF solution. To define the suspended areas, etch holes have to be used to perforate the device layer. The reason for the perforations is to allow access of the HF solution to the oxide layer through the device layer. The oxide layer underneath the areas without perforation is not exposed to HF and remains intact to serve as mechanical anchors. The process flow overview is outlined in Figure 3.7, and a prototype gyroscope fabricated in this process is presented in Figure 3.8.

In this example process, $15\mu m \times 15\mu m$ holes were used to define suspended areas, and $10\mu m$ gaps were used in the sensing and actuation electrodes. The lowest etch rates were observed for the $15\mu m \times 15\mu m$ holes, at approximately $1.25\mu m$/min, and 85min DRIE time was used to assure complete through-etch of the $100\mu m$ thick device layer, while minimizing excessive undercutting in larger areas. The anchors were designed as unperforated areas larger than $40\mu m \times 40\mu m$ for 25 min release in 49% HF solution. The details of process flow are given next as an example recipe:

- The process starts with a Silicon-on-Insulator (SOI) wafer with a $4\mu m$ oxide layer between a $100\mu m$ thick single-crystalline silicon device layer with (100) crystal orientation, and a $400\mu m$ e-crystalline silicon handle layer. In order to eliminate the organic residue and film on the device layer, RCA cleaning is performed, which decontaminates the wafer by sequential oxidative desorbtion. A 1 hour dehydration step follows cleaning to remove moisture in the device layer.

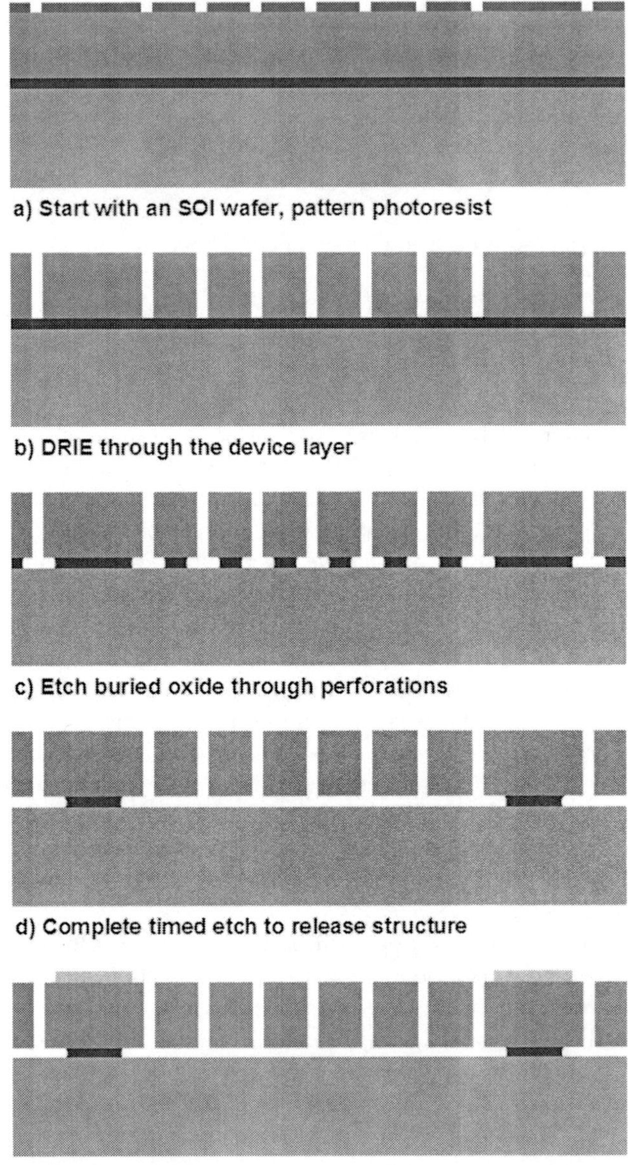

Fig. 3.7 The process flow of a metallized one-mask SOI-based bulk micromachining process.

- An adhesion promoter (HDMS) is spun on the device layer, and soft-baked for 1 min. Then, $7\mu m$ thick Shipley 1827 photoresist is spun with a spin time of 30 sec, spin rate of 1500 rpm, and an acceleration of 510 rpm/sec. The resist thickness is determined to be sufficient to withstand the deep-reactive ion etching (DRIE). For evaporating the photoresist solvents, samples are soft baked on a hot plate at 90°C for 1.5 minutes.
- The sample is exposed to UV-light in a Karl-Suss aligner, using a Chrome-Soda lime mask. The mask defines the features of the whole mechanical structure. For the $7\mu m$ thick Shipley 1827 photoresist, 35 sec exposure time is used.
- The exposed photoresist is developed using MF-319 photodeveloper for 1 minute, and rinsed with flowing DI water to stop developing. Then, the samples are hard baked in an oven at 90°C.

Fig. 3.8 An SOI-based bulk-micromachined gyroscope, diced and released.

- The deep-reactive ion etching (DRIE) step is performed in an STS Inductively Coupled Plasma (ICP) chamber, using 8sec etch step cycle with 130sccm SF_6 and 13sccm O_2, 600W coil power and 15W platen power; and 5sec passivation step cycle time with 85sccm C_4F_6, 600W coil power, and 0W platen power. The etching time was optimized to assure complete through-etching in the smallest trench areas, which are $15\mu m \times 15\mu m$ perforation holes in the proof-masses. 85min DRIE time was observed to suffice for through-etching in these holes, at approximately $1.25\mu m$/min etch rate.
- The photoresist layer is stripped after DRIE, using first a 10 minute acetone dip, and then oxygen plasma with a base pressure of 15 mTorr, oxygen pressure of 200 mTorr and 400 Watt power for 15 minutes.

- Finally, the sample is soaked in a 49% HF solution to etch the oxide layer underneath the perforated areas. The HF etchant can access the oxide layer only underneath the perforated areas through the $15\mu m \times 15\mu m$ perforation holes. For a 25min release, the unperforated areas larger than $40\mu m \times 40\mu m$ remain attached to the substrate, being utilized as anchors. Etching is stopped with a DI water dip and then isopropyl alcohol soak, which is immediately vaporized on a $100°C$ hot plate.

3.2.2 Silicon-on-Glass Bulk Micromachining

In the one-mask process presented above, the process simplicity also imposes limitations on the device design. The etching perforations required in the suspended areas could be undesirable due to reduced mass or parasitic modes. One way to eliminate this design constraint is to bond the device layer onto a substrate with pre-existing cavities underneath the suspended areas.

The basic silicon-on-glass (SOG) process is a very common technology to attain suspended devices above pre-etched cavities [127]. The process starts with a glass wafer. After cleaning, photoresist is coated on the wafer, and patterned using the cavity mask. The cavities in the glass wafer are etched. Then the silicon device layer wafer is anodic bonded to the glass substrate. Photoresist is coated on the device layer, and patterned using the device mask aligning to the cavity pattern on the substrate. The device layer is through-etched using DRIE. It is common to deposit metal in the cavities prior to wafer bonding, in order to dissipate charge accumulation on the glass surface during DRIE and minimize device layer undercut. To allow wirebonding, the bonding pads are metallized by shadow masking, which involves temporarily bonding a mask wafer with through-etched openings, and sputtering or evaporating metal through the mask wafer openings. The overall process flow is outlined in Figure 3.9.

Another method to minimize backscattered ions during DRIE other than cavity metallization is to use an SOI wafer for the device layer (Figure 3.10). The SOI wafer is coated with photoresist, and patterned using the device layer mask. Then the device layer is through-etched in the DRIE step. In parallel with preparation of the device layer SOI, the cavities in the glass substrate are etched. Then the device layer pattern is aligned with the substrate cavity pattern, and the two wafers are anodically bonded. The handle layer and the buried oxide layer are dissolved, exposing the device layer. That is the reason this type of process is also called the *dissolved wafer process*. Similarly, the bonding pads are metallized by shadow masking to complete the processes.

The same fundamental process can be realized using silicon substrates instead of glass. One of the primary process differences would be the use of a bonding technique compatible with two silicon wafers such as fusion bonding. The other difference is the need for an insulating layer between the two wafers, when a conductive silicon substrate is used. This can easily be achieved by thermally oxidizing

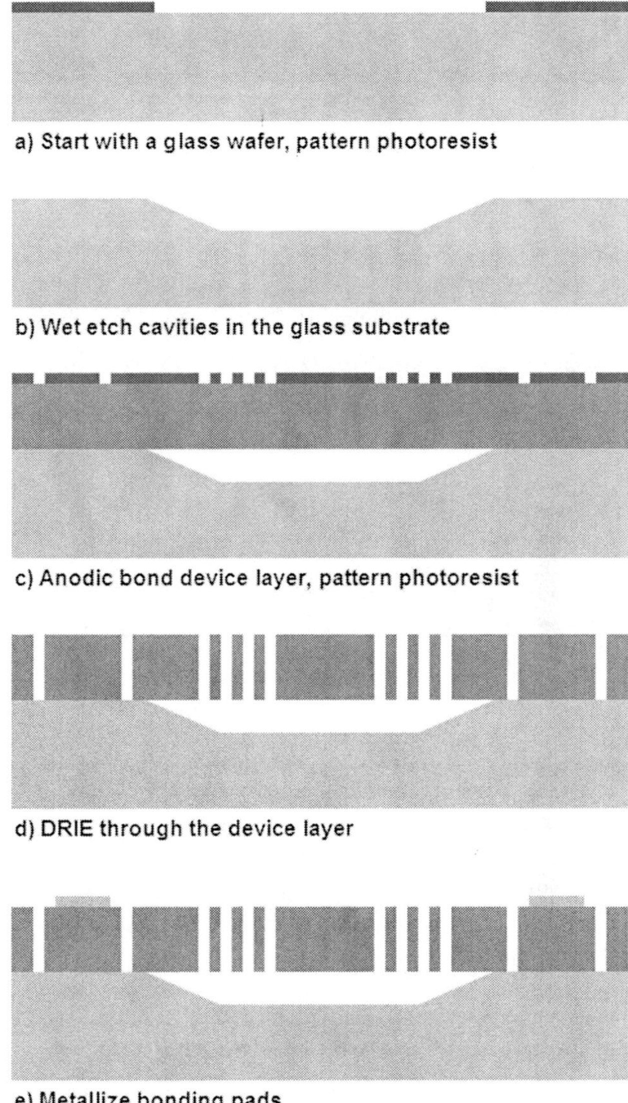

Fig. 3.9 The process flow of the basic Silicon-on-Glass (SOG) process.

the substrate surface. The major advantage in using silicon wafers as a substrate is perfect matching of thermal expansion coefficients between the device layer and the substrate, which enhances temperature stability of gyroscopes.

The illustrative example process flows presented above are representative of the basics in bulk micromachining. Many different bulk micromachining processes have

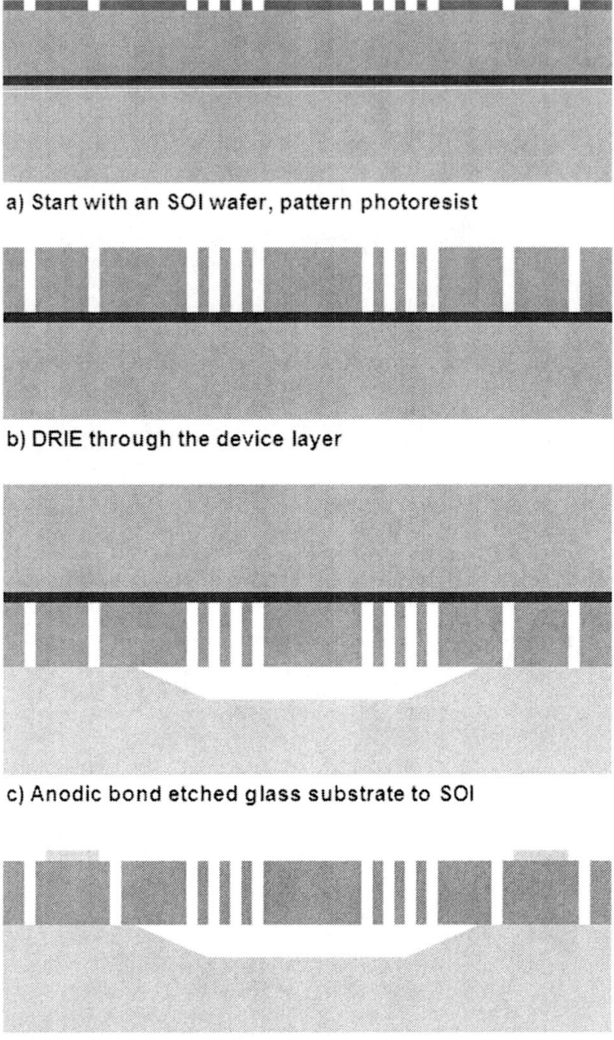

Fig. 3.10 Silicon-on-Glass process that utilizes SOI wafers to improve DRIE process.

been utilized for inertial sensors, using various process and material combinations, and at various functionality versus complexity and cost trade-off points.

3.3 Surface-Micromachining Processes

In contrast to subtractive bulk-micromachining processes, surface micromachining is in essence an additive technique. Surface micromachining relies on successive deposition and patterning of thin structural layers on the surface of a substrate, rather than etching thick bulk layers.

Development of surface micromachining started in the late 1980's to achieve a micromachining process compatible with CMOS fabrication technologies. Readily available in IC fabrication, polycrystalline silicon has been widely used as the primary mechanical structure material. The sacrificial layer technique has been the enabling technology to realize a movable free-standing structure out of a conformally deposited thin layer.

In surface micromachining, complex three-dimensional devices are built by depositing multiple stacks of alternating structural layers and sacrificial layers. Each sacrificial layer supports the structural layer above it during fabrication, and separates it from the other layers below. At the end of the process, the sacrificial layers are selectively etched away, releasing the structural layers.

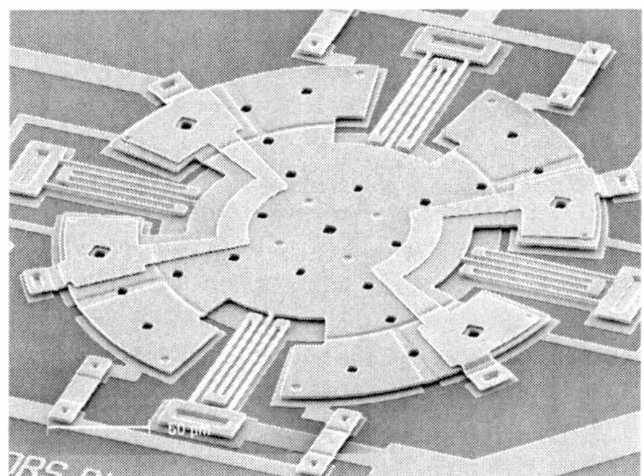

Fig. 3.11 Scanning electron microscope image of a torsional surface-micromachined gyroscope that utilizes three Polysilicon structural layers.

The most common surface micromachining processes use polysilicon as the structural material, LPCVD deposited phosphosilicate glass (PSG) oxide as the sacrificial layer, and silicon nitride as electrical isolation between the polysilicon and the substrate.

In a standard three polysilicon layer surface micromachining process, the moving parts of the device are formed using the second structural polysilicon layer (Poly1) or the third (Poly2). The electrical connections are formed using the first structural polysilicon layer (Poly0) deposited on the nitride-covered substrate.

Multi-user surface micromachining processes are commercially available. For example, PolyMUMPs is the surface micromachining Multi-User MEMS Processes (MUMPs) offered by MEMSCAP Inc. PolyMUMPs is a standard three-layer

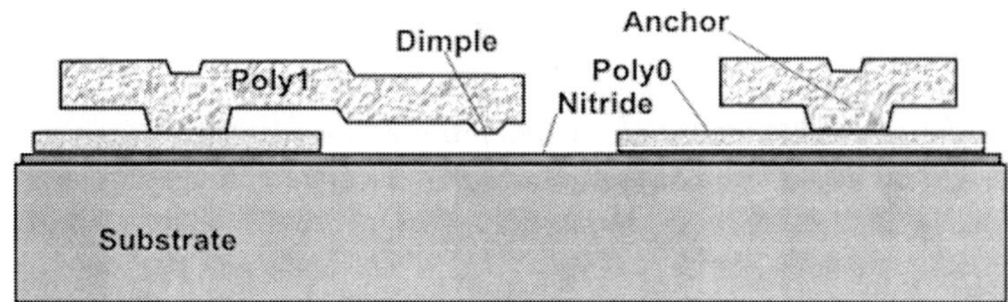

Fig. 3.12 Cross-section of a device fabricated using the first two structural layers of a general surface micromachining process [57].

polysilicon surface micromachining technology. The process flow can be summarized as follows:

The first step of the fabrication process is deposition of a 600 *nm* low-stress Silicon Nitride layer on the silicon n-type (100) wafers as an electrical isolation layer. This is followed directly by the deposition of the first structural polysilicon film, Poly0, which is 500 *nm* thick. Poly0 is then photolithographically patterned by etching in an RIE (Reactive Ion Etch) system. The remaining photoresist is stripped away.

A 2.0 *μm* phosphosilicate glass (PSG) sacrificial layer is then deposited by LPCVD. This sacrificial layer of PSG, known as the Oxide layer, is removed at the end of the process to free the first mechanical layer of polysilicon. The sacrificial layer is lithographically patterned with the dimples mask and the dimples are transferred into the sacrificial PSG layer by RIE. The wafers are then patterned with the third mask layer, the anchor mask, and reactive ion etched. This step provides anchor holes that will be filled by the second polysilicon layer (Poly1). After etching anchors, the second structural layer of polysilicon is deposited. This structural layer has a thickness of 2.0 *μm*, and the moving structures including the proof mass, suspension system and the capacitors are formed in this layer. The polysilicon is lithographically patterned using a mask designed to form the second structural layer Poly1. After etching the polysilicon, the photoresist is stripped.

The same procedure is followed to form the second sacrificial layer Oxide2 and the third structural layer Poly2. Throughout these steps, corresponding masks are used to form Anchor2 and Poly1-Poly2-Via structures. Finally, the wafer is diced, and the structures are released in HF Solution.

Device Implementation in Surface Micromachining

In PolyMUMPs, the proof masses and the flexures of the devices could be formed in Poly1 with a $2\mu m$ thickness. The mass values of the proof masses are calculated as $m = At\rho$, where the total footstep area A is extracted from the layout, t is the structural thickness, and $\rho = 2330 kg/m^3$ is the density of Polysilicon. Since the first sacrificial layer Oxide1, has a thickness of $2\mu m$, the masses are suspended over the substrate with a $2\mu m$ clearance. Interdigitated comb-drives structures, air-gap sense

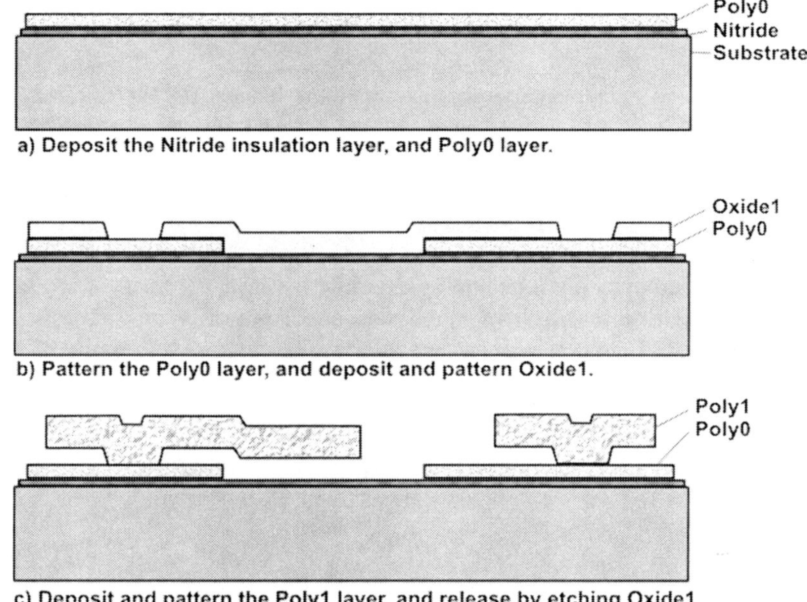

Fig. 3.13 The process flow of a standard surface-micromachining process.

capacitors, parallel-plate capacitors, are formed in Poly1, and are connected to the bonding-pads by the interconnect lines formed in Poly0. Out-of-plane actuation and detection electrodes underneath the Poly1 structure can also be formed in the Poly0 layer.

Fig. 3.14 Scanning Electron Microscope (SEM) photograph of (a) the comb-drives, and (b) the air-gap sense capacitors.

Utilizing Stacked Polysilicon Layers in Standard Surface-Micromachining Technologies

The reliability and sensitivity of micromachined inertial sensors such as accelerometers and gyroscopes are primarily determined by their structural properties. One of the major failure modes of surface-micromachined inertial sensors is the stiction of the proof mass to the substrate due to a shock applied on the device. The out-of-plane motion of the proof mass can be suppressed by increasing the out-of-plane stiffness of the suspension beams, which is proportional to the thickness cubed. Bulk micromachining technologies allow fabrication of very thick suspension beams resulting in orders of magnitude larger out-of-plane stiffness values, however structural thicknesses in surface micromachining are very limited.

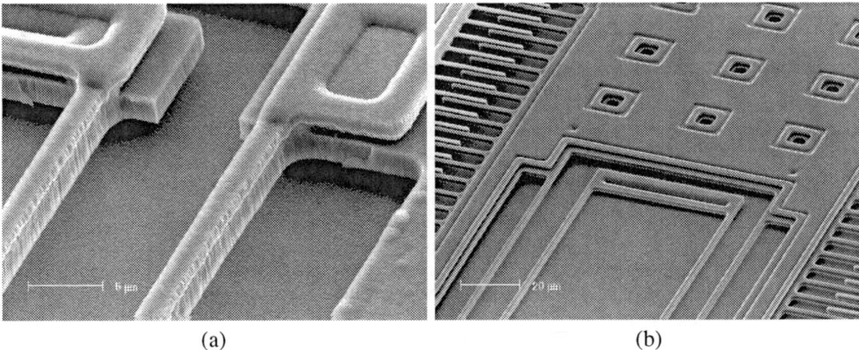

(a)   (b)

Fig. 3.15 SEM photographs of (a) the device with two-polysilicon-layer suspension beams, and (b) the device with proof mass made up of stacked two polysilicon layers and one trapped silicon oxide layer.

By utilizing stacked polysilicon layers, it is possible to achieve suspension beams with increased thickness in a standard surface-micromachining technology, even though the thickness of each structural layer is fixed. The same approach can be extended to increase the proof mass inertia of an inertial sensor fabricated in a standard surface-micromachining technology. It is often essential to have a large proof mass in inertial sensors such as accelerometers and gyroscopes for increasing sensitivity, since the detected phenomena is generally proportional to the proof mass of the device, which increases linearly with thickness. In order to increase the inertia of the proof mass without changing the footstep area, two or more polysilicon layers are stacked, together with the sacrificial silicon oxide layers trapped in between the layers. By designing an enclosed polysilicon via window on the proof mass, access of the etchant to the sacrificial oxide layer is prevented, preserving the oxide layer between the polysilicon layers for added mass.

3.4 Combined Surface-Bulk Micromachining

Many process flows have been reported to combine characteristics and advantages of both surface and bulk micromachining. An excellent example is the combined surface-bulk micromachining process by Bosch, developed for automotive gyroscopes (Figure 3.16).

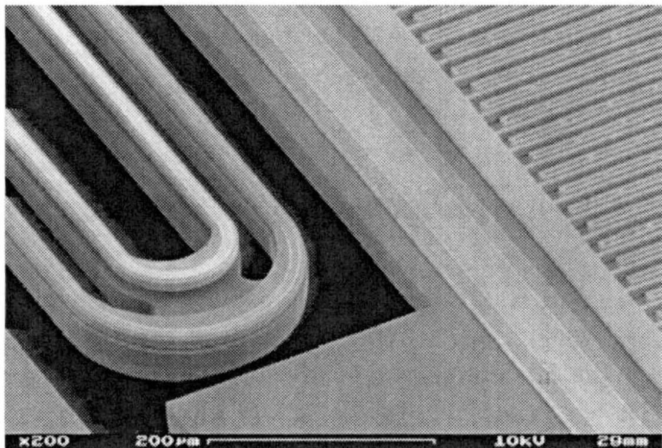

Fig. 3.16 SEM of gyroscope fabricated in the combined surface-bulk micromachining process by Bosch [135].

The process starts with a 150μm thick single-crystal silicon wafer. Both for electrical isolation and use as a sacrificial layer, 2.5μm thick thermal oxide layer is grown. Then a 12μm thick epitaxial polysilicon layer is grown, and aluminum interconnect layer is sputtered. The backside of the silicon wafer active area is thinned down to 50μm with a KOH wet etching. The polysilicon and single-crystal silicon device layers are through-etched by DRIE, and sacrificial oxide is vapor etched. Glass substrate and cap wafers are used to encapsulate the device (Figure 3.17).

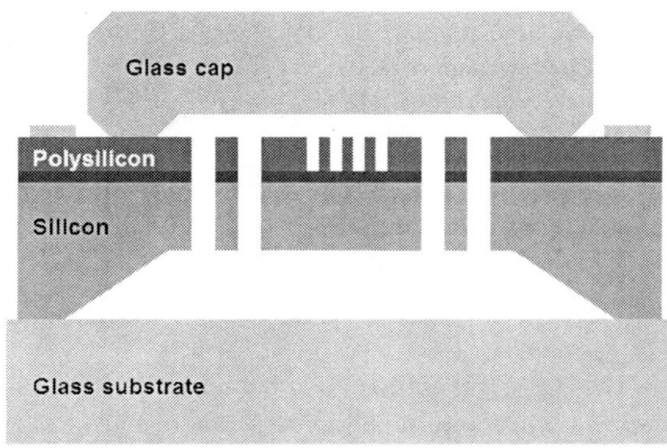

Fig. 3.17 Cross-section of the combined surface-bulk micromachining process by Bosch [135].

3.5 CMOS Integration

Inertial MEMS devices, and especially gyroscopes, require quite complicated electronics for control and signal conditioning. In stand-alone gyroscope units, all interface electronics are usually built into an Application-specific integrated circuit (ASIC), connected directly to the MEMS sensing element. Complementary Metal-Oxide Semiconductor (CMOS) technology and its derivatives have been extensively used for fabricating gyroscope interface ASICs. Since the signal levels between the gyroscope and the ASIC are extremely low, the integration method of these two components is of utmost importance.

3.5.1 Hybrid Integration

The most popular and least demanding approach for MEMS-ASIC integration is hybrid integration. In hybrid integration, the MEMS sensing element die and the ASIC die are fabricated completely separately, and brought together in a single package. The dice can either be stacked, or mounted side-by-side as in Figure 3.18.

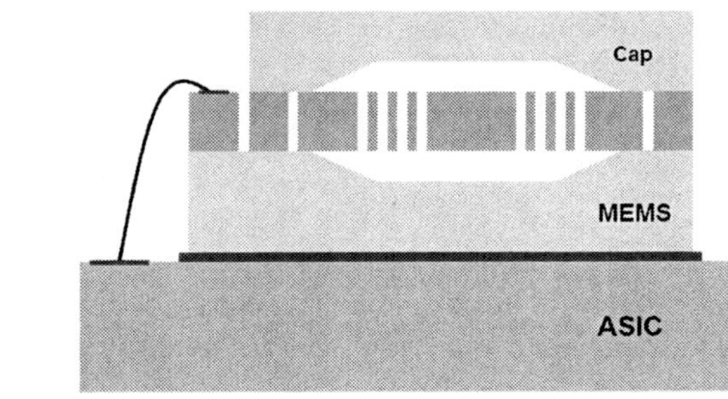

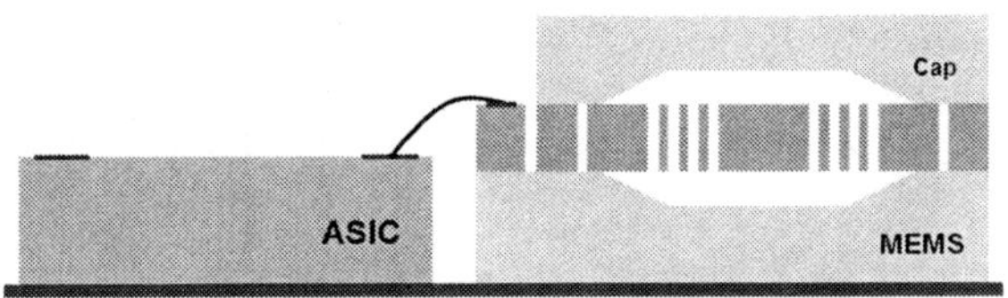

Fig. 3.18 Illustration of hybrid integration of the MEMS and ASIC dice, by die stacking and side-by-side mounting. In either case, the MEMS device is connected to the ASIC by direct wire-bonding.

The electrical connections between the MEMS and ASIC dice can be made either by direct wire-bonding or flip-chip ball-grid arrays. Wire bonding is the most versatile connection method, allowing the most flexibility in the design of MEMS and ASIC. For flip-chip BGA bonding, the bumped pad locations on the MEMS die has to match the pad locations on the ASIC die. However, this method provides less parasitics than wirebonding.

Hybrid integration has become the mainstream integration approach for bulk-micromachined devices. These devices usually provide an order of magnitude larger signal levels than surface-micromachined devices. Thus, parasitic capacitances at the ASIC interface becomes less of a concern.

3.5.2 Monolithic Integration

Integrating the MEMS sensing element with the electronics on the same die provides the lowest parasitics and best possible noise performance. Thus, development of fabrication processes that allow monolithic integration has been a major focus area in the MEMS community.

Surface micromachining technologies are inherently more compatible with CMOS. In integrated surface micromachining processes, very close integration with electronics compensates for the limitations of surface micromachining such as limited structural thickness and limited device area. Surface and bulk micromachining processes can also be combined to improve the MEMS device characteristics, while maintaining very low interconnection parasitics.

For monolithic integration, micromachining steps can be interjected into a CMOS process before (Pre-CMOS), during (Intra-CMOS) or after (Post-CMOS) the CMOS processing steps. Sandia National Laboratories IMEMS process is a Pre-CMOS process, in which the MEMS device is formed in a shallow trench on the substrate and completely covered with sacrificial oxide, only exposing polysilicon vias after planarization [129].

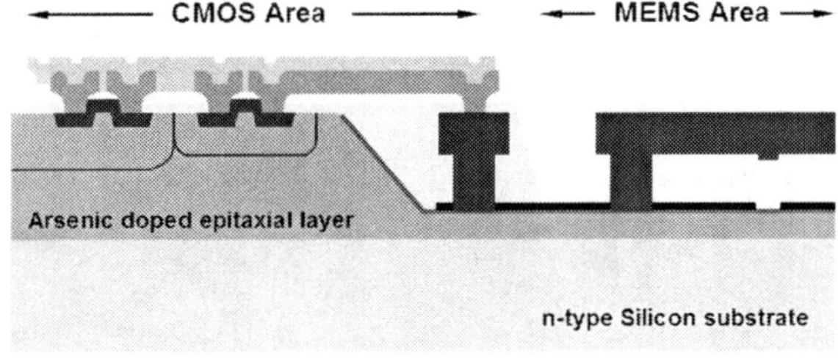

Fig. 3.19 Sandia National Laboratories IMEMS technology, a Pre-CMOS process integrated micromachining steps prior to CMOS [129].

The iMEMS process by Analog Devices is an Intra-CMOS process based on a standard BiCMOS process, where the device layer is deposited and patterned before metallization. Intra-CMOS processes require very tight compatibility of MEMS and CMOS process steps [130].

Pre-CMOS and Intra-CMOS processes are often difficult to realize in external ASIC foundries, and require a dedicated production line. Post-CMOS processes

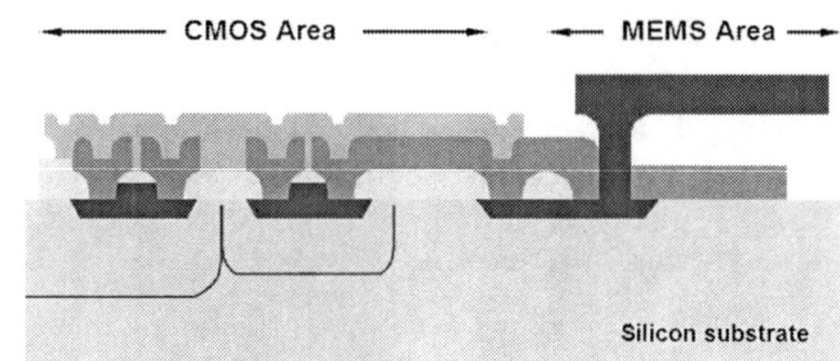

Fig. 3.20 Intra-CMOS iMEMS micromachining process by Analog Devices [130].

provide the most flexibility for ASIC foundry selection, while imposing temperature constraints on the micromachining steps, especially limiting new material deposition. Front-side thin-film etch and back-side DRIE processes developed at Carnegie Mellon University utilize existing CMOS layers to define micromachined structures in completed CMOS devices. An excellent review of monolithic integration methods are presented in detail by Xie and Fedder in [5].

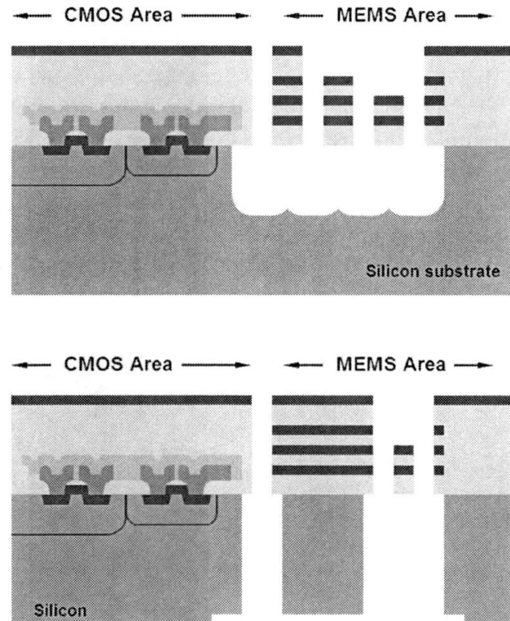

Fig. 3.21 Post-CMOS micromachining processes developed at CMU. The first process utilizes front-side etching, while the second process combines bulk-micromachining steps with CMOS by back-side DRIE to use CMOS layers as structural materials [5].

Regardless of the processing approach, monolithic integration technologies consume valuable die area on the CMOS devices. On the other hand, a two-chip hybrid integration approach allows more flexibility on the design and fabrication of both

MEMS and ASIC dice. Many factors have to be considered in the selection of electronics integration approach to optimize the cost and performance.

3.6 Packaging

The package is the interface of MEMS devices with the environment, and performs many critical functions. It provides the mechanical and electrical interface in the final application platform, as well as during test and calibration. Most importantly, it protects the MEMS device from environmental factors.

Packaging is one of the primary cost-drivers in the MEMS. Apart from cost, many technical design criteria have to be met, such as mechanical and thermal stresses, electrical characteristics, hermeticity, package resonances, heat flow, sensor orientation, and final board assembly compatibility.

Inertial sensors do not require interaction with the environment, unlike pressure sensors or optical MEMS devices. Thus, standard IC packaging technologies such as plastic, ceramic or metal have been successfully applied to inertial MEMS.

Plastic packaging is the lowest cost packaging method, but requires devices to be hermetically sealed at the wafer level. The sealed die is attached on a leadframe, wirebonded, and the thermoset plastic material is transfer-molded over the assembly. The molding pressure and the resulting packaging stresses are known to cause problems in stress-sensitive inertial sensors. Pre-molded cavity packages or gel die-coatings have been developed to alleviate stress issues in plastic packages.

Fig. 3.22 A micromachined gyroscope vacuum-packaged in a ceramic quad flat package (CQFP) with J-leads. The ceramic package is sealed under vacuum using a kovar lid and a fluxless solder preform.

Ceramic packages provide excellent mechanical and electrical characteristics. The most important advantage of ceramic packages is that the die is placed in a cavity which is hermetically sealed, eliminating packaging stresses and hermeticity problems. Ceramic packages are typically more expensive but more robust. Metal packages can also be hermetically sealed, and they offer excellent heat conduction and electrical shielding.

3.6.1 Wafer-Level Packaging

Wafer-level packaging (WLP) suggests to incorporate the packaging process into the micromachining process at the wafer level. WLP aims to minimize or even completely eliminate the added cost and increased size of an external package.

The mainstream WLP approach is encapsulation of the MEMS device at the wafer level using a cap wafer. The cap layer and the substrate perform most of the packaging functions, such as hermeticity, and mechanical and electrical interface.

Fig. 3.23 Wafer-level encapsulated micromachined gyroscope [95]. Courtesy of Bosch.

Common wafer-bonding methods such as fusion bonding, eutectic bonding, anodic bonding or glass-frit bonding are widely used for bonding a silicon or glass cap wafer to achieve wafer-level encapsulation. Alternative methods have also been reported, such as the thick-film encapsulation process (Figure 3.24) by Stanford and Bosch using epitaxial polysilicon [136]. Regardless of the capping method, wafer-level packaging has many advantages over an external package. Contamination is minimized, since the device is completely sealed prior to dicing, and never leaves the clean-room environment before sealing. Most importantly, a large fraction of the overall device cost due to packaging is reduced.

Fig. 3.24 Thick-film encapsulation using epitaxial polysilicon by Stanford and Bosch [136].

Through-wafer via technology is essential to achieve electrical connections with minimal die size in wafer-level packaging. Many approaches have been reported based on through etching the substrate wafer using DRIE, insulating the sidewalls and filling the via holes with conductive materials. Another alternative is to etch trenches through the wafer that form isolated via connections in the bulk silicon and filling the trenches with dielectric material, as reported by Silex (Figure 3.25).

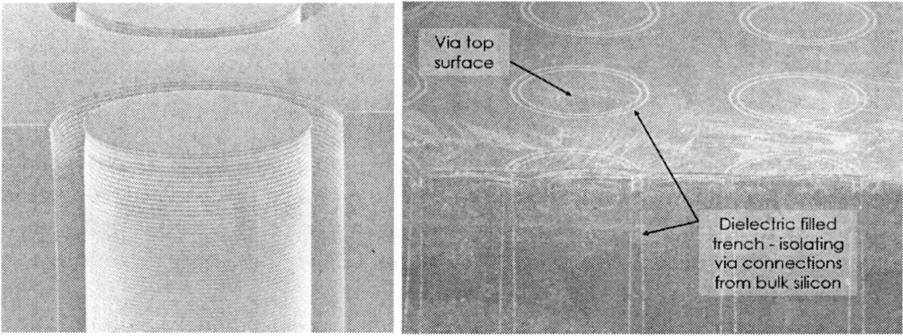

Fig. 3.25 Through-wafer via technology based on etching trenches through the wafer using DRIE to form via connections in the bulk silicon and filling the trenches with dielectric material. Courtesy of Silex.

3.6.2 Vacuum Packaging

Gyroscopes often require vacuum environment to minimize damping and increase the Q factors in drive and sense directions. Maintaining a high and stable Q factor over the lifetime of the device is extremely critical. Furthermore, higher vacuum lev-

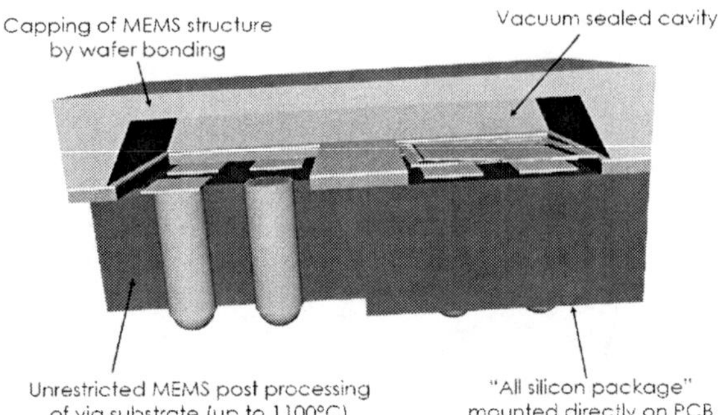

Fig. 3.26 Wafer-level packaging with through-wafer via technology by Silex.

els help ensuring stability of Q factor with temperature, improving the bias stability over temperature.

Vacuum packaging can be implemented at the die level or at the wafer level. Die-level vacuum packaging is commonly realized by sealing the die in a ceramic package in a high-vacuum environment. Wafer-level vacuum packaging involves sealing the capping wafer in vacuum. In both approaches, two primary factors have to be controlled to achieve vacuum: outgassing and leaking.

The leak rates required for over 10 years lifetime are on the order of 10^{-13} standard cc/sec, well below the capability of any leak test method. The Q factors of gyroscopes are usually monitored, as a good indicator of vacuum. Lid sealing of ceramic packages using fluxless solder preforms is known to provide sufficiently low leak rates. However, outgassing over time have been proven to be a limiting factor.

To minimize outgassing, material selection and decontamination is extremely important. Inside the cavity any material that could outgas such as polymers should be avoided. Hydrogen atoms trapped in certain alloys should be extracted. A common practice is to degas all components by baking in ultra high vacuum for extended times. Die attach adhesives and epoxies can not be used, and only eutectic or thermocompression die attach should be considered.

To prevent degradation of vacuum over time, getters are widely used in MEMS vacuum packaging. Getters usually consist of zirconium, vanadium and iron alloys

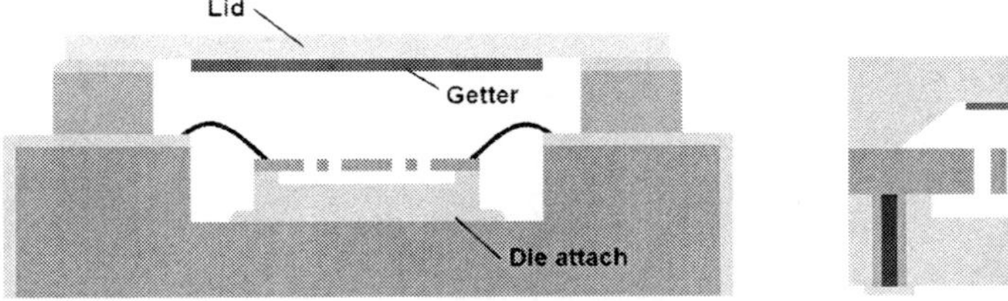

Fig. 3.27 Getters are used in discrete packaging and wafer-level vacuum packaging for maintaining a high and stable Q factor over the lifetime of micromachined gyroscopes.

and titanium. After being activated at around 400°C, they remove by chemical sorption active gases, including H_2O, O_2, CO, CO_2, N_2 and H_2 [118]. Getters are available in the form of discrete films, or could be deposited on package lids. For wafer-level vacuum packaging, getters are generally sputtered inside cap wafer cavities.

3.7 Summary

In this chapter, the most common micromachining processes and fabrication process flows used for MEMS gyroscopes were summarized, along with discussions on electronics integration and packaging approaches. A detailed description of microfabrication technologies highlighted in this chapter, and other techniques related to MEMS devices in general can be found in [107].

Chapter 4
Mechanical Design of MEMS Gyroscopes

This chapter describes the fundamental mechanical elements in the MEMS implementation of vibratory gyroscopes. Common mechanical structures are presented and analyzed for both linear and torsional gyroscopes, discussing primary vibratory system design issues to realize the gyroscopic dynamical system. Analysis of various flexure systems is followed by discussions on anisoelasticity and quadrature error due to mechanical imperfections. Finally, damping related issues are addressed, important material properties of silicon are highlighted, and mechanical design considerations to achieve a robust sensing element are discussed.

4.1 Mechanical Structure Designs

Various vibratory MEMS gyroscopes have been reported in the literature based on a wide range of mechanical structures. The common goal of all vibratory gyroscope structures is to realize a drive oscillator that generates and maintains a constant momentum, and a sense-mode accelerometer that measures the sinusoidal Coriolis force.

Vast majority of micromachined rate gyroscopes form the drive oscillator and the sense-mode accelerometer out of a mass or a combination of masses suspended by flexible beams above a substrate. The Coriolis force induced on the masses due to the drive vibration and the angular rate input has to be transferred to the sense-mode accelerometer in the orthogonal direction. Thus, at least one proof mass is required to be common to both the drive oscillator and the sense-mode accelerometer.

The primary objective of mechanical structure is to form the coupled orthogonal drive and sense dynamical systems by providing the required degree-of-freedom (DOF) for the masses. Usually both the drive and sense dynamical systems are 1-DOF oscillators. Thus, the resulting overall gyroscope can be modeled as a 2-DOF dynamical system as in Figure 4.1, where the modes are coupled by the Coriolis effect.

The drive-mode oscillator and the sense-mode accelerometer can be based on either linear or torsional motion. In the case of a linear vibratory gyroscope, conservation of linear momentum results in energy transfer from the drive axis to the sense axis, while in a torsional gyroscope conservation of angular momentum results in energy transfer. The following sections outline the basics of both linear and torsional gyroscope designs.

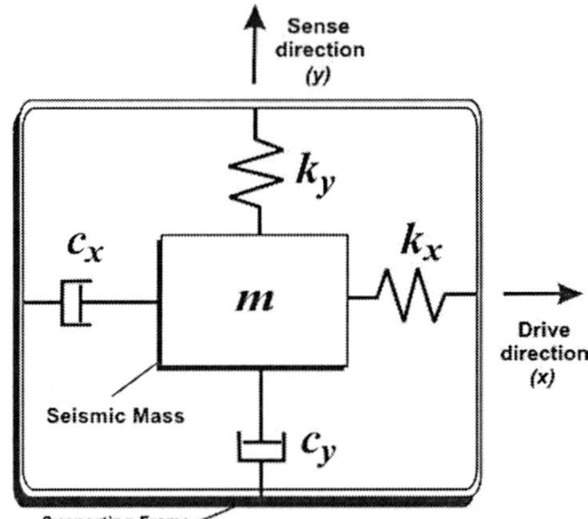

Fig. 4.1 The proof mass, which is free to oscillate in the drive and sense directions, forms the 2-DOF gyroscope system.

4.2 Linear Vibratory Systems

Linear or translational micromachined vibratory gyroscopes are based on sustaining a linear drive oscillation, and detecting a linear sense-mode response to the sinusoidal Coriolis force in the presence of an angular-rate input. Since the induced Coriolis force is orthogonal to the drive-mode vibration, the proof-mass is required to be free to oscillate in two orthogonal linear directions, and desired to be constrained in other vibrational modes. The suspension system design becomes critical in achieving these objectives.

The drive and sense axes are determined primarily by the desired angular rate detection axis. For example, a z-Axis gyroscope (Figure 4.2) requires the proof mass to be free to oscillate in the two in-plane orthogonal directions: the drive direction along the x-Axis and the sense direction along the y-Axis. The proof mass that provides the Coriolis coupling is allowed to oscillate along both x-Axis and y-Axis, becoming a 2-DOF system.

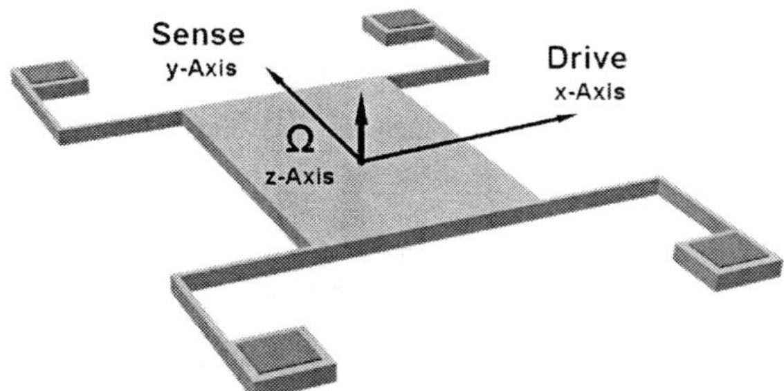

Fig. 4.2 Z-Axis gyroscope, the drive direction is along the x-Axis, and the sense direction is along the y-Axis.

An in-plane y-Axis gyroscope (Figure 4.3) proof mass is required to oscillate in one in-plane direction along the x-Axis and one out-of-plane direction along the z-Axis. The drive and sense axes can be interchanged depending on the actuation and detection schemes. An x-Axis gyroscope requires either the drive or the sense direction to be along the y-Axis. Thus, in many cases a y-Axis gyroscope design can be used as x-Axis gyroscope by rotating 90° in plane.

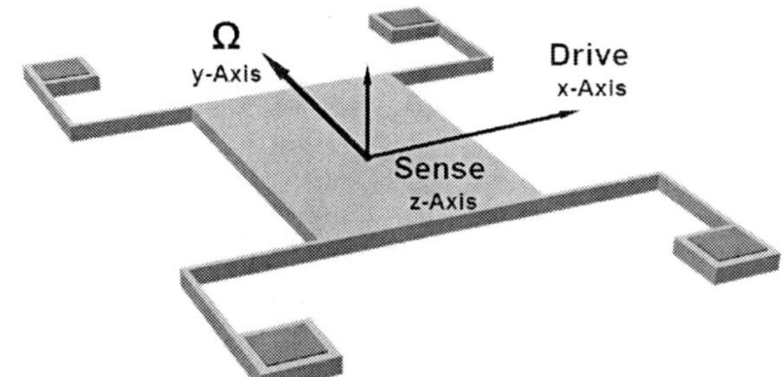

Fig. 4.3 Y-Axis gyroscope, the drive direction is along the x-Axis, and the sense direction is along the z-Axis.

4.2.1 Linear Suspension Systems

The flexure system that suspends the proof-mass above the substrate usually consists of thin flexible beams, formed in the same structural layer as the proof-mass. The thin beams have to be oriented to be compliant in both the drive and sense motion directions. Common suspension structures utilized in z-Axis micromachined gyroscopes include crab-leg suspensions (Figure 4.4), serpentine suspensions (Figure 4.5), hairpin suspensions (Figure 4.6), H-type suspensions (Figure 4.7), and U-beam suspensions (Figures 4.9 and 4.11).

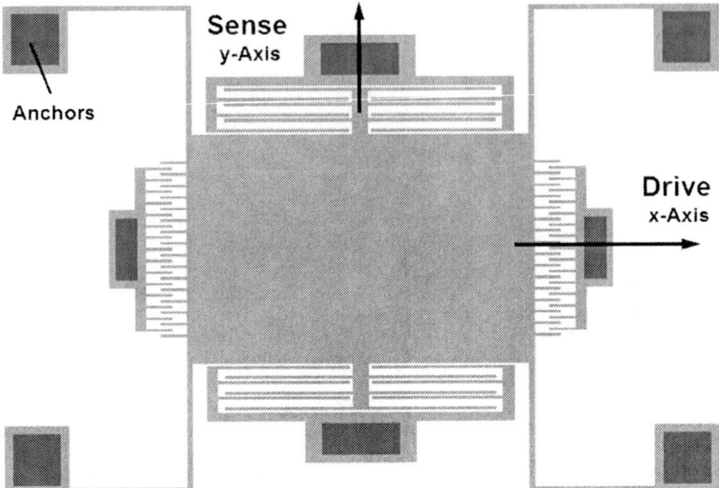

Fig. 4.4 Crab-leg suspensions.

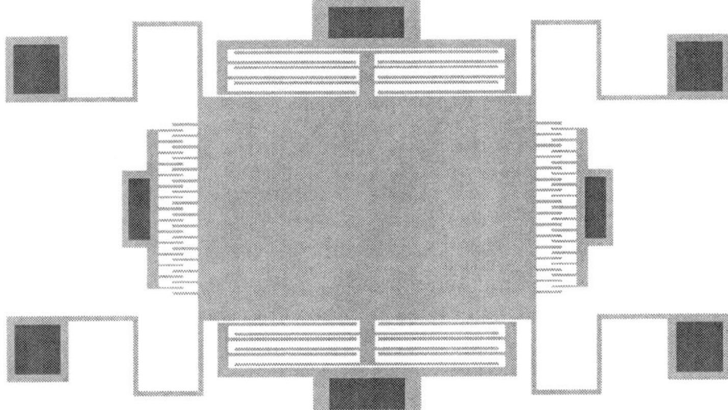

Fig. 4.5 Serpentine suspensions.

These suspension systems are widely used for both z-Axis gyroscopes with in-plane drive and sense modes, and x/y-Axis gyroscopes with one in-plane and one out-of-plane mode. Gyroscopes with an out-of-plane mode are usually fabricated with a thin structural layer to allow out-of-plane deflections of the beams, while gyroscopes with in-plane drive and sense modes are preferably fabricated with a thicker structural layer to suppress out-of-plane modes.

The crab-leg and H-type suspensions are known to provide better symmetry among the drive and sense-modes, allowing to easily locate the drive and sense

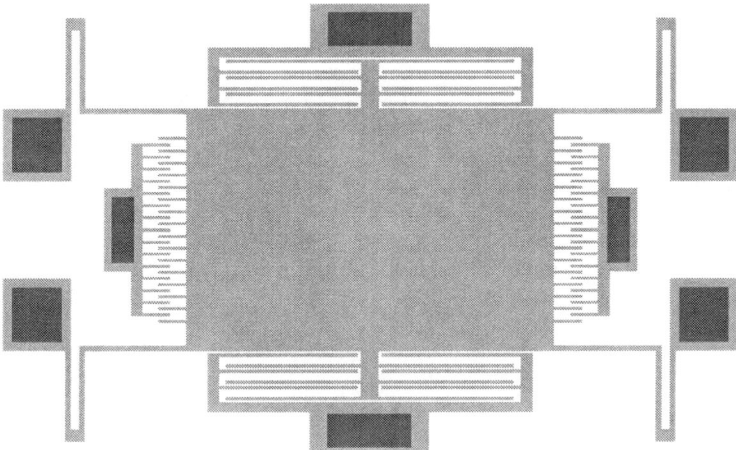

Fig. 4.6 Hairpin suspensions.

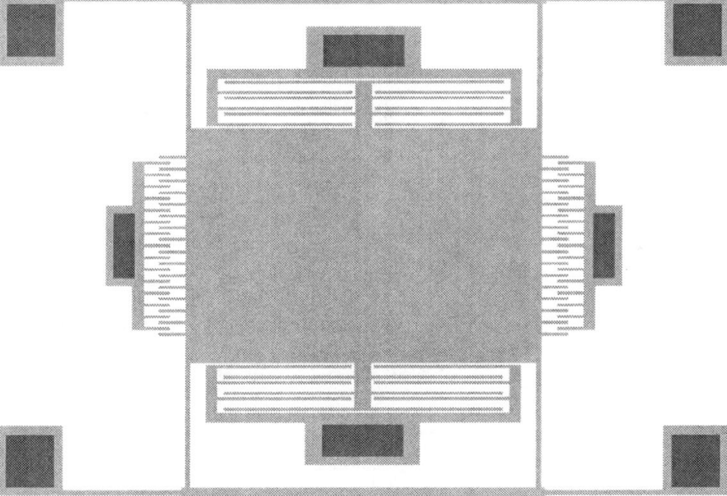

Fig. 4.7 H-type suspensions.

modes closer. In the crab-leg, serpentine and hairpin suspensions, drive motion results in deflections also in the sense-mode beams, which often causes undesired energy transfer into the sense-mode. Thus, H-type suspension and especially U-beam suspensions with decoupling frames provide better mode-decoupling, which will be discussed further next.

4.2.1.1 Frame Structures

Suspension systems similar to crab-leg, serpentine or hairpin suspensions are compliant in two orthogonal directions. The same beams experience deflections in both modes, resulting in undesired coupling between the drive and sense modes. Since the drive-mode amplitude is orders of magnitude larger than sense-mode, it is often required to isolate the drive motion from the sense motion. It is also desired to limit the deflection direction of the drive and sense electrodes, so that drive electrodes deflect only in drive direction, and sense electrodes deflect only in sense direction. This enhances the precision and stability of the drive actuation and sense detection electrodes.

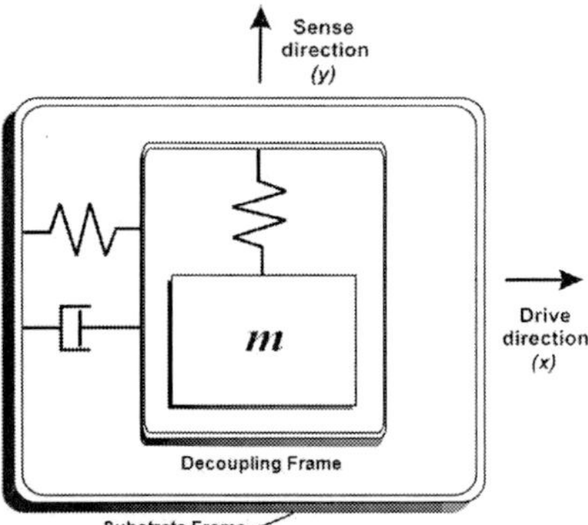

Fig. 4.8 Lumped model of the drive frame implementation with U-beam suspensions.

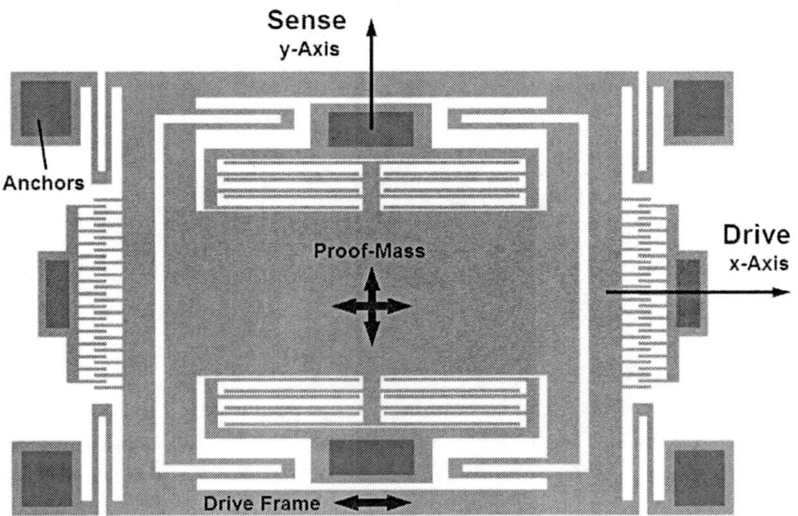

Fig. 4.9 Drive frame implementation with U-beam suspensions, minimizing the component of the actual drive motion along the sense detection axis.

To decouple the drive motion and sense motion, it is common to implement a frame structure that nests the proof-mass. Two basic approaches in frame implementation are using a drive frame as in Figure 4.9, or a sense frame as in Figure 4.11. In the drive frame implementation, the proof mass is nested inside a frame that is constrained to move only in the drive direction. This approach assures that the drive motion is very well aligned with the designed drive axis, and minimizes the component of the actual drive motion along the sense detection axis. It also provides improved side stability and minimal parasitic sense-direction forces in the drive actuators.

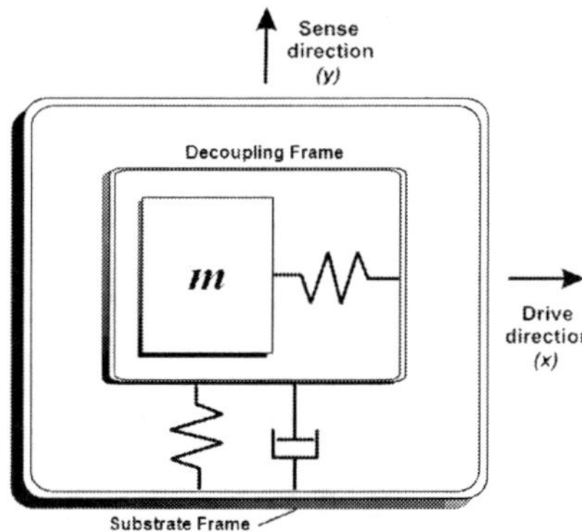

Fig. 4.10 Lumped model of the sense frame implementation with U-beam suspensions.

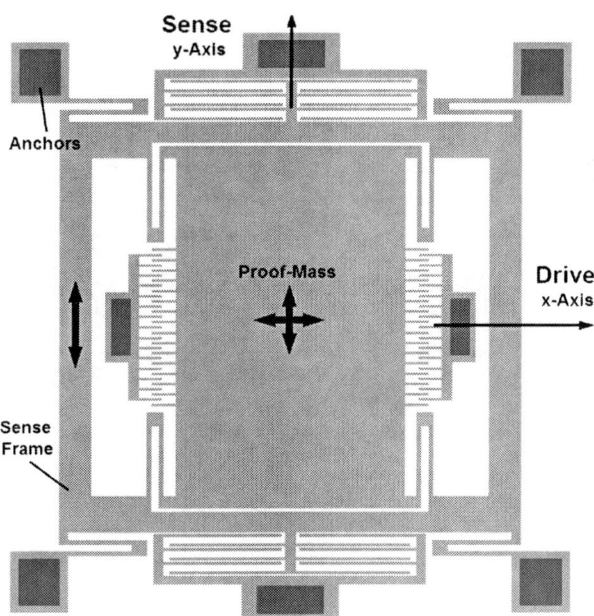

Fig. 4.11 Sense frame implementation with U-beam suspensions, minimizing the undesired capacitance change in the sense electrodes due to the drive motion.

The sense frame implementation is based on nesting the proof mass inside a frame that is constrained to move only in the sense direction. The sense electrodes are attached to the frame, and relative motion in the sense electrodes along the drive direction is prevented. This approach minimizes the undesired capacitance change in the sense electrodes due to the drive motion.

More sophisticated frame structures that provide the advantages of both drive and sense frame implementations are also possible. For example, the z-axis gyroscope in [94] by Bosch utilizes a drive frame around the proof-mass which oscillates only in the drive direction, and a sense frame inside the proof-mass which is fixed in the drive direction and oscillates with the proof-mass in the sense direction (Figure 4.12). With this double frame structure, the drive oscillations are very well aligned with the drive axis, parasitic components of the drive forces in the sense direction are suppressed, and the motion of the sense-electrodes in the drive direction is eliminated.

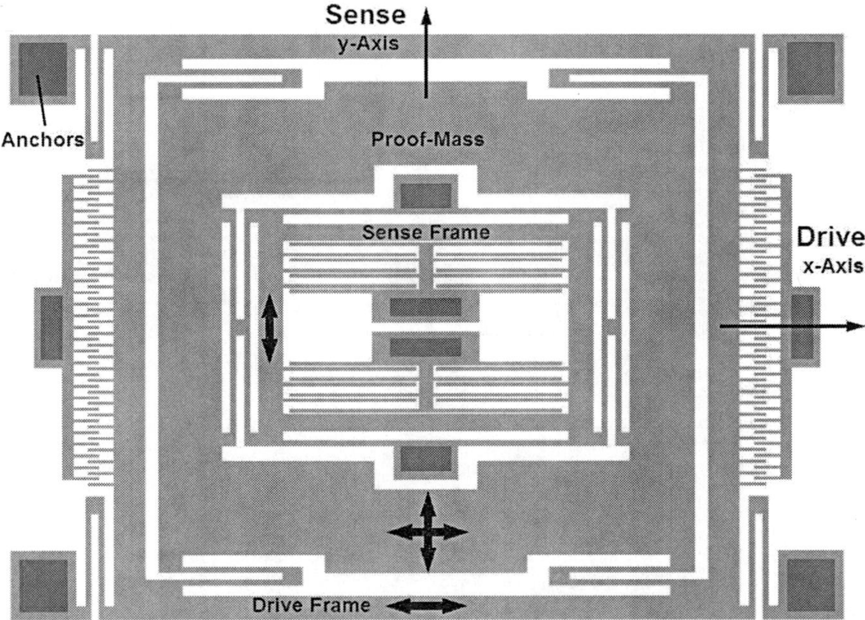

Fig. 4.12 A double-frame implementation example based on [94], which combines the advantages of drive and sense frames.

4.2.1.2 Anti-Phase Devices

Gyroscopes are inherently sensitive to external inertial inputs such as ambient vibrations and shock. Many applications require gyroscopes to function under a certain degree of vibration environment. Given the small amplitudes of the Coriolis response, the response to external accelerations and vibrations could easily disrupt the rate measurements.

Anti-phase systems, also known as tuning fork gyroscopes (TFG), aim to cancel common-mode inputs. In the tuning fork architecture, two identical masses are driven in opposite directions (anti-phase), which causes the Coriolis forces induced on the two masses to be in opposite directions also (Figure 4.13). When the sense-mode response of the two masses are detected in a differential mode, their response to Coriolis forces are added, but their common-mode response in the same direction are canceled out. Thus, common-mode rejection is achieved while the rate signal is preserved.

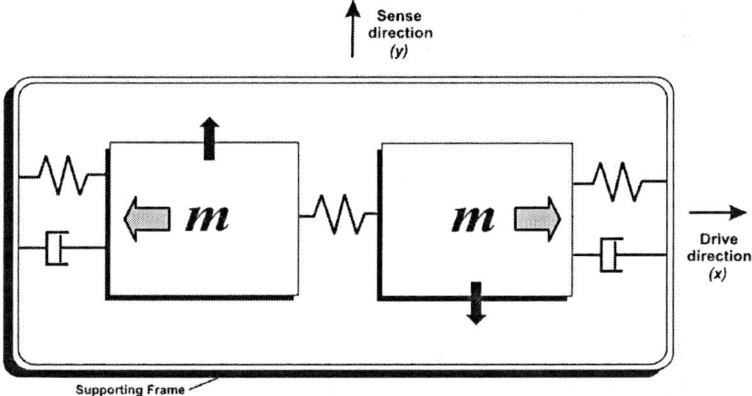

Fig. 4.13 Anti-phase tuning fork gyroscopes (TFG), which provide common-mode rejection by utilizing two anti-phase vibrating masses in a differential mode.

To achieve an anti-phase oscillation in the drive-mode, the two masses are coupled with a coupling spring as in Figures 4.13 and 4.14. The coupling spring results in a 2-DOF drive dynamical system, with an in-phase and an anti-phase mode. The device is operated at the anti-phase drive frequency, which excites the masses in opposite directions as desired.

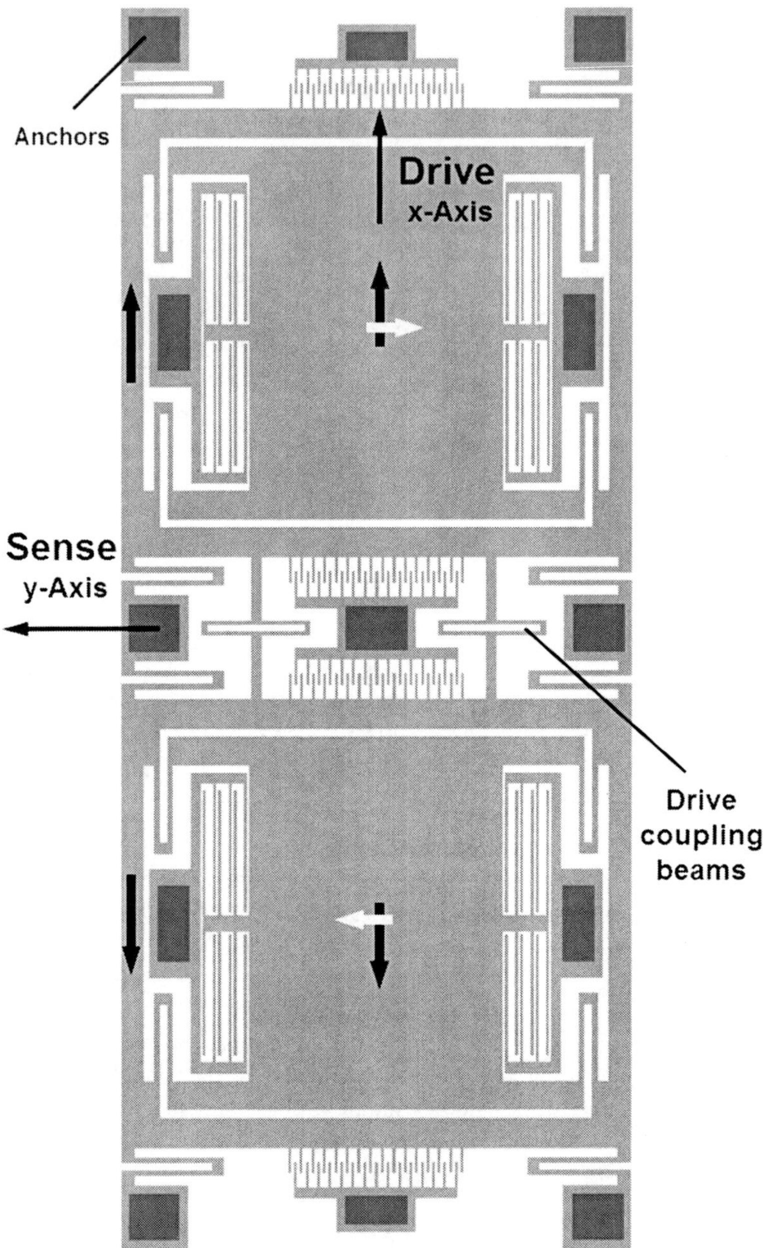

Fig. 4.14 An example anti-phase tuning fork gyroscope system, which consists of two drive-frame gyroscopes coupled in the drive mode by a spring.

4.2.2 *Linear Flexure Elements*

In linear micromachined gyroscopes, the suspension systems are usually designed to be compliant along the desired motion direction, and stiff in other directions. Most suspension systems utilize narrow beams as the primary flexure elements, aligning the narrow dimension of the beam normal to the motion axis (Figure 4.15).

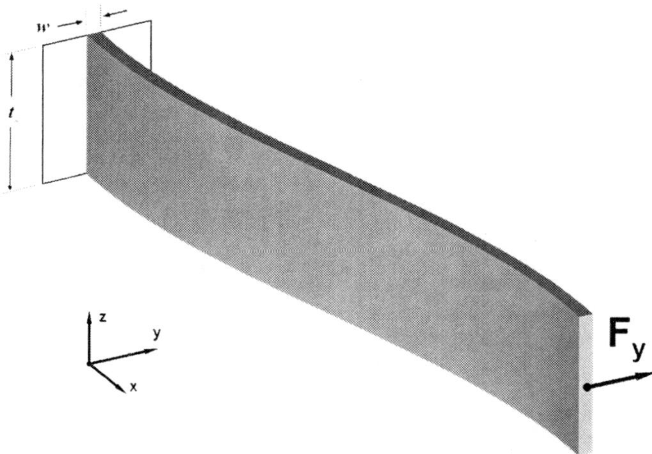

Fig. 4.15 The fixed-guided end beam under translational deflection.

In purely translational modes, the boundary conditions of the beams that connect the components of the gyroscopes are most commonly the fixed-guided end configuration (Figure 4.15), in which the moving end of the beam remains parallel to the fixed end. Many complete gyroscope suspension systems can be modeled as a combination of fixed-guided end beams.

If we define the length of a beam L as the x-axis dimension, width w as the y-axis dimension, and the thickness t as the z-axis dimension, the area moments of inertia of the beam in the y and z directions become

$$I_y = \frac{1}{12} t w^3 \tag{4.1}$$

$$I_z = \frac{1}{12} w t^3 \tag{4.2}$$

For a single fixed-guided beam (Figure 4.15), the translational stiffness for motion in the orthogonal direction to the beam axis is given by [106]

$$k_{y,z} = \frac{1}{2} \frac{3EI_{y,z}}{\left(\frac{L}{2}\right)^3} \tag{4.3}$$

where E is the Young's Modulus. Thus, the stiffness values of the fixed-guided beam along the three principle axes become

$$k_x = E\frac{wt}{L} \tag{4.4}$$

$$k_y = E\frac{tw^3}{L^3} \tag{4.5}$$

$$k_z = E\frac{wt^3}{L^3} \tag{4.6}$$

It should be noticed that the ratio of the stiffness values along the z and y axes is $\frac{k_z}{k_y} = (\frac{t}{w})^2$. Thus, large thickness is a key factor in suppressing the out-of-plane deflections. The ratio of the axial stiffness to the y-axis stiffness is $\frac{k_x}{k_y} = (\frac{L}{w})^2$. This ratio could be quite large depending on the beam design, providing excellent suppression of orthogonal motion.

Even though theoretical expressions of the beam elements could be a practical guide in design, finite element analysis (FEA) simulations are essential in estimation of the flexure characteristics. For example, let us analyze a fixed-guided beam with the dimensions $L = 500\mu$m, $w = 10\mu$m, and $t = 100\mu$m. Assuming an elastic modulus of $E = 130$ GPa, FEA results indicate the y-axis reaction force for a 1μm purely y-axis deflection to be 142μN, yielding $k_y = 142$N/m. However, the reaction force for 10μm deflection increases to 1980μN, resulting in $k_y = 198$N/m. This illustrates the non-linearity of the beam for increased deflections.

FEA results on the fixed-guided beam also reveals one of the major limitations of this suspension type. The x-axis reaction force for a 1μm y-axis deflection is 410μN. For 10μm deflection, the x-axis reaction force increases to over 40400μN. This extraordinarily large force is due to the fact that the beam starts to be axially loaded as the lateral force increases. Thus, single fixed-guided beams should not be used for large deflection flexures such as the drive-direction suspensions.

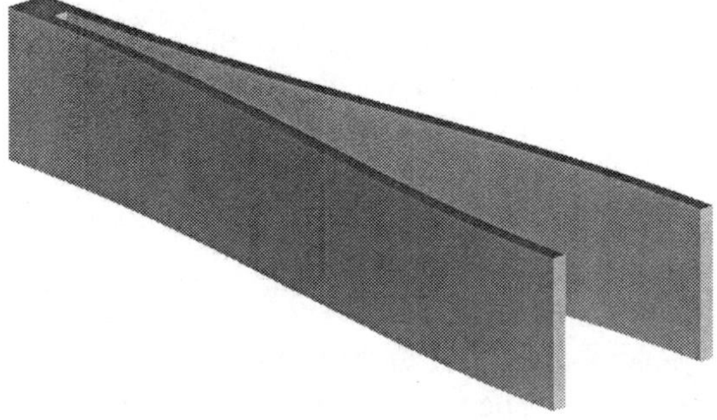

Fig. 4.16 Folded beam (U-shaped) suspensions consist of two fixed-guided beams in series, and eliminate the nonlinearity and axial-loading limitations of single fixed-guided beams.

Folded beam (U-shaped) suspensions eliminate the nonlinearity and axial-loading limitations of single fixed-guided beams. By connecting two fixed-guided beams in series in the folded beams (Figure 4.16), the two connection points of the suspension are on the same side, and lateral deflections do not result in axial loading. Since a folded beam consists of two fixed-guided beams of stiffness k_y in series, the stiffness of a folded beam of length L, width w and thickness h becomes

$$\frac{1}{k_{folded}} = \frac{1}{k_y} + \frac{1}{k_y} \tag{4.7}$$

$$k_{folded} = E\frac{tw^3}{2L^3} \tag{4.8}$$

One limitation of the folded beams is the reduced axial stiffness. The distance between the two beams results in a moment arm under an axial load, and causes bending as in Figure 4.17. This could become a disadvantage in designs that require substantial suppression of axial motion.

Fig. 4.17 The folded beam (U-shaped) suspensions under lateral and axial loading. The compliance of the folded beam under axial loading due to bending could be undesirable.

The axial compliance problem of the folded beams could be solved by providing symmetry in the axial direction. The double-folded suspension beam contains two folded beams symmetrically connected (Figure 4.18). Since it could be modeled as two folded beams in parallel, the stiffness of a double-folded beam is

$$k_{double-folded} = 2k_{folded} = E\frac{tw^3}{L^3} \tag{4.9}$$

In summary, for small deflections the folded beam provides half the stiffness of a fixed-guided beam while the double-folded beam has the same stiffness as a fixed-guided beam. As the deflections get larger compared to the beam dimensions, the fixed-guided beam stiffness becomes non-linear, as seen in Figure 4.19.

In complete suspension systems, a number of flexure elements are connected to the proof-mass. The total stiffness in a certain direction could be approximated by the sum of all flexure stiffness values in that direction. However, this approximation assumes that the compliance of the proof-mass, frame structures, and flexures

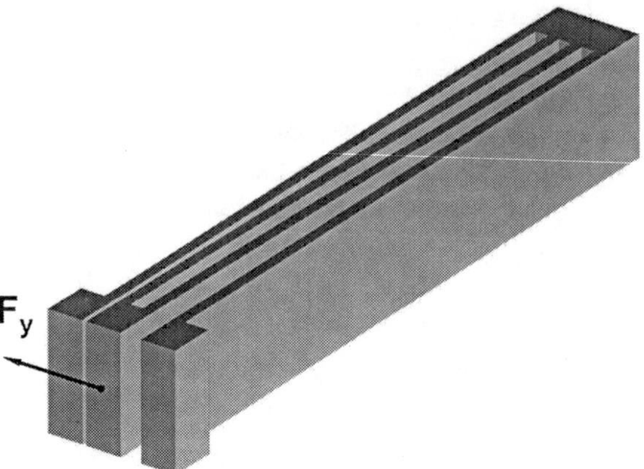

Fig. 4.18 Double-folded beam suspension, which contains two symmetrically connected folded beams, provide excellent axial stiffness and linearity.

in other directions are negligible. In reality, these factors dramatically reduce the overall stiffness. Thus, modal analyses in FEA software is absolutely necessary for accurate estimation and design of resonant frequencies.

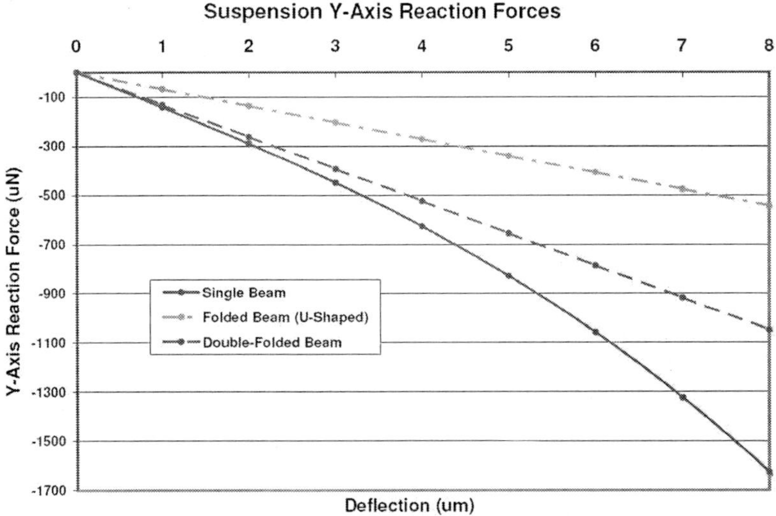

Fig. 4.19 Comparison of the y-axis reaction forces in fixed-guided, folded, and double-folded beams.

4.3 Torsional Vibratory Systems

Torsional or rotation-based micromachined gyroscopes utilize rotational vibratory motion in their drive and sense modes. The operation principle is based on conservation of primarily angular momentum, instead of linear momentum as in translational vibratory gyroscopes.

Torsional gyroscope structures consist of a rotational drive oscillator that generates and maintains a constant angular momentum, and a sense-mode angular accelerometer that measures the sinusoidal Coriolis moment. Thus, similar to linear vibratory gyroscopes, a 2-DOF oscillatory system is formed.

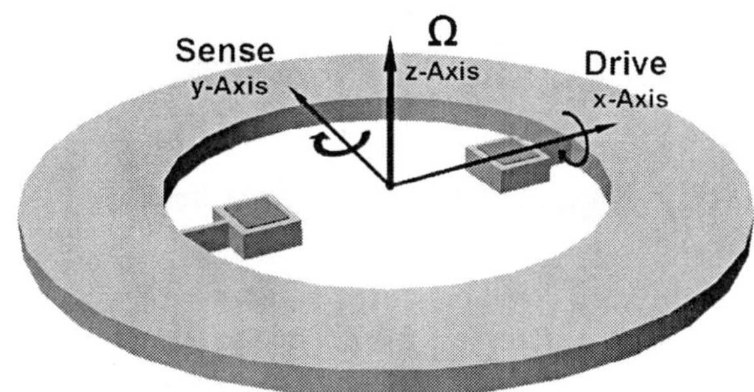

Fig. 4.20 Torsional Z-Axis gyroscope: The drive oscillation is about the x-Axis, and the sense oscillation is about the y-Axis.

In a z-Axis torsional gyroscope, the proof mass rotates about the two in-plane orthogonal directions: the drive direction rotation about the x-Axis and the sense direction rotation about the y-Axis as in Figure 4.20. The Coriolis moment is about the axis cross-product of the input angular rate and the drive angular velocity vectors.

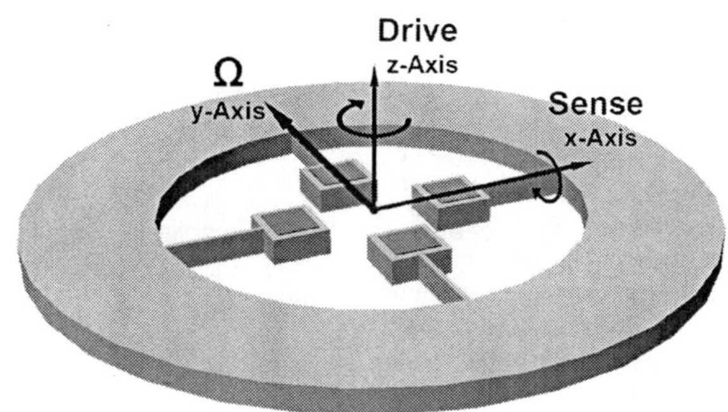

Fig. 4.21 Torsional Y-Axis gyroscope: The drive oscillation is about the z-Axis, and the sense oscillation is about the x-Axis.

An in-plane y-Axis torsional can be implemented with either an in-plane drive (about the z-Axis) and out-of-plane sense (about the x-Axis) configuration, or an out-of-plane drive (about the x-Axis) and in-plane sense (about the z-Axis) configuration as in Figure 4.21.

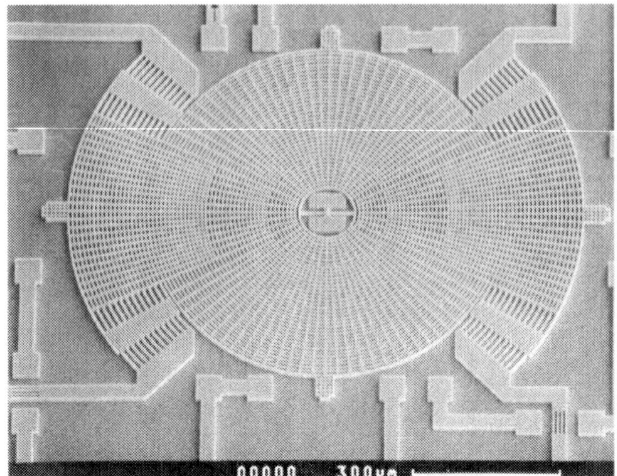

Fig. 4.22 Torsional gyroscope by Bosch, with a drive-mode about the z-Axis [95]. SEM courtesy of Bosch.

4.3.1 Torsional Suspension Systems

Gimbals are commonly used in torsional gyroscope suspension systems to decouple the drive and sense modes, and to suppress undesired modes. Many suspension system and gimbal configurations are possible in torsional vibratory gyroscopes. Similar to linear gyroscope systems, the suspension system that supports the masses and gimbals usually consists of thin flexible beams, formed in the same structural layer as the proof-mass.

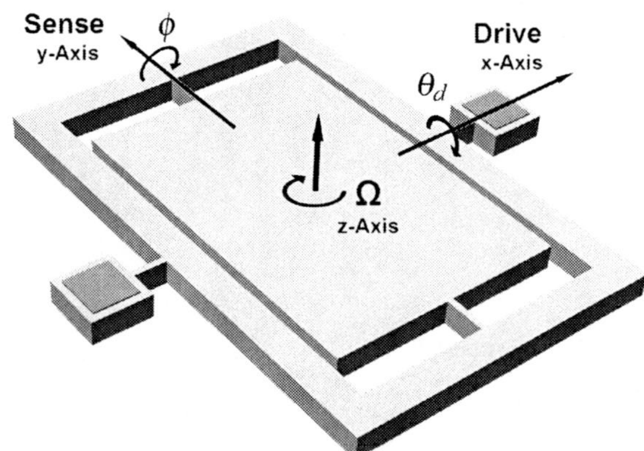

Fig. 4.23 Torsional z-Axis gyroscope with drive gimbal structure. The drive-mode deflection angle of the gimbal is θ_d, and the sense-mode deflection angle of the sensing mass is ϕ.

An example gimbal system for a z-Axis torsional gyroscope based on [97] was shown in Figure 4.23. In the drive-mode, the outer drive gimbal is excited about the x-axis. In the presence of an angular rate input about z-axis, the sinusoidal Coriolis torque is induced about the y-axis, which causes the sense-mode response of the inner mass (Figure 4.24).

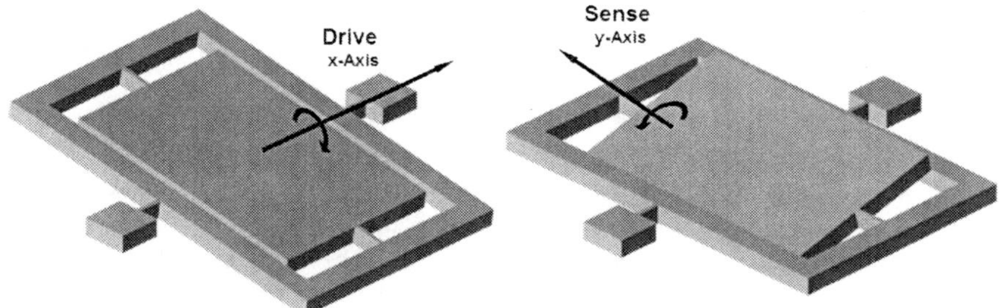

Fig. 4.24 The drive and sense modes of a typical torsional z-axis gyroscope, similar to [97].

A representative gimbal implementation in y-Axis torsional gyroscopes based on [99] is presented in Figure 4.25. The system consists of an inner gimbal that can deflect torsionally in-plane about the z-Axis, and an the outer mass attached to the inner gimbal. The drive-mode is in-plane about the z-Axis, and the sense-mode is out-of-plane about the x-Axis. In the drive-mode, the inner gimbal and the outer mass oscillate together, and the angular rate input about the y-Axis generates a Coriolis torque about the x-Axis. The outer mass responds to the Coriolis torque by deflecting torsionally about the x-Axis relative to the drive gimbal. The sense-mode response is detected by the out-of-plane electrodes located underneath the outer mass structure.

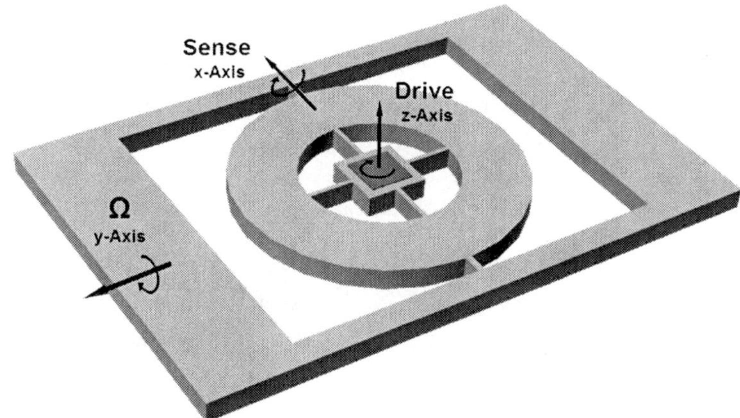

Fig. 4.25 The gimbal system in a y-axis torsional gyroscope, based on [99].

4.3.2 Torsional Flexure Elements

4.3.2.1 Out-of-Plane Torsional Hinges

Out-of-plane deflections, which are about the x-axis or y-axis, are most commonly achieved by torsional beams (Figure 4.26) aligned along the deflection axis. In the purely torsional mode, the boundary conditions of the beam are such that the moving end of the beam remains parallel to the fixed end, but rotates about the axis normal to the end plane.

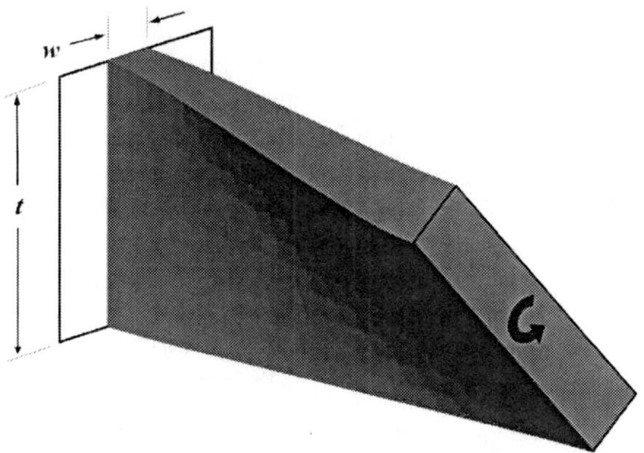

Fig. 4.26 Typical torsional beams used for out-of-plane deflections, about the x or y axes.

Assuming each torsional beam is straight with a uniform cross-section, and the structural material is homogeneous and isotropic; the torsional stiffness of each beam with a length of L can be modeled as

$$K = \frac{SG}{L} \tag{4.10}$$

where $G = \frac{E}{2(1-v)}$ is the shear modulus with the elastic modulus E and Poisson's ratio v. For a beam with a rectangular cross-section of width w and thickness t, given that $w \leq t$, the cross-sectional coefficient S can be expressed as [106]

$$S = tw^3 \left[\frac{1}{3} - 0.21 \frac{w}{t} \left(1 - \frac{t^4}{12t^4} \right) \right] \tag{4.11}$$

4.3.2.2 In-Plane Torsional Flexures

In-plane torsional deflections about the z-axis are usually achieved by a combination of fixed-guided end beams (Figure 4.27). The beams are configured such that the center lines along their lengths intersect at the rotation center of the mass. The boundary conditions of the beams are different from linear gyroscope systems, in that the moving end of the beam does not remain parallel to the fixed end, and deflects with an angle equal to the total rotation angle.

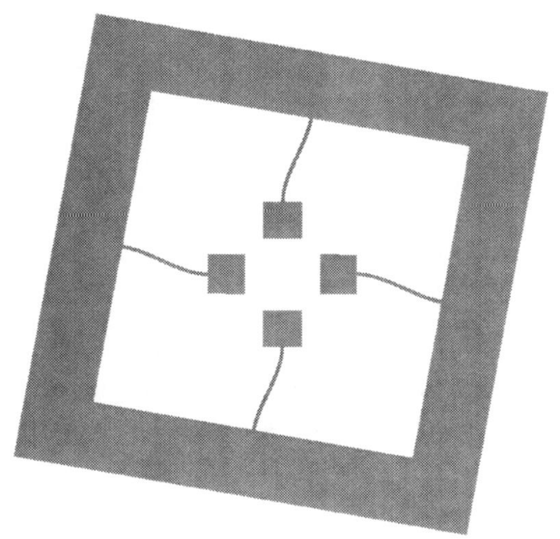

Fig. 4.27 Fixed-guided flexure beams commonly used for in-plane (about the z-Axis) torsional suspensions.

The suspension beams can be located inside or outside the mass. The direction of the moving end deflection angle, and thus the boundary condition depends on the location of the beams. Let us consider a single fixed-guided beam, and define the length of a beam L as the y-axis dimension, width w as the y-axis dimension, and the thickness t as the z-axis dimension. The area moment of inertia of the beam in the y direction is

$$I_y = \frac{1}{12} t w^3 \tag{4.12}$$

When the beam is located inside the mass as in Figure 4.28 and the moving end is at a distance R from the rotation center, the torsional spring constant of the single beam about the z-axis becomes

$$K_{zz} = 4\frac{EI_y}{LN}\left[3\left(\frac{R}{L}\right)^2 - 3\frac{R}{L} + 1\right]$$

(4.13)

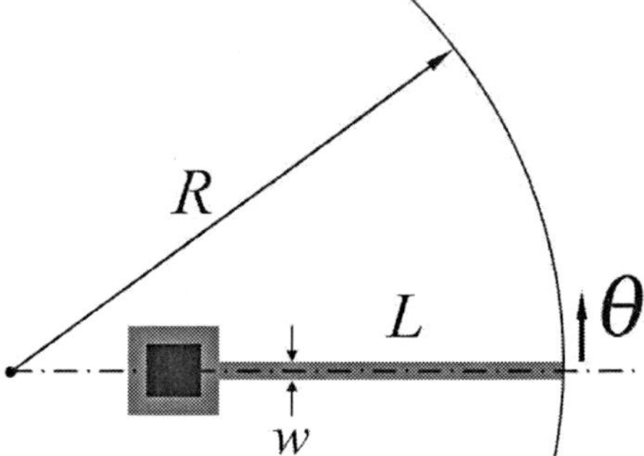

Fig. 4.28 Interior beam configuration for in-plane torsional suspensions.

where N is the number of folds in the beam.

When the beam is located outside the mass as in Figure 4.29 and the moving end is at a distance R from the rotation center, the torsional spring constant of a single beam is

$$K_{zz} = 4\frac{EI_y}{LN}\left[3\left(\frac{R}{L}\right)^2 + 3\frac{R}{L} + 1\right]$$

(4.14)

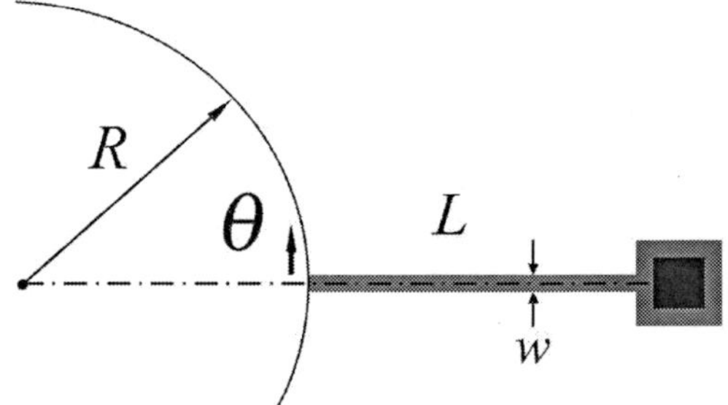

Fig. 4.29 Exterior beam configuration for in-plane torsional suspensions.

Derivations of the torsional spring constants are presented in detail in [115]. The interior beam configuration allows to achieve more compliant suspension systems with compact dimensions. However, large residual stresses in the structural layer could cause excessive curling. Exterior beam configuration minimizes curling, while consuming more die area. Thus, interior beams could be more suitable for bulk micromachined low-stress devices, and exterior beams more suitable for surface micromachined devices.

4.4 Anisoelasticity and Quadrature Error

Most of the major challenges in vibratory gyroscopes arise because of the fact that the magnitude of the sense-mode response amplitude is extremely small. Let us illustrate common order of magnitudes on an example gyroscope system. In Chapter 2, the sense-mode response amplitude of a matched mode gyroscope system with $\omega_s = \omega_d$ was derived as

$$y_{0matched} = \Omega_z \frac{2Q_s x_0 m_C}{m_s \omega_s} \tag{4.15}$$

If we consider a mode-matched gyroscope system with the total sense-mode mass equal to the mass that generates the Coriolis force, i.e. $m_s = m_C$, the ratio of the sense amplitude to the drive amplitude becomes

$$\frac{y_{0matched}}{x_0} = 2Q_s \frac{\Omega_z}{\omega_s} \tag{4.16}$$

Typical operation frequencies of vibratory gyroscopes are around $f_s = 10$ kHz, yielding $\omega_s = 2\pi \cdot 10,000$ rad/s. Assuming a sense-mode quality factor $Q_s = 1000$, the sense to drive amplitude ratio for a $1°/s = \pi/180$ rad/s angular rate input is 556 ppm. If the drive-mode amplitude is 10 μm, the sense-mode amplitude is 5.56 nm. This example illustrates a best case scenario, since it is based on a high-Q and mode-matched system. If the drive and sense modes are separated with $\Delta f = 100$ Hz, the sense response amplitude drops to 0.16 nm.

In reality, fabrication imperfections result in non-ideal geometries in the gyroscope structure, which in turn causes the drive oscillation to partially couple into the sense-mode. Even though several cross-coupling mechanisms such as elastic, viscous and electrostatic couplings exist, in most cases the elastic coupling in the suspension elements is the largest in magnitude. Considering the relative magnitudes of the drive and sense oscillations, even extremely small undesired coupling from the drive motion to the sense-mode could completely mask the Coriolis response.

To investigate the dynamical effects of cross-coupling, let us start with the ideal system dynamics. The simplified dynamics of an ideal z-axis gyroscope system with the drive-mode along the x-axis and the sense-mode along the y-axis in vector form can be expressed as

$$\begin{bmatrix} m_d & 0 \\ 0 & m_s \end{bmatrix} \begin{bmatrix} \ddot{x} \\ \ddot{y} \end{bmatrix} + \begin{bmatrix} c_x & 0 \\ 0 & c_y \end{bmatrix} \begin{bmatrix} \dot{x} \\ \dot{y} \end{bmatrix} + \begin{bmatrix} k_x & 0 \\ 0 & k_y \end{bmatrix} \begin{bmatrix} x \\ y \end{bmatrix} = \begin{bmatrix} F_d \\ -2m_C\Omega_z\dot{x} \end{bmatrix} \tag{4.17}$$

where m_C is the portion of the driven proof mass that contributes to the Coriolis force, m_d is the total drive-mode mass, and m_s is the total sense-mode mass. The total stiffness matrix is equal to the sum of the stiffness matrices of each suspension element in the system.

Almost all suspension elements in real implementations of vibratory gyroscopes have elastic cross-coupling between their principal axes of elasticity. This phenomenon is called *anisoelasticity*, and is the primary cause of mechanical quadrature error in gyroscopes. The anisoelastic forces that result in elastic coupling between the x and y axes are modeled through the off-diagonal springs constants k_{xy} and k_{yx} in the system stiffness matrix. When these anisoelastic elements are included in the dynamics, the equations of motion become

$$\begin{bmatrix} m_d & 0 \\ 0 & m_s \end{bmatrix} \begin{bmatrix} \ddot{x} \\ \ddot{y} \end{bmatrix} + \begin{bmatrix} c_x & 0 \\ 0 & c_y \end{bmatrix} \begin{bmatrix} \dot{x} \\ \dot{y} \end{bmatrix} + \begin{bmatrix} k_x & k_{xy} \\ k_{yx} & k_y \end{bmatrix} \begin{bmatrix} x \\ y \end{bmatrix} = \begin{bmatrix} F_d \\ -2m_C\Omega_z\dot{x} \end{bmatrix} \tag{4.18}$$

Since the oscillation amplitudes in the sense-mode are orders of magnitude smaller than the drive-mode, the coupling spring k_{xy} in the drive dynamics is negligible. The impact of anisoelasticity is primarily on the sense-mode dynamics due to k_{yx}, which couples the drive-mode displacement into the sense-mode oscillator. The simplified sense-mode equation of motion with anisoelasticity can be expressed as

$$m_s\ddot{y} + c_y\dot{y} + k_y y = -2m\Omega_z\dot{x} - k_{yx}x \tag{4.19}$$

The total cross-coupling stiffness k_{yx} in the suspension system is equal to the sum of the k_{yx} values of each suspension beam. In an ideal gyroscope system with identical springs located symmetrically, even if the cross-axis stiffness values of each individual spring are not zero, the off-diagonal cross-axis stiffness values exactly cancel out when added. Thus, the total k_{xy} and k_{yx} become zero, and the total stiffness matrix becomes diagonal. With a diagonal stiffness matrix, system eigenmodes align perfectly with the drive and sense axes.

However, fabrication imperfections and variations are inevitable, and exist to a certain degree in every actual gyroscope structure. Non-uniform variations within the die result in slight differences among the suspension elements. Therefore, the off-diagonal terms do not exactly cancel out in real suspension systems, and yield residual off-diagonal terms in the total stiffness matrix.

For z-axis gyroscopes, slight variation in the average widths of the suspension beams is the basic cause of suspension asymmetries which ultimately leads to anisoelasticity. The nominal cross-axis coupling values for folded beams (assuming $L_1 \sim L_2$ and $L_C \ll L_1$) and crab-leg suspensions are derived in [98] as

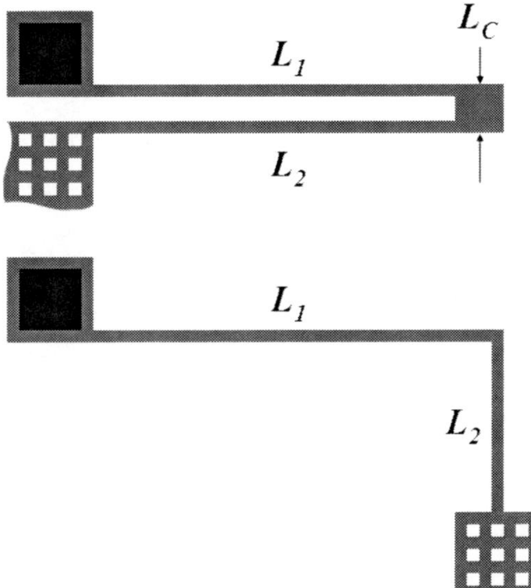

Fig. 4.30 Models of folded beams and crab-leg suspensions for calculation of cross-axis coupling stiffness values.

$$\text{Folded beam: } k_{yx} = \frac{3EI_C(L_1 - L_2)}{L_C L_1^3} \tag{4.20}$$

$$\text{Crab-leg: } k_{yx} = \frac{9EI_1 I_2}{L_1 L_2 (I_1 L_2 + I_2 L_1)} \tag{4.21}$$

where I_C and L_C are the moment of inertia and the length of the connecting beam in the folded beam, and $I_{1,2}$ and $L_{1,2}$ are the moment of inertia and the length of the first and second springs in the folded beam and the crab-leg suspension as shown in Figure 4.30.

Cross-axis stiffness of the folded beam nominally becomes zero when the length of the two beams are equal. Even though in reality geometrical mismatches between the two beams of a folded suspension could result in a net cross-axis stiffness, this value is relatively small compared to crab-leg or similar suspensions, especially when the length of the connecting beam is kept to a minimum. The resulting net cross-axis stiffness of each suspension element is canceled out further by symmetrically placing suspension beams, but will not be nulled completely due to slight variations among the suspension elements.

In gyroscope systems with out-of-plane operational modes, cross-axis stiffness between the in-plane and out-of-plane directions, i.e. k_{zx} or k_{zy}, becomes critical. The primary factor that results in this type of anisoelasticity is the deviation of the cross-section of the beams from a perfect rectangle. Sidewall tilt in DRIE is known to result in a non-rectangular cross-section, causing the principle axes of elasticity of the suspension beams to deviate from perfectly parallel and orthogonal to the device surface.

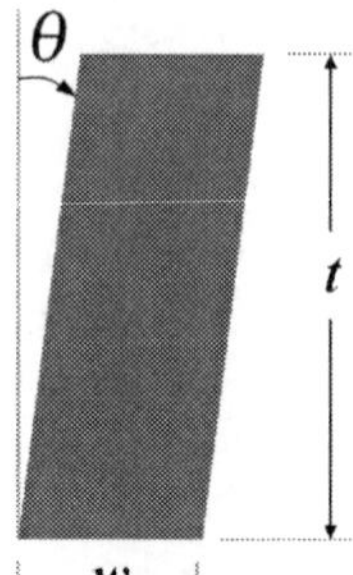

Fig. 4.31 Sidewall angle in suspension beams, resulting in the cross-axis stiffness k_{zx} or k_{zy} between the in-plane and out-of-plane directions.

If we consider a y-axis gyroscope system with drive direction along the x-axis and sense direction along the z-axis as an example, the cross-axis coupling stiffness term k_{zx} excites the sense-mode. For a single suspension beam with a parallelogram cross section, the cross-axis coupling stiffness due to the sidewall angle θ is [101]

$$k_{zx} = k_x \frac{t^2}{w^2} \theta \qquad (4.22)$$

Similarly, the k_{zx} values of symmetrically designed beams ideally cancel out in the total stiffness matrix, while non-uniform variations among the beams result in a net cross-axis coupling from the in-plane drive motion to the out-of-plane sense-mode.

Having explained the fundamental causes of cross-axis stiffness in suspension beams, let us investigate the impact of these off-diagonal stiffness terms on the system dynamics. The two forces that excite the sense-mode oscillator in a z-axis gyroscope system were previously shown to be the Coriolis force F_C and the quadrature force F_Q

$$F_C = -2m_C \Omega_z \dot{x} \qquad (4.23)$$
$$F_Q = -k_{yx} x \qquad (4.24)$$

The rate-equivalent quadrature error in the sense-mode response can be found by taking the ratio of the quadrature force F_Q to the Coriolis force per unit angular rate (F_C/Ω_z)

$$\Omega_Q = \frac{F_Q}{\frac{F_C}{\Omega_z}} = \frac{k_{yx}|x_0 \sin \omega_d t|}{2m_C|\omega_d x_0 \cos \omega_d t|} \qquad (4.25)$$

Assuming that the mass that generates the Coriolis force is equal to the drive-mode mass, i.e. $m_C \simeq m_d$, the rate equivalent quadrature becomes

$$\Omega_Q = \frac{k_{yx}}{k_x} \frac{\omega_d}{2} \tag{4.26}$$

This expression illustrates how very small imbalances in the suspension structure can lead to high quadrature error values. For example, if the cross coupling stiffness k_{yx} is only 1% of the drive stiffness in a gyroscope system with $\omega_d = 10\text{kHz}$, the resulting quadrature error is $18000°/\text{sec}$.

Simply due to the fact that the Coriolis force F_C is proportional to the drive velocity $\dot{x}$ and the quadrature force F_Q is proportional to the drive position x, there is always a $90°$ phase difference between the Coriolis response and the mechanical quadrature. Since both F_C and F_Q are excitation forces applied on the sense-mode oscillator, both the relative amplitude and phase of the quadrature and Coriolis response are independent of the sense-mode dynamics (Figure 4.32). This means that the rate-equivalent quadrature remains constant for varying mode mismatch Δf and sense quality factor Q_s, even though the actual quadrature signal magnitude varies.

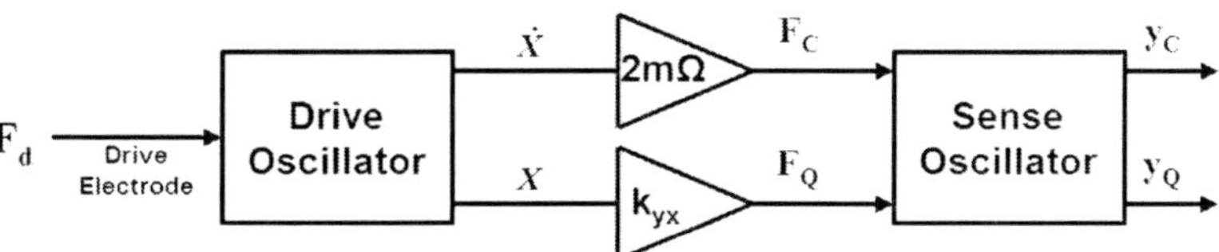

Fig. 4.32 Block diagram of the mechanical quadrature model, showing that both the Coriolis force F_C and the quadrature force F_Q are simultaneously applied on the sense-mode oscillator. Both the amplitude and phase of the quadrature signal relative to the Coriolis response are independent of the sense-mode dynamics.

Having explained the underlying reason behind the constant $90°$ phase difference between the Coriolis response and quadrature, let us now investigate the phase relations between the drive motion and quadrature error for the two main system types: mode-matched system with $\Delta f = 0$ and mode-mismatched system with $\Delta f > 0$. It is assumed that in the steady state the drive-mode displacement is of the form

$$x(t) = x_0 \sin(\omega_d t + \phi_d) \tag{4.27}$$

where ϕ_d is the drive-mode position phase relative to the drive AC signal. Since the drive oscillator is usually operated at resonance, in most systems $\phi_d = -90°$.

For a system with matched drive and sense modes, i.e. $\Delta f = 0$, the phases of the Coriolis response and quadrature relative to the drive signal are

$$\phi_{Coriolis} = \phi_d - 180° \tag{4.28}$$

$$\phi_{Quadrature} = \phi_d - 270° \tag{4.29}$$

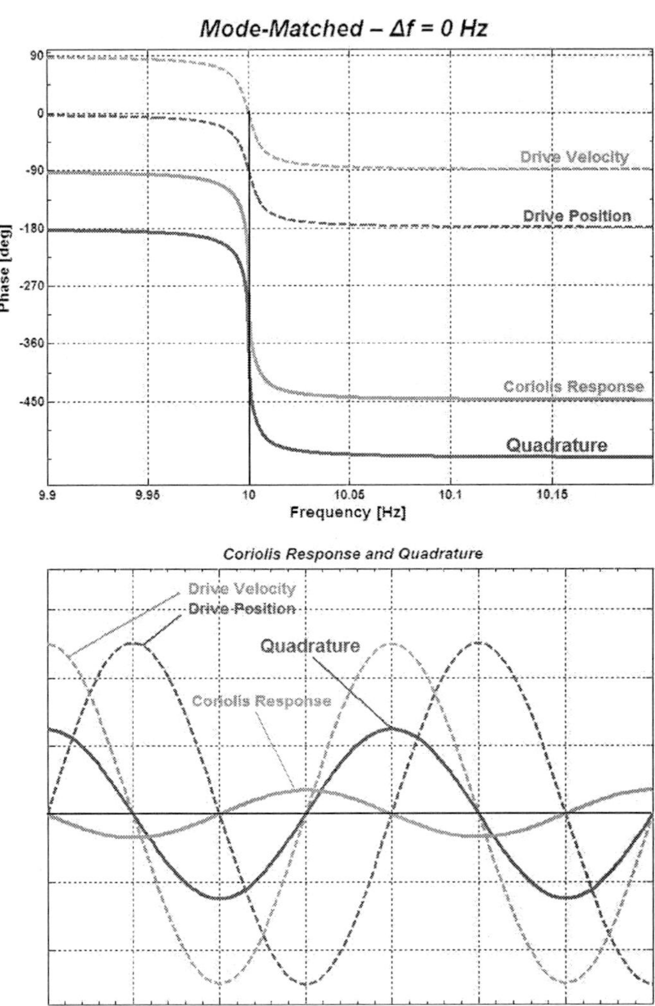

Fig. 4.33 The drive position, drive velocity, Coriolis response and quadrature phase relations for a mode-matched system, when $\Delta f = 0$ Hz.

For a mismatched system with the sense-mode sufficiently higher than the drive-mode, i.e. $\Delta f > 0$, the Coriolis response phase converges to the Coriolis force phase, which is $-90°$ from drive position, and the quadrature response lags the Coriolis response by $90°$

$$\phi_{Coriolis} = \phi_d - 90° \tag{4.30}$$

$$\phi_{Quadrature} = \phi_d - 180° \tag{4.31}$$

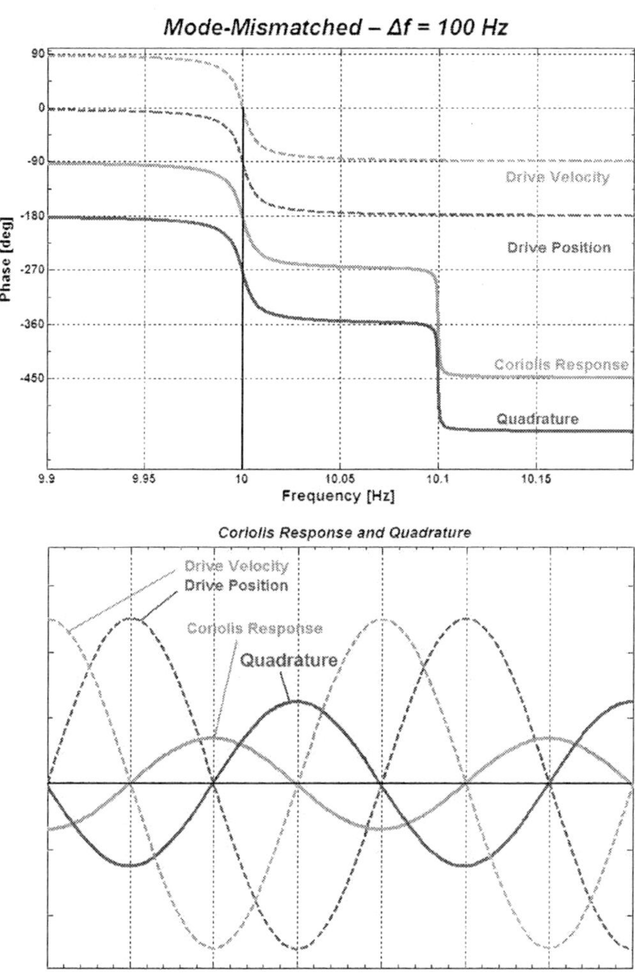

Fig. 4.34 The drive position, drive velocity, Coriolis response and quadrature phase relations for a mode-mismatched system, when $\Delta f = 100$ Hz.

In a mode-matched system, the proof-mass oscillation trajectory due to mechanical quadrature becomes an ellipse, since the sense-mode quadrature response has a $90°$ phase difference with the drive position (Figure 4.35).

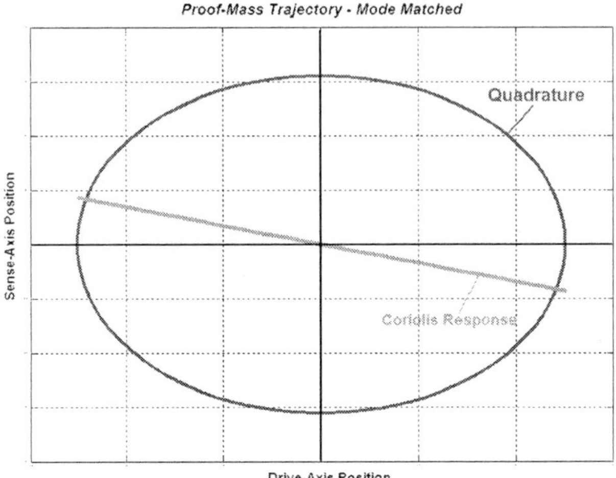

Fig. 4.35 Coriolis response and quadrature trajectories for a mode-matched system.

When the drive and sense frequencies are mismatched, the oscillation trajectory is already an ellipse without anisoelasticity. For a system with $\Delta f > 0$, the quadrature response has a $180°$ phase difference with the drive position, and the oscillation trajectory due to quadrature becomes a straight line as in Figure 4.36.

4.4.1 Quadrature Compensation

Since there is always a $90°$ phase difference between the Coriolis response and the mechanical quadrature, the quadrature signal can be separated from the Coriolis signal during amplitude demodulation at the drive frequency. However, the substantially large relative magnitude of the quadrature signal has many implications on the detection electronics:

- First and foremost, the dynamic range of front-end electronics has to be designed to accommodate the large levels of quadrature signal, which could be quite larger than the dynamic range required for the full range of the gyroscope. This could result in lower resolution and signal to noise ratio for the Coriolis signal.
- To be able to discriminate large levels of quadrature signal, the phase accuracy of synchronous demodulation also has to be high. Lower phase accuracy results in more quadrature signal to mix into the rate signal, which could result in large rate bias for high quadrature levels.

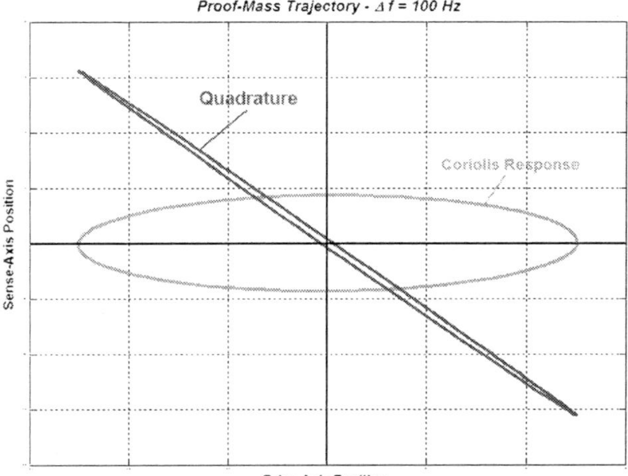

Fig. 4.36 Coriolis response and quadrature trajectories for a mode-mismatched system. The straight line of oscillation due to quadrature turns into a narrow ellipse in the presence of an angular rate input.

- The stability of quadrature over temperature and over time is extremely important when the quadrature level is high. If the part of the quadrature signal that mixes into the rate signal varies over temperature and time, then the temperature stability and long term stability of the gyroscope deteriorate.

Even though the quadrature signal can be electrically nulled, it is desirable to cancel the actual mechanical quadrature motion at the sensing element level. Several approaches could be implemented to compensate for mechanical quadrature:

- Mechanically balancing the sensing element by altering the mechanical system. Post-fabrication trimming by laser ablation can be used to remove mass from proper areas of the proof-mass until mechanical imbalance is eliminated [100]. Although this process can be automated, it is not compatible with wafer-level packaged devices.
- Electrostatically canceling quadrature force via DC bias. This approach requires quadrature compensation electrodes that vary overlap area as a result of the drive motion. The electrostatic force is modulated by the drive motion at the drive frequency, proportional to the drive amplitude. By adjusting the DC bias level, a quadrature cancellation force $F_{comp} = k_{yx}x_0 \sin \omega_d t$ equal and opposite to the quadrature force is achieved.
- Electrostatically canceling quadrature force by directly applying an AC compensation signal. In this approach, an AC signal in-phase with the drive position is extracted from the drive oscillator, and applied together with a DC bias on an electrode that exerts force in the sense direction. The amplitude of the AC signal and the DC bias are controlled to achieve a quadrature cancellation force $F_{comp} = k_{yx}x_0 \sin \omega_d t$ that exactly nulls the quadrature force. Since the compen-

sation force does not depend on the drive motion, precise phase control and an AGC loop that regulates the drive amplitude are required.

The quadrature error can reach thousands of $°/s$ signals in actual gyroscope systems. Usually a combination of different measures have to be taken to minimize the yield loss due to excessive quadrature. Tight process control, robust gyroscope design, and electrical/electrostatic quadrature cancellation capabilities have to work together to achieve acceptable quadrature levels over the entire wafer area.

4.5 Damping

Damping is the energy dissipation effect in an oscillatory system. In micromachined vibratory systems such as gyroscopes, many dissipation mechanisms contribute to the total damping. The following is an overview of prominent damping phenomena such as viscous damping and structural damping.

4.5.1 Viscous Damping

In the presence of a gas surrounding the vibratory structure of a gyroscope, the primary damping mechanism in the gyroscope dyanmical system is the viscous effects of the gas confined between the proof mass surfaces and the stationary surfaces. The damping of the structural material is usually orders of magnitude lower than the viscous damping except under high-vacuum conditions.

Viscous damping in the gyroscope dynamical system is dominated by the internal friction of the gas between the proof-mass and the substrate, and between the comb-drive and sense capacitor fingers. These viscous damping effects can be captured by using two general damping models: slide film damping and squeeze film damping.

4.5.1.1 Slide Film Damping

Slide film damping, or lateral damping, occurs when two plates of an area A, separated by a distance y_0, slide parallel to each other (Figure 4.37). At low pressures and when the mean free path of the gas is comparable to the gap, gas rarefaction effects can be modeled by the effective viscosity of the gas μ_{eff}. Assuming a Newtonian gas, the lateral damping coefficient can be expressed as

$$c_{slide} = \mu_{\text{eff}} \frac{A}{d} \tag{4.32}$$

where A is the overlap area of the plates, and d is the plate separation. The effective viscosity μ_{eff} is approximated in [109] as

$$\mu_{\text{eff}} = \frac{\mu}{1 + 2K_n + 0.2K_n^{0.788}e^{-K_n/10}} \tag{4.33}$$

The Knudsen number K_n is the measure of the gas rarefaction effect, which is a function of the gas mean free path λ and the gap d:

$$K_n = \frac{\lambda}{d} \tag{4.34}$$

Pressure dependence of the gas viscosity is captured in the mean free path, which can be calculated for air as

$$P\lambda = 5.1 \times 10^{-5}\text{Torr.m} \tag{4.35}$$

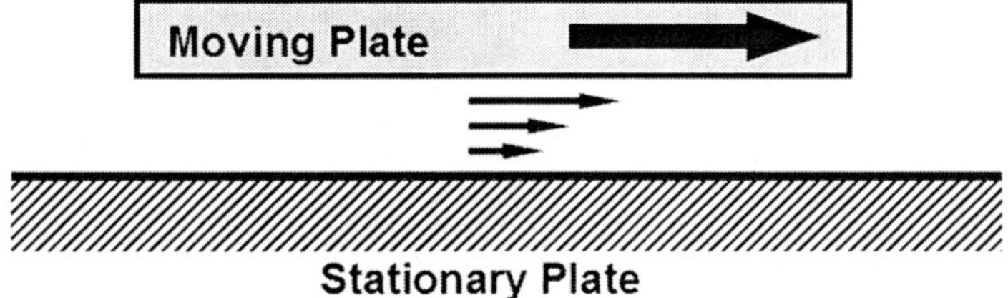

Fig. 4.37 Slide-film damping between two plates, which occurs when the plates slide parallel to each other.

4.5.1.2 Squeeze Film Damping

Squeeze film damping occurs when two parallel plates move toward each other and squeeze the fluid film in between (Figure 4.38). Squeeze film damping effects are quite complicated, and can exhibit both damping and stiffness effects depending on the compressibility of the fluid.

The effective viscosity μ_{eff} to model gas rarefaction effects in squeeze film damping is given in [110] as

$$\mu_{\text{eff}} = \frac{\mu}{1 + 9.638K_n^{1.159}} \tag{4.36}$$

Solving the linearized Reynolds equation yields one force in-phase with displacement, and one force out-of-phase. The in-phase force due to squeeze-film effect is the spring force, and the out-of-phase force is the damping force. The squeeze-film damping force F_c and spring force F_k are reported in [111] as

$$\frac{F_c}{z} = \frac{64\sigma P_a A}{\pi^6 d} \sum_{m,n \text{ odd}} \frac{m^2 + c^2 n^2}{(mn)^2[(m^2 + c^2 n^2)^2 + \sigma^2/\pi^4]} \qquad (4.37)$$

$$\frac{F_k}{z} = \frac{64\sigma^2 P_a A}{\pi^8 d} \sum_{m,n \text{ odd}} \frac{1}{(mn)^2[(m^2 + c^2 n^2)^2 + \sigma^2/\pi^4]} \qquad (4.38)$$

where z is the plate deflection, P_a is the ambient pressure, m and n are odd integers, $c = w/l$ and $A = wl$ for a plate with width w and length l. The squeeze number σ as a function of frequency ω is

$$\sigma = \frac{12\mu_{\text{eff}} w^2}{P_a d^2} \omega \qquad (4.39)$$

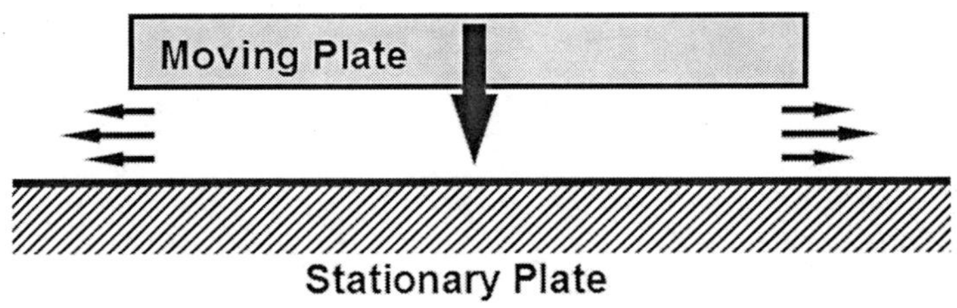

Fig. 4.38 Squeeze-film damping between two plates, which occurs when the plates move towards each other.

Detailed discussions on slide-film and squeeze-film damping in micromacined structures along with more advanced phenomena such as non-linear effects, kinetic gas models, plate motions that propagate into the fluid with rapidly diminishing amplitude, and computational fluid dynamics simulations are presented in [108–110, 112, 113].

4.5.2 Viscous Anisodamping

Mechanical cross-coupling sources between the two oscillation axes of vibratory gyroscopes is not limited to the elastic coupling due to anisoelasticity. Depending on the design, viscous coupling due to hydrodynamic forces could be a major error mechanism.

Hydrodynamic lift, also known as the *surfboard effect,* is the primary source of anisodamping, which couples the drive motion into the sense-mode [101]. This phenomenon occurs when a plate slides over a viscous medium and the hydrodynamic lift due to the plate velocity generates a force orthogonal to the motion direction.

The anisodamping forces appear as the off-diagonal damping terms c_{xy} and c_{yx} in the system damping matrix:

$$\begin{bmatrix} m_d & 0 \\ 0 & m_s \end{bmatrix} \begin{bmatrix} \ddot{x} \\ \ddot{y} \end{bmatrix} + \begin{bmatrix} c_x & c_{xy} \\ c_{yx} & c_y \end{bmatrix} \begin{bmatrix} \dot{x} \\ \dot{y} \end{bmatrix} + \begin{bmatrix} k_x & k_{xy} \\ k_{yx} & k_y \end{bmatrix} \begin{bmatrix} x \\ y \end{bmatrix} = \begin{bmatrix} F_d \\ -2m_C\Omega_z\dot{x} \end{bmatrix} \quad (4.40)$$

Similar to anisoelasticity, the coupling in the drive dynamics is negligible, and the impact is primarily on the sense-mode dynamics. The simplified sense-mode equation of motion including the anisodamping effect is

$$m_s\ddot{y} + c_y\dot{y} + k_y y = -2m\Omega_z\dot{x} - c_{yx}\dot{x} \quad (4.41)$$

It should be noticed that, unlike anisoelasticity, anisodamping is proportional to the drive velocity, and causes a coupling exactly in phase with the Coriolis response. Thus, it is indistinguishable from the rate response of the gyroscope.

Even though the bias due to anisodamping could be canceled from the output as on offset, viscous forces are highly temperature dependent, and could contribute greatly to the temperature bias of the gyroscope. Vacuum packaging becomes crucial in minimizing this effect. The design of the gyroscope is also critical. Avoiding the use of parallel plate sense electrodes that move with the drive motion, or providing symmetry about the drive axis alleviate the effects of anisodamping.

4.5.3 Intrinsic Structural Damping

Although viscous damping is the dominating damping mechanism in the presence of a gas in the gyroscope ambient, the total damping in the gyroscope system is a combination of multiple effects. The damping components other than viscous damping start limiting the quality factor as the pressure inside the gyroscope cavity approaches high vacuum.

Under vacuum conditions, thermoelastic damping is one of the primary damping mechanisms. Thermoelastic damping is the intrinsic material damping that occurs as a result of thermal energy dissipation due to elastic deformation. In a vibrating beam, alternating tensile and compressive strains across the width cause irreversible heat flow, which in turn results in an effective damping due to dissipation of vibrational energy [103]. Thermoelastic damping has been reported to limit the Q factor of vacuum packaged gyroscopes to values from 100,000 to 200,000.

Many other factors from anchor losses to die attach methods contribute to the total damping in vibratory gyroscopes. The total quality factor can be expressed as a combination of these effects as [102]

$$\frac{1}{Q_{total}} = \frac{1}{Q_{viscous}} + \frac{1}{Q_{TED}} + \frac{1}{Q_{anchor}} + \frac{1}{Q_{electronics}} + \frac{1}{Q_{other}} \quad (4.42)$$

where Q_{TED} is due to thermoelastic damping, $Q_{electronics}$ is due to electronics damping which is in the order of 10^{11}, Q_{other} captures remaining damping effects estimated around 250,000 [102]. Q_{anchor} is due to the anchor losses which could be as low as 10,000 depending on the anchor type and material. Anti-phase devices that locally cancel vibration injection into the anchors provide much higher Q_{anchor} values.

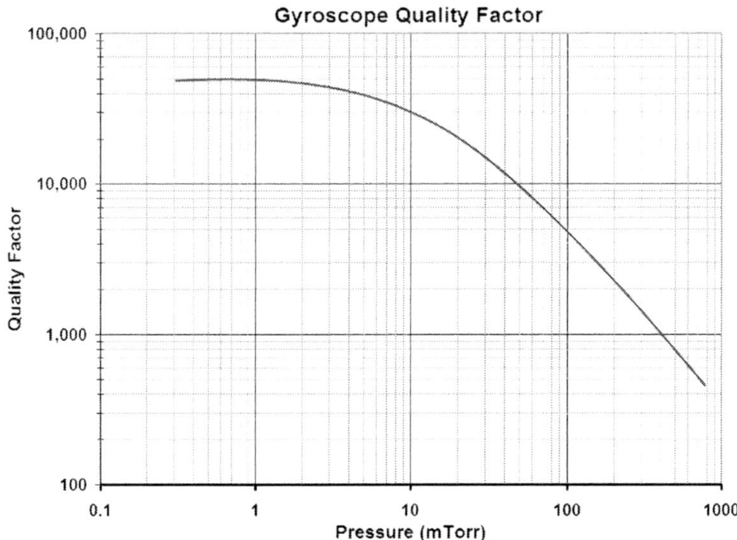

Fig. 4.39 Typical quality factor versus pressure curve, where the structural damping becomes the Q limit as viscous damping diminishes.

Usually viscous damping and other intrinsic damping components are very difficult to estimate theoretically for complicated gyroscope systems. It is common to empirically measure the overall Q factor of drive and sense modes by a frequency response or ring-down test.

Q factor versus pressure curves as in Figure 4.39 are usually obtained to guide the packaging pressure requirements of gyroscopes. At sufficiently low pressures, the quality factors usually start converging to a limit value set by the total structural damping, which depends on the specifics of the device geometry and anchor structure. To minimize effects of pressure variations, it is desirable to operate the device in the flat region of the curve.

4.6 Material Properties of Silicon

Single crystal silicon is one of the most common materials used in inertial sensors. Since the cubic nature of the single crystal silicon lattice results in orthotropic material properties, it requires special attention in design and modeling. The following table summarizes the important material properties of single crystal silicon and polysilicon for reference.

Table 4.1 Material properties of Silicon

Material Property	Single Crystal Si		Polysilicon
	[100]	[111]	
Young's modulus	131 GPa	190 GPa	161 GPa
Poisson ratio	0.28	0.26	0.23
Density	2330 kg/m^3		

When suspension beams with arbitrary angles are designed in a gyroscope system, anisotropy of Young's modulus has to be taken into account. For example, in a (100) silicon wafer, elastic modulus is 131 GPa parallel to the flat and 169 GPa at a 45° angle to the flat. Variation of Young's modulus in the <100> plane for single crystal silicon is presented in Figure 4.40 [96].

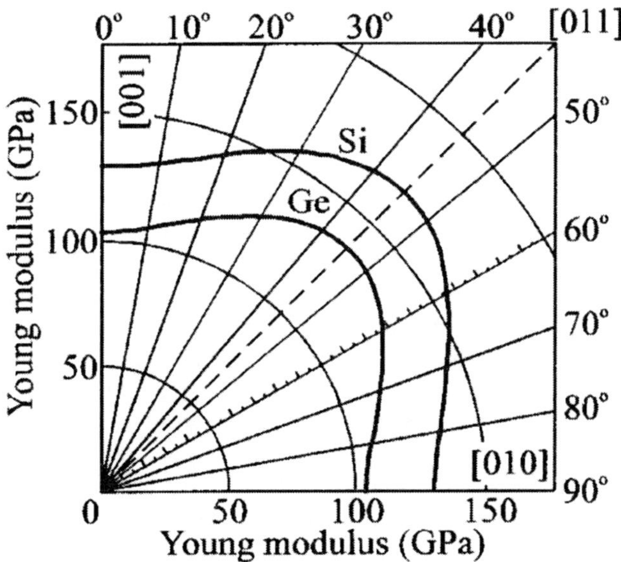

Fig. 4.40 Young's modulus vs. direction in the <100> plane for Silicon and Germanium. Reprinted with permission from [96]. Copyright 1965, American Institute of Physics.

It should be noted that silicon is an excellent structural material since it exhibits no plastic deformation or creep below 500 °C. Since it is impervious to fatigue, it can withstand millions of cycles without failure, which is essential for vibratory gyroscopes.

4.7 Design for Robustness

The micromachined gyroscope sensing element is a highly complex dynamic mechanical system. In Chapter 2 we illustrated that the dynamic response of the gyroscope is very sensitive to variations in system parameters that shift the drive or sense resonant frequencies. Both structural variations and environmental effects are known to result in quite large variations in the resonant frequencies. Even though mode-mismatching is a common practice to reduce sensitivity to variations by operating away from the sense resonant frequency, many other factors and failure modes have to be considered to achieve robustness. Some of the most important design aspects are summarized below.

4.7.1 Yield

Numerous types of fabrication process variations have direct effect on the device performance. Variations in structural thickness, critical dimensions such as beam widths and capacitive gaps, sidewall angles and cross-section profiles result in differences in mechanical and electrical characteristics from die to die on a wafer, from wafer to wafer, and from lot to lot.

The sensing element design must meet specifications such as scale factor and quadrature error with a given process window to maximize yield. Many critical features can often be determined with a top-down approach. For example, the scale-factor compensation capability of the electronics determines the acceptable range of capacitive gap variations and relative variations of the drive and sense resonant frequencies. Then in turn, acceptable relative tolerance on the suspension beam widths is estimated. With a given the critical dimension control capability of the fabrication process, the minimum beam widths allowed in the gyroscope design are determined.

Especially in low-cost and high-volume applications, design for yield is of utmost importance. Corner analysis and Monte Carlo methods are widely used to estimate the impact of the combination of the variation in many fabrication parameters. In corner analysis, every combination of maximum and minimum values of each variable is simulated. In Monte Carlo simulations, each variable is randomly sampled and combined based on their statistical distribution, and the resulting statistical distribution of the system parameters is obtained.

4.7.2 Vibration Immunity

Gyroscopes are typically complicated vibratory systems with many high-Q resonant modes. Severe vibration environments in many applications could easily excite an undesired mode that could lead to a large bias error, saturation and even catastrophic failure.

In the gyroscope design cycle, it is crucial to define the vibration spectrum in the target application. Thorough FEA simulations and design optimization have to be performed to locate the operational modes and parasitic modes of the system away from the high-energy region of the vibration spectrum.

FEA simulations also allow to estimate the deflections in the gyroscope structure under acceleration. Simulation of deflections under acceleration reveal many important design aspects, such as maximum deflection points to guide the shock-stop design, the level of acceleration that cause contact, and capacitance changes due to acceleration. Common-mode rejection methods, such as anti-phase devices previously presented in this chapter, help alleviate the adverse effects of acceleration and vibration.

4.7.3 Shock Resistance

In micromachined gyroscopes, the most common failure mechanism due to shock is fracture. Crystalline silicon is a purely brittle material, which means that it deforms elastically until the maximum stress level reaches the yield strength. Even though the yield strength of Silicon is 7 GPa [107], a very wide range of yield strength values have been measured in the literature.

To minimize the probability of fracture at a given shock level, a practical design criteria is to keep the maximum von-Mises stress lower than 1 GPa. Almost all FEA software packages have the capability to model an acceleration on the structure and calculate the von-Mises stress distribution. In gyroscopes, the highest stress concentration points are usually suspension beam connections. In drive suspension beams, the superposition of the stresses due to the drive motion together with the shock stress has to be taken into consideration.

4.7.4 Temperature Effects

Stability of the gyroscope scale factor and bias over temperature are among the most challenging performance specifications. Temperature affects the electro-mechanical system parameters in many ways. Due to thermal expansion, the device geometry and the characteristics of the actuation and detection electrodes are altered. Thermally induced stresses due to expansion also affect the suspension stiffness values. Stress relief mechanisms and central anchors help minimize the thermal stresses.

Avoiding bi-morph effects by the use of materials with matched thermal expansion coefficients is crucial in process design. FEA simulations are essential in identifying and controlling thermal effects in a particular design.

Temperature also directly affects the Young's modulus of the structural material. It should also be noted that the viscous damping is also temperature dependent, which could have a drastic influence on scale factor and bias in devices with closely matched modes.

4.8 Summary

In this chapter, we presented fundamental mechanical design aspects of linear and torsional micromachined vibratory gyroscopes. Basic gyroscope dynamical system structures and their corresponding flexure systems were covered. The root cause and system-level implications of anisoelasticity and quadrature error were explained. Damping in vibratory systems, material properties of silicon, and critical design aspects for a robust sensing element were discussed.

Chapter 5
Electrical Design of MEMS Gyroscopes

In this chapter, we discuss the electrical design issues in a generic micro-electro-mechanical vibratory system to realize the complete gyroscopic system. First, the fundamentals of electrostatic actuation and capacitive sensing methods are covered. Then the details on complete characterization of MEMS gyroscopes, including frequency response acquisition, system identification, Coriolis signal detection and amplification, and rate signal extraction are presented.

5.1 Introduction

Micromachined gyroscopes are active devices, which require both actuation and detection mechanisms. Various vibratory MEMS gyroscopes have been reported in the literature employing a wide range of actuation and detection methods. For exciting the gyroscope drive mode oscillator, the most common actuation methods are electrostatic, piezoelectric, magnetic and thermal actuation. Most common Coriolis response detection techniques include capacitive, piezoelectric, piezoresistive, optical, and magnetic detection.

In many MEMS applications, capacitive detection and electrostatic actuation are known to offer several benefits compared to other sensing and actuation means, especially due their ease of implementation. Capacitive methods do not require integration of a special material, which makes them compatible with almost any fabrication process. They also provide good DC response and noise performance, high sensitivity, low drift, and low temperature sensitivity [4, 42].

5.2 Basics of Capacitive Electrodes

Various configurations of parallel-plate capacitors have been widely used for both electrostatic actuation and capacitive sensing in MEMS devices. In most micromachined gyroscopes, the actuation and sensing electrodes can be modeled as a combination of moving parallel-plate capacitors.

The term *capacitor* refers to a device that stores charge when a voltage is applied across its terminals. For linear capacitors such as parallel-plates, the *capacitance C* is the proportionality constant between the stored charge Q and the applied voltage V, governed by

$$Q = CV \tag{5.1}$$

For a generic parallel-plate electrode as in Figure 5.1, the capacitance between two parallel plates is expressed as

$$C = \varepsilon_0 \frac{A_{overlap}}{y_0} = \varepsilon_0 \frac{x_0 z_0}{y_0}. \tag{5.2}$$

where $\varepsilon_0 = 8.854 \times 10^{-12}$ F/m is the permittivity of free space, $A_{overlap} = x_0 z_0$ is the total overlap area, y_0 is the electrode gap.

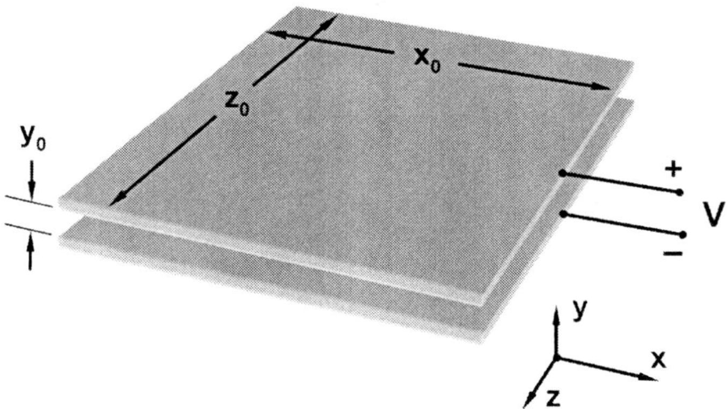

Fig. 5.1 Parallel-plate capacitor model.

The energy $W(Q)$ stored in capacitor as a function of charge and the co-energy $W^*(V)$ as a function of voltage applied across the capacitor are

$$W(Q) = \frac{Q^2}{2C} \tag{5.3}$$

$$W^*(V) = \frac{CV^2}{2} \tag{5.4}$$

which are equal in a linear capacitor.

5.3 Electrostatic Actuation

In parallel-plate actuation electrodes, the electrostatic force is generated due to the electrostatic conservative force field between the plates. In the case where the voltage across the capacitor is controlled, the electrostatic force can be expressed as the gradient of the co-energy $W^*(V)$ stored in the capacitor

$$\mathbf{F} = \nabla W^*(V) = \frac{\nabla C(x,y,z)V^2}{2} = \nabla\left(\frac{x_0 z_0}{y_0}\right)\frac{\varepsilon_0 V^2}{2} \qquad (5.5)$$

Parallel-plate actuation electrodes are usually designed to exert an electrostatic force in one particular direction, aligned with the degree of freedom and the motion direction of the actuated mass. The electrostatic force expression could be simplified by taking the derivative of capacitance with respect to the motion direction. The most general expression of electrostatic force in a specific motion direction vector $\hat{e}$ denoting the displacement along that direction as e is

$$\mathbf{F}_e = \frac{\partial W^*(V)}{\partial e} = \frac{1}{2}\frac{\partial C}{\partial e}V^2\hat{e} \qquad (5.6)$$

The two primary implementation types of parallel-plate actuation electrodes are variable-gap and variable-area actuators. The respective characteristics of each electrode type are discussed below.

5.3.1 Variable-Gap Actuators

In variable-gap actuators, the electrostatic actuation force of interest is the force component generated in the direction normal to the electrode plane. If we denote the normal direction as the y-axis and the nominal gap as y_0 as in Figure 5.2, the force in the y-direction becomes

$$F_y = \frac{1}{2}\frac{\partial C}{\partial y}V^2 = -\frac{1}{2}\frac{\varepsilon_0 z_0 x_0}{(y_0 + y)^2}V^2 \qquad (5.7)$$

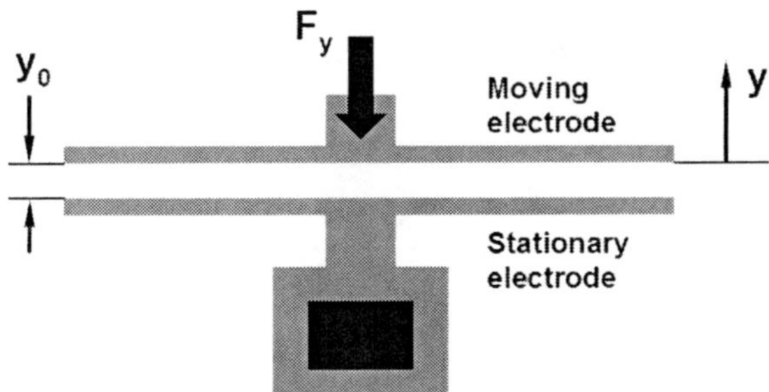

Fig. 5.2 Variable-gap electro-static actuator model.

It should be noticed that in variable-gap actuation, which is also known as parallel-plate actuation, force is a nonlinear function of the displacement y. This means that the electrostatic force increases as the gap between the plates decreases, which is widely known to result in instability and snap-down. The advantage of parallel-plate actuation is that it provides much larger force per area compared to variable-area actuators.

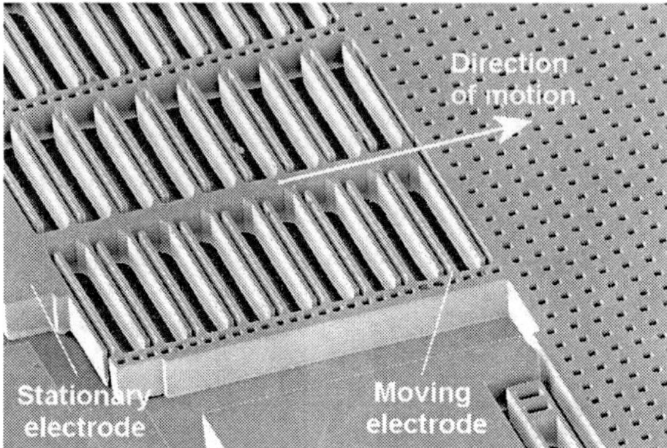

Fig. 5.3 Variable-gap actuators, also known as parallel-plate actuators, in a bulk micromachined vibratory gyroscope.

The nonlinear electrostatic force profile in parallel-plate actuation electrodes actually also becomes one of the major capabilities of electrostatic action, which is resonance frequency tuning. The electrostatic spring constant due to the force nonlinearity can be found by taking the derivative of the electrostatic force F_y with respect to displacement

$$k_{el} = -\frac{\partial F_y}{\partial y} = -\frac{\varepsilon_0 A}{(y_0 + y)^3} V^2 \tag{5.8}$$

The electrostatic spring constant of parallel-plates is a negative spring constant, and always reduces the resonant frequency with increasing net DC bias across the electrodes.

5.3.2 Variable-Area Actuators

Variable-area actuators aim to linearize the capacitance change versus displacement, in order to achieve constant electrostatic force with respect to displacement. The interdigitated comb-drive structure is based on generating the actuation force through a series of parallel plates sliding parallel to each other, without changing the gap between the plates. The electrostatic force generated in the x-direction for two parallel plates as in Figure 5.4 is

$$F_x = \frac{1}{2}\frac{\partial C}{\partial x}V^2 = \frac{1}{2}\frac{\partial(x_0 - x)}{\partial x}\frac{\varepsilon_0 z_0}{y_0}V^2 \tag{5.9}$$

$$F_x = -\frac{1}{2}\frac{\varepsilon_0 z_0}{y_0}V^2 \tag{5.10}$$

It should be noticed that this force is independent of displacement in the x-direction and the overlap length of the capacitor plates, x_0.

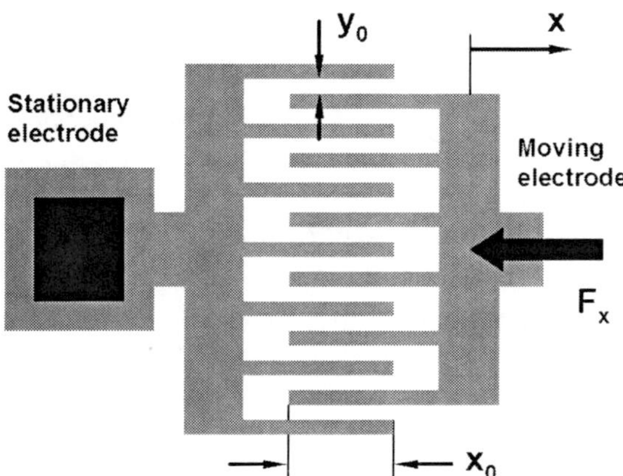

Fig. 5.4 Variable-area electrostatic actuator model.

Interdigitated comb-drives (Figure 5.4) based on variable-area actuation are one of the most common actuation structures used in MEMS devices. The primary advantages of comb-drives are long-stroke actuation capability and the ability to apply displacement-independent forces, which provide highly stable actuation.

In a comb-drive structure made of N fingers, each finger forms two parallel-plate pairs, and the total electrostatic force generated in the x-direction becomes

$$F_{comb} = -\frac{\varepsilon_0 z_0}{y_0}NV^2 \tag{5.11}$$

where z_0 is the structure thickness, and y_0 is the distance between the fingers. Since the force is independent of the initial overlap length, a good practice in comb-drive design is to keep the overlap length minimum, while greater than the expected actuation peak amplitude. This minimizes the parasitic lateral forces, which occur in the direction normal to the fingers, i.e. y-direction, due to the parallel-plate effect, and improves lateral stability.

Since the comb-drive force in the x-direction is not a function of the displacement in the x-direction, its partial derivative with respect to displacement is zero. This means that comb-drives do not result in a negative electrostatic spring constant, when fringing field effects are negligible.

Fig. 5.5 Bulk micromachining implementation of interdigitated comb-drives.

5.3.3 Balanced Actuation

The electrostatic force generated by any capacitor structure is proportional to the square of the potential difference. When a sinusoidal net actuation force is desired, the drive force can be linearized with respect to the actuation voltages by appropriate selection of voltages applied to the opposing electrode sets. The net electrostatic force generated by two opposing capacitors C_1 and C_2 is

$$F_{net} = \frac{1}{2}\frac{\partial C_1}{\partial x}V_1^2 - \frac{1}{2}\frac{\partial C_2}{\partial x}V_2^2 \tag{5.12}$$

A balanced actuation scheme is a common method to linearize the force with respect to a constant bias voltage V_{DC} and a time-varying voltage v_{AC}. The method is based on applying $V_1 = V_{DC} + v_{AC}$ to one set of electrodes, and $V_2 = V_{DC} - v_{AC}$ to the opposing set (Figure 5.6). Assuming two electrodes are identical, the net electrostatic force reduces to

$$F_{balanced} = \frac{1}{2}\frac{\partial C}{\partial x}\left[(V_{DC} + v_{AC})^2 - (V_{DC} - v_{AC})^2\right] = 2\frac{\partial C}{\partial x}V_{DC}v_{AC} \tag{5.13}$$

For an interdigitated comb-drive structure with N fingers on each side, structural thickness t, and finger distance d, the total drive force in a balanced actuation scheme becomes

$$F_{bal-comb} = 4\frac{\varepsilon_0 t N}{d}V_{DC}v_{AC} \tag{5.14}$$

For a parallel-plate actuator structure with N plates on each side, thickness t, overlap length L and plate gap d, the total drive force in a balanced actuation scheme assuming small deflections is

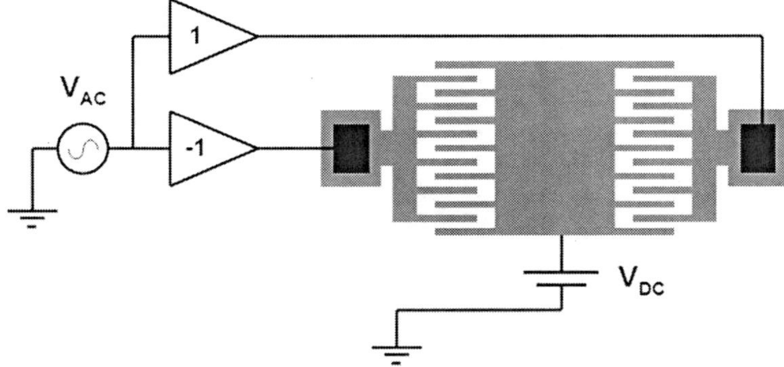

Fig. 5.6 The balanced driving scheme, based on applying $V_1 = V_{DC} + v_{AC}$ to one set of electrodes, and $V_2 = V_{DC} - v_{AC}$ to the opposing set.

$$F_{bal-pp} = 2\frac{\varepsilon_0 L t N}{d^2} V_{DC} v_{AC} \tag{5.15}$$

5.4 Capacitive Detection

Parallel-plate capacitors can be mechanized in several ways to detect deflection. For a generic parallel-plate electrode plate with a gap d and overlap area $A_{overlap}$, the capacitance is

$$C = \varepsilon_0 \varepsilon_r \frac{A_{overlap}}{d} \tag{5.16}$$

where ε_r is the dielectric constant of the material between the plates.

Each parameter in this expression can be modulated by a deflection to result in a capacitance change. In variable-gap capacitors, the motion is normal to the plane of parallel plates, and the gap d changes with deflection. In variable-area capacitors, the motion is parallel to the plane, which results in a change in $A_{overlap}$. By placing a moving media between the parallel plates, the dielectric constant ε_r can be modulated by deflection. The most common electrode types in inertial sensors are variable-gap and variable-area capacitors, which are summarized below.

5.4.1 Variable-Gap Capacitors

Variable-gap capacitors are the most widely used electrode type for detection of small displacements. When the parallel plates are oriented normal to the motion direction, deflections cause a change in the gap d. The electrode gap is usually determined by the minimum gap requirement of the fabrication process, and could be anywhere from a few microns to sub-micron dimensions. Thus, small gap changes could result in high capacitance changes, providing very large capacitance sensitivity.

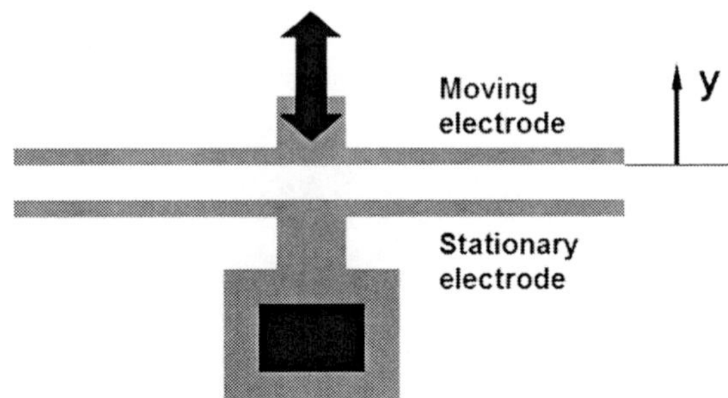

Fig. 5.7 Variable-gap capacitor electrode model, in which the deflections cause a change in the gap d.

It should be noticed that capacitance is a nonlinear function of displacement in variable-gap capacitors. However, for very small deflections relative to the initial gap, the capacitance change is linearized. Denoting the displacement in the motion direction as y and assuming $y \ll d$, the capacitance change in a variable-gap capacitor with an overlap length L becomes

$$\Delta C = \varepsilon_0 \frac{tL}{d-y} - \varepsilon_0 \frac{tL}{d} \approx \varepsilon_0 \frac{tL}{d^2} y \qquad (5.17)$$

5.4.2 Variable-Area Capacitors

Variable-area capacitors are ideal when the detected motion magnitudes are larger, especially either when variable-gap capacitors become significantly nonlinear, or deflections are larger than a minimum gap. Since the overlap area is proportional to both dimensions in the plate plane, capacitance change is purely linear with respect to motion parallel to the plates. Denoting x as the displacement in the motion direction parallel to the plates, the capacitance change becomes

$$\Delta C = \varepsilon_0 \frac{t(L+x)}{d} - \varepsilon_0 \frac{tL}{d} = \varepsilon_0 \frac{t}{d} x \qquad (5.18)$$

The sensitivity ratio of variable-gap capacitors to variable-area capacitors with similar feature sizes is L/d. This ratio illustrates that much higher sensitivities can be achieved by variable-gap capacitors by using long electrode fingers, with the expense of linearity.

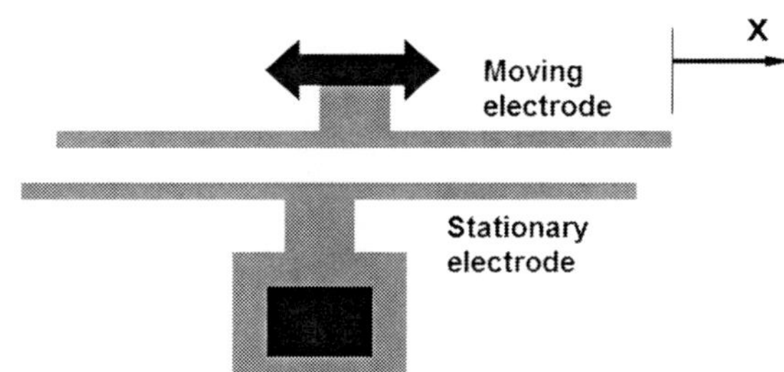

Fig. 5.8 Variable-area capacitor electrode model, in which
the deflections cause a change
in the overlap area.

5.4.3 Differential Sensing

Differential capacitance sensing (Figure 5.9) is generally employed to linearize the capacitance change with deflection. Differential detection is achieved by symmetrically placing capacitive electrodes on two opposing sides of the proof mass, such that the capacitance change in the electrodes are in opposite directions.

Depending on the fabrication process, differential electrodes could be designed in several ways. Most common design method for surface micromachining is to have two separate stationary electrodes on each side of a moving finger, as in Figure 5.9. Since anchor size requirements are small, the two stationary fingers can be anchored separately, and the set of fingers on the same detection side can be connected via an interconnect line.

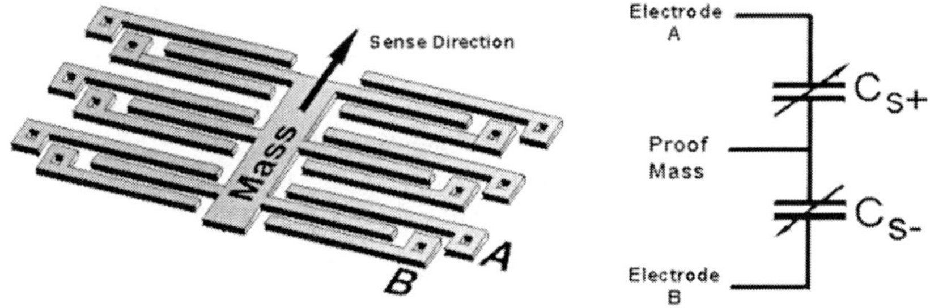

Fig. 5.9 Differential configuration of variable-gap capacitors to achieve linearity.

In bulk micromachining, individual fingers have to be attached to a branch and share a common anchor as in Figure 5.10. Moving fingers are also attached to a branch connected to the proof-mass. The opposing set of electrodes are placed on the other side, connected to a separate anchor.

Regardless of the design, functionality of differential detection is identical. For a positive displacement, the finger attached to the mass approaches finger A increasing the capacitance C_{s+}, and moves away from finger B decreasing the capacitance C_{s-}. Thus, a differential capacitive bridge is formed. Defining d as the finger separation, L as the length of the fingers, and t as the thickness of the fingers; the differential capacitance for an electrode set with N fingers on each side can be calculated as

$$C_{s+} = N\frac{\varepsilon_0 t L}{d-y}, \qquad C_{s-} = N\frac{\varepsilon_0 t L}{d+y}$$

$$\Delta C = C_{s+} - C_{s-} \approx 2N\frac{\varepsilon_0 t L}{d^2}y \qquad (5.19)$$

It is seen that the capacitance change is inversely proportional to the square of the initial gap. Thus, the performance of the sensor (i.e. sensitivity, resolution, and signal to noise ratio) increases with minimizing the electrode gaps.

Fig. 5.10 Bulk-micromachining implementation of the differential sensing capacitors.

For the bulk-micromachining implementation, the large gap on the opposing side of detection gap also becomes critical. It is desired to have minimum capacitance sensitivity to displacement on the large gap size, while keeping each set as compact as possible. A practical rule could be to use at least four times the sensing gap for the large gap.

5.5 Capacitance Enhancement

The performance of micromachined sensors employing capacitive detection is determined primarily by the sensitivity of detection electrode capacitance to displacement. Even though increasing the overall sensing area provides improved sensing capacitance, it was shown in the previous section that the sensing electrode gap is the foremost factor that defines the capacitance sensitivity.

In electrostatically actuated devices such as micromachined gyroscopes, actuation capacitance also directly affects performance. The actuation capacitance determines the required drive voltages. For a small actuation capacitance, large voltages are needed to achieve sufficient forces, which in turn results in a large drive signal feed-through. The drive signal feed-through is generally a major noise source, and often a larger signal than the measured Coriolis signal. Gyroscopes are generally designed with largest possible actuation capacitance, and operated in vacuum to achieve large amplitudes with low actuation voltages to minimize the drive feed-through.

Consequently, it is desired to minimize gap in both actuation and detection electrodes. Several approaches have been explored to reduce electrode gaps as outlined below.

5.5.1 Gap Reduction by Fabrication

Various advanced fabrication technologies have been reported to decrease electrode gaps beyond the capabilities of a typical etching process. Most of these approaches are based on deposition of thin layers on electrode sidewalls [123–125]. For example, in one reported approach, high aspect-ratio polysilicon structures are created by refilling deep trenches with polysilicon deposited over a sacrificial oxide layer. Thick single-crystal silicon structures are released from the substrate through the front side of the wafer by means of a combined directional and isotropic silicon dry etch and are protected on the sides by refilled trenches. This process involves one layer of low-pressure chemical vapor deposited (LPCVD) silicon nitride, one layer of LPCVD silicon dioxide, and one layer of LPCVD polysilicon [124].

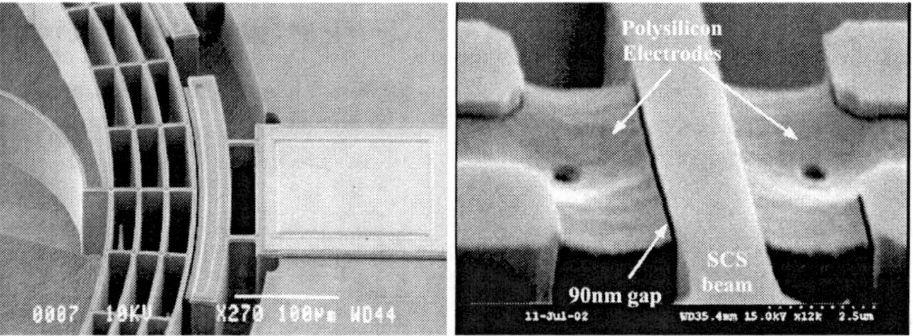

Fig. 5.11 Prior advanced fabrication techniques for capacitance enhancement [124, 125]. In this approach, high aspect-ratio polysilicon structures are created by refilling deep trenches with polysilicon deposited over a sacrificial oxide layer.

5.5.2 Post-Fabrication Capacitance Enhancement

An alternative to process modification to shrink electrode gaps is device geometry modification after fabrication. Post-release assembly techniques allow to increase the sensing and actuation capacitances in micromachined gyroscopes without the cost of complex fabrication steps, in order to enhance the performance and noise characteristics beyond the fabrication process limitations.

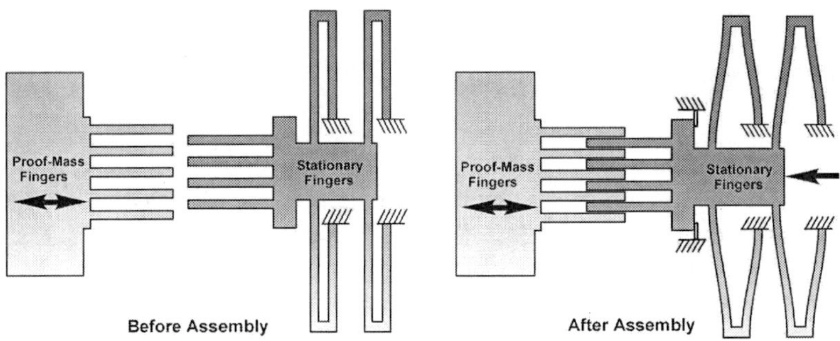

Fig. 5.12 Illustration of the post-release assembly of comb-drives.

The post-release assembly approach is based on attaching the stationary electrodes of the device to a moving stage that locks into the desired position to minimize the electrode gap after completion of the fabrication process. Thus, the initial gap before assembly can be designed as the minimum gap of the etching process, and the final gap after assembly is determined by design, independent of process requirements [65]. Thermal actuators could be used for assembling the moving stages. The concept can be applied both on variable-area capacitors such as comb-drives, and variable-gap capacitors.

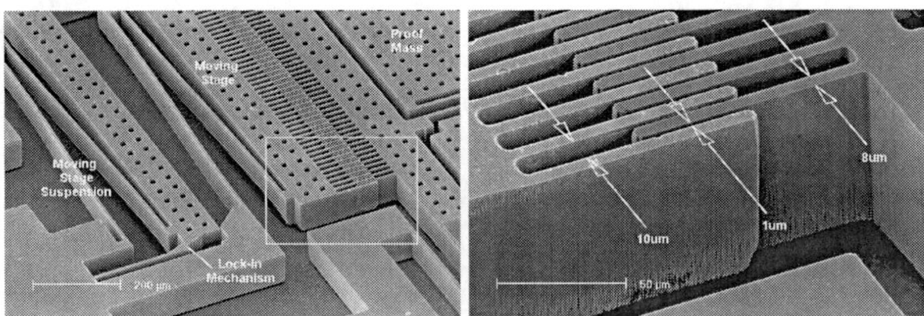

Fig. 5.13 The scanning-electron-microscope images of post-release positioned comb-drives integrated in a micromachined gyroscope. The minimum gap requirement of the fabrication process is $10\mu m$. The resulting gap between the stationary and moving fingers after the assembly is $1\mu m$ [65].

Scanning-electron-microscope images of post-release positioned comb-drives integrated in a micromachined gyroscope are presented in Figure 5.13. In this example device, the minimum gap requirement is $10\mu m$, which would be the gap between conventional comb-drive fingers. With the post-release positioning approach, $8\mu m$ wide fingers attached to opposing electrodes are designed initially apart with a gap of $10\mu m$. When interdigitated after release, the resulting gap between the stationary and moving fingers is $1\mu m$. This results in 10 times increase in the force per finger, and the number of fingers per unit area is increased 2 times by allowing smaller gaps. Figure 5.14 shows a gyroscope system successfully driven into resonance with this approach. Even smaller gaps could be achieved if the lateral stability of the moving stage is designed to be sufficient.

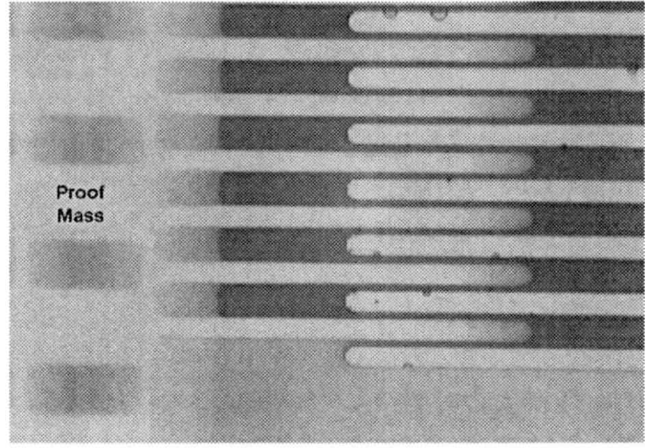

Fig. 5.14 Microscope photograph of the assembled post-release positioning comb-drives integrated in a micromachined gyroscope, showing resonance in the drive-mode with the assembled comb fingers.

Figure 5.15 presents SEM images of the post-release positioning variable-gap sensing electrodes integrated in a micromachined gyroscope, before and after assembly. It is observed that uniform gaps on the order of $1\text{-}2\mu m$ have been achieved with $100\mu m$ thick structures with minimum-gap requirement of $10\mu m$.

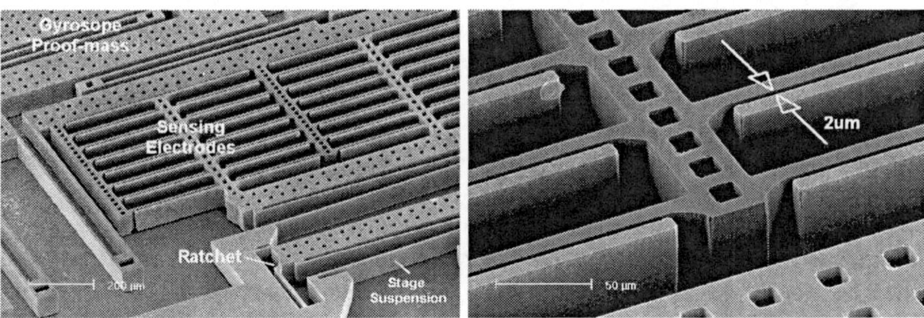

Fig. 5.15 Electrode gaps in the order of $1\text{-}2\mu m$ have been achieved with $100\mu m$ thick structures, while the minimum-gap requirement of the fabrication process is $10\mu m$.

5.6 MEMS Gyroscope Testing and Characterization

Various experimental characterization techniques are available for evaluation of micromachined gyroscope designs. Optical detection methods, including laser-Doppler vibrometry, confocal imaging, or computer vision with stroboscopic illumination provide very reliable in-plane and out-of-plane measurements, but require expensive optical characterization setups. Electrical characterization allows lower cost and simpler setups. Below is an overview of the most common optical and electrical methods used in experimental characterization of gyroscopes.

Optical Methods

The in-plane oscillation amplitude of vibrating micro-structures can be measured by the motion blur envelope with low accuracy. However, measuring the phase of the vibration is not possible. Utilizing stroboscopic illumination at an appropriate frequency, the position of the vibrating structure can be captured clearly, providing high accuracy amplitude and phase measurements. For automated accurate optical characterization frequency response, computer controlled stroboscopic illumination and image processing tools such as Polytec MMA-300 Micro Motion Analyzer [119] are available.

For out-of-plane vibration measurements, laser-Doppler-vibrometry (LDV) is the most accurate and effective technique. LDV is a non-contact vibration measurement method using the Doppler effect, based on the principle of detecting the Doppler shift of coherent laser light, that is scattered from a small area of the test object. Scanning Laser Doppler Vibrometers can measure the response of a dense array of points on the whole vibrating structure, allowing to detect the mode-shapes of the structure at specific frequencies.

For static in-plane or out-of plane measurements, non-contact optical profilers are ideal, and could be utilized for measuring static deflections, or obtaining the structural parameters, such as layer thickness, elevation, suspension beam geometry, and any possible curling in the structure. For example, the Confocal Imaging Profilers acquire confocal images using a confocal arrangement, fast scanning devices and high contrast algorithms.

Electrical Methods

When capacitive actuation and detection electrodes are available in the device, electrostatic actuation and capacitive detection is the most convenient characterization technique. The frequency response can be acquired under varying pressure values with a vacuum chamber, and at varying temperatures with a temperature controlled chamber. Using a signal analyzer in sine-sweep mode, the device can be excited over a desired frequency range, and capacitance change due to the response is amplified and converted into a voltage signal by transimpedance amplifiers. To obtain accurate results, suppressing the parasitic capacitances and electromagnetic interference is critical. Thus, it is often desired to package the device together with preamplifiers in the same package by direct wire bonding (Figure 5.16). This approach is usually implemented for Coriolis response detection. Fabricating the preamplifiers and signal conditioning electronics on the same chip with the device by an integrated fabrication process potentially provides the highest accuracy. Using a high-frequency carrier signal with subsequent synchronous demodulation also efficiently amplifies and separates the signal from electrical feedthroughs.

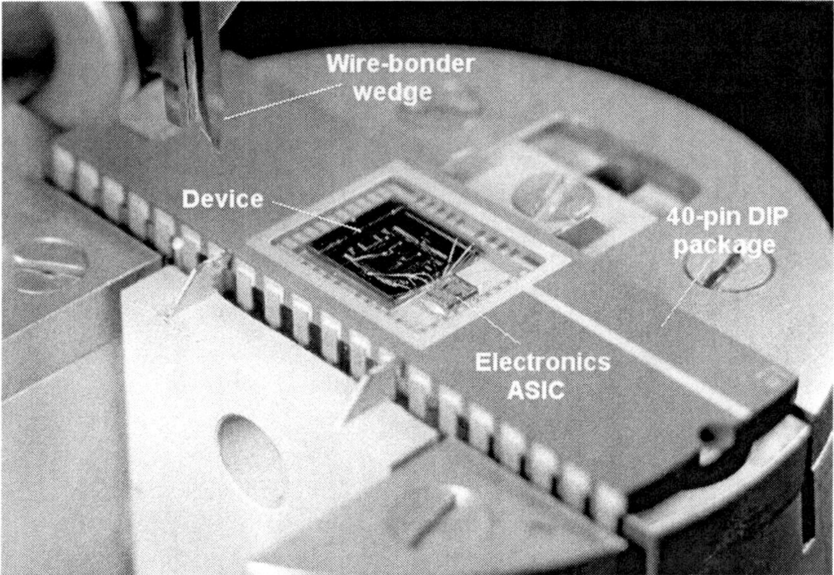

Fig. 5.16 A prototype gyroscope, packaged together with preamplifiers in the same package by direct wire bonding.

5.6.1 Frequency Response Extraction

The most effective way to identify the dynamical parameters of the resonant gyroscope system is to experimentally acquire the frequency response, usually by electrostatically exciting the system with a sine wave in frequency sweep mode, and capacitively detecting the response simultaneously.

When two electrodes are available in the device for electrostatic characterization, two-port actuation and detection can be utilized. In this approach, one probe is used to impose the DC bias voltage on the gyroscope structure through the anchor, one probe is used to apply the AC drive voltage on the actuation port, and the detection port is directly connected to the transimpedance amplifier (Figure 5.17).

If only one capacitive port is available in the device, one-port actuation and detection can be applied, which requires only two electrical contacts on the device (Figure 5.18). In the one-port actuation and detection approach, a single electrode is used for both driving and sensing at the same time. The driving AC signal plus the

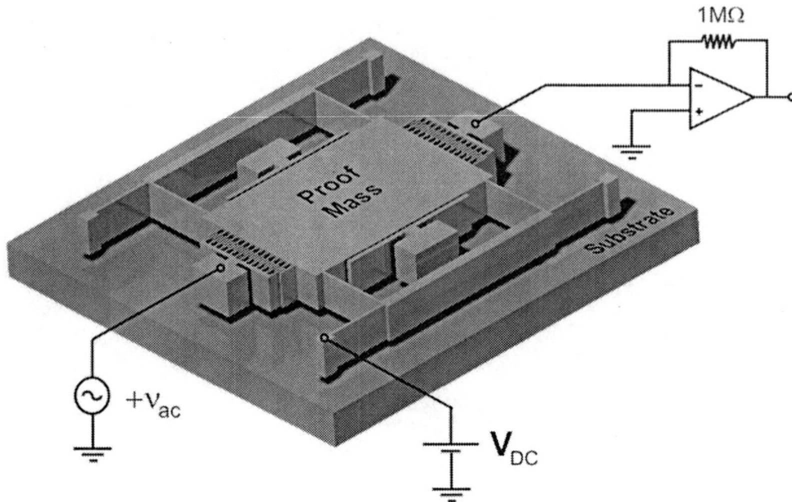

Fig. 5.17 The 2-port actuation and detection scheme for frequency response extraction.

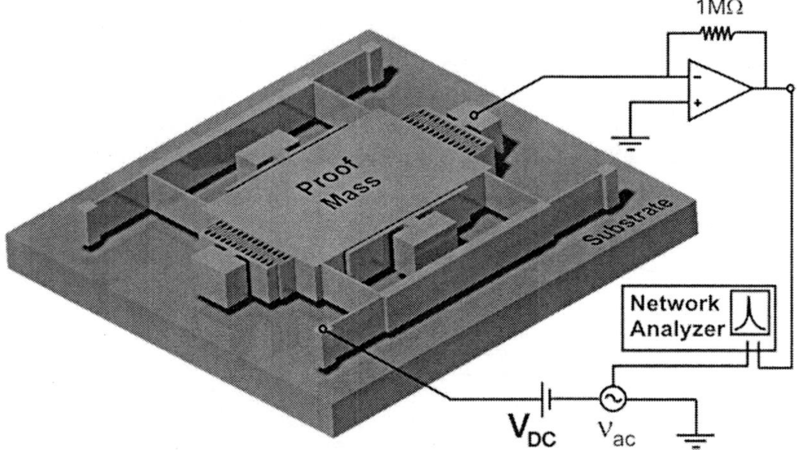

Fig. 5.18 The 1-port actuation and detection scheme for frequency response extraction.

DC bias voltage is imposed on the gyroscope structure through the anchor, and the actuation and detection port is directly connected to the transimpedance amplifier.

In both configurations, since one electrode is used for excitation of the resonator, and the net drive voltage between the drive electrode C_{drive} and the proof-mass is $V_{drive} = V_{DC} + v_{AC} \cos \omega t$, the drive force can be expressed as

$$F_{drive} = \frac{1}{2} \frac{\partial C_{drive}}{\partial x} (V_{DC} + v_{AC} \cos \omega t)^2 \tag{5.20}$$

When the DC terms and the double frequency terms are ignored, the forcing function with one electrode becomes

$$F_{drive} = \frac{\partial C_{drive}}{\partial x} V_{DC} \upsilon_{AC} \cos \omega t \qquad (5.21)$$

On the sensing electrode side, the motion of the proof-mass modulates the sense capacitor, which in turn causes a motional current. Since the net potential between the proof-mass and the sense electrode C_s in the two-port scheme is simply V_{DC}, the motional current is

$$i_s = \frac{d}{dt}\left[V_{DC} C_{sense}(t) \right] = V_{DC} \frac{\partial C_{sense}}{\partial x} \dot{x} \qquad (5.22)$$

The transimpedance amplifier converts the motional current into voltage, with a gain of K equal to the feedback resistor value. Thus, the output voltage from the transimpedance amplifier becomes

$$V_o = K V_{DC} \frac{\partial C_{sense}}{\partial x} \dot{x} \qquad (5.23)$$

In the two-port scheme, the overall transfer function of the two-port resonator with a mass m, spring constant k and damping c, from the input AC voltage $V_{AC} = \upsilon_{AC} \cos \omega t$ to the output voltage V_o can be expressed as

$$\frac{V_o}{V_{AC}} = K \frac{\partial C_{drive}}{\partial x} \frac{\partial C_{sense}}{\partial x} V_{DC}^2 \frac{s}{m s^2 + c s + k} \qquad (5.24)$$

In the one-port scheme, the excitation force is the same as the two-port scheme. However, the net potential between the proof-mass and the sense electrode is $V_{DC} + \upsilon_{AC} \cos \omega t$, which results in the motional current

$$i_s = \frac{\partial C_{sense}}{\partial x}\left[\dot{x}(V_{DC} + \upsilon_{AC} \cos \omega t) - x \upsilon_{AC} \sin \omega t \right] \qquad (5.25)$$

5.6.1.1 Frequency Response Setups

The previously discussed frequency response extraction methods can be realized using a DC power source, an on-chip or off-chip transimpedance amplifier with a feedback resistor, and a Dynamic Signal Analyzer. The device under characterization may either be packaged with wire-bonded permanent electrical connections as in Figure 5.16, or tested using a probe-station under microscope for temporary electrical connections.

The source of the Dynamic Signal Analyzer can be used as the driving AC signal, while the DC power supply applies the DC bias voltage on the structure. The frequency response is acquired by connecting the transimpedance amplifier output to the input of the Dynamic Signal Analyzer in sine-sweep mode (Figure 5.19).

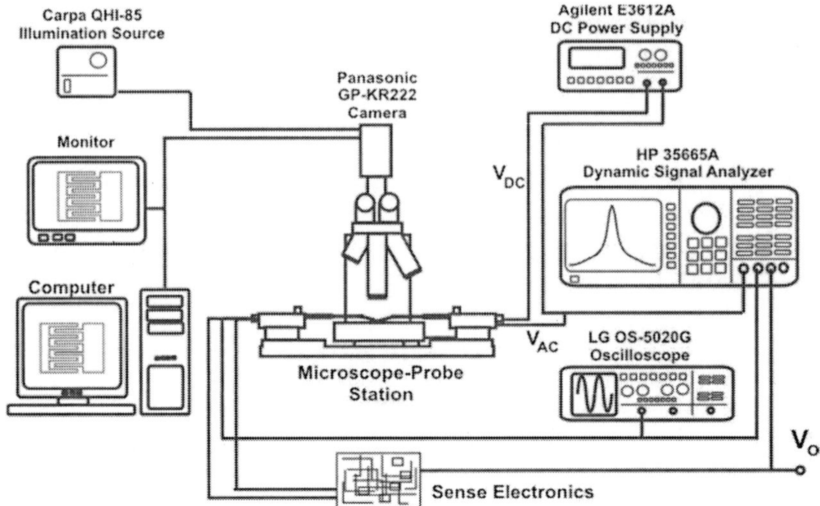

Fig. 5.19 An example experimental setup for electrostatic characterization of gyroscopes.

5.6.1.2 System Identification

Electrostatically acquiring the drive and sense mode frequency responses of a gyroscope system is the most common method to identify the dynamical parameters. However, the output signal is generally corrupted by the feed-through of the excitation signal to the detected signal over the parasitic electrical components in parallel to the ideal system dynamics. Although this corrupted signal could be used to give a rough approximation of the system parameters, accurate estimation of parameters, even the resonance frequency, is not possible. However, the ideal system response can be extracted from the corrupted response with analytical investigation of the real and imaginary components of the response.

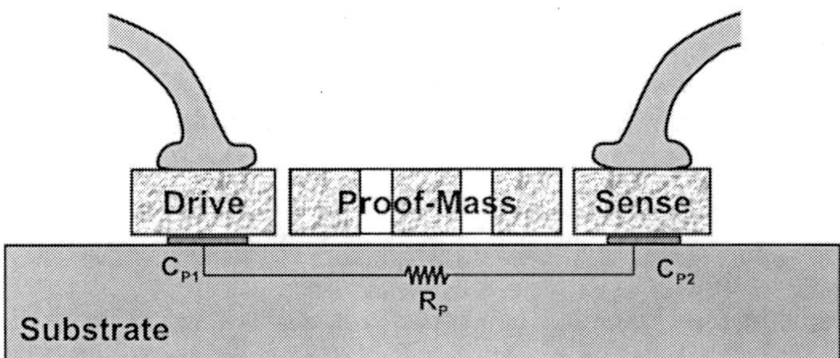

Fig. 5.20 The parasitic signal runs over the parasitic capacitances through the substrate.

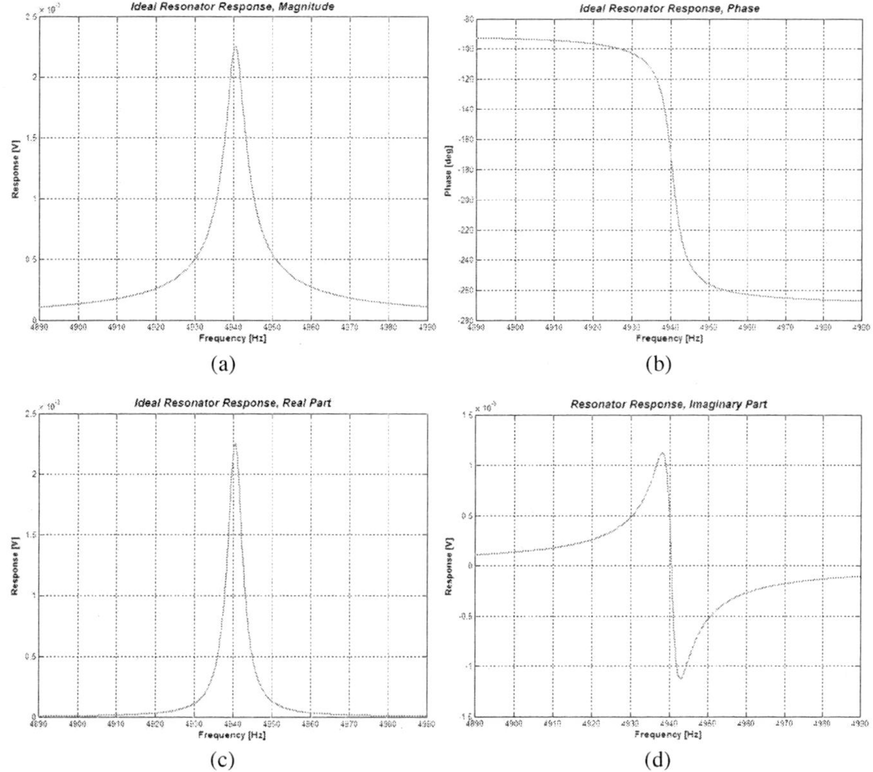

Fig. 5.21 The frequency sweep response of an ideal electro-mechanical resonator, with $C_p = 0$ and $R_p = 0$. (a) Magnitude, (b) Phase, (c) Real part, (d) Imaginary part. Notice that, the imaginary part of the response is zero at the resonance point. Away form resonance, the real part becomes zero.

In a device with a conductive substrate and no shielding, the overall feed-through current runs over a parasitic capacitance at the driving port pad, then through the substrate which has a finite resistance, and finally over another parasitic capacitance at the sensing port pad. This parasitic model can be lumped into one parasitic capacitance C_p and one parasitic resistance R_p in series, parallel to the electro-mechanical resonator, as seen in (Figure 5.20). If we include the parasitic signal that couples the output voltage to the input voltage in parallel to the electro-mechanical resonator dynamics, the transfer function of the overall system introduced earlier in this section but including the parasitic capacitance and resistances becomes

$$\frac{V_o}{V_{AC}} = K\alpha \frac{s}{m s^2 + c s + k} + K \frac{C_p s}{R_p C_p s + 1} \tag{5.26}$$

where K is the transimpedance amplifier gain, and the constant α, which contains the coefficients for conversion of the input sine wave to the mechanical force and the mechanical displacement to the motion induced current, is

$$\alpha = \frac{\partial C_{drive}}{\partial x} \frac{\partial C_{sense}}{\partial x} V_{DC}^{2} \tag{5.27}$$

With the parasitic feedthrough, the real and imaginary parts of the total transfer function frequency response become

$$Re = K \left(\frac{\alpha c \omega^{2}}{(k - m\omega^{2})^{2} + c^{2}\omega^{2}} + \frac{R_{p}C_{p}^{2}\omega^{2}}{(R_{p}C_{p}\omega)^{2} + 1} \right) \tag{5.28}$$

$$Im = K \left(\frac{\alpha(k - m\omega^{2})\omega}{(k - m\omega^{2})^{2} + c^{2}\omega^{2}} + \frac{C_{p}\omega}{(R_{p}C_{p}\omega)^{2} + 1} \right) \tag{5.29}$$

It should be noticed that, the real part of the response includes a term that is purely mechanical with none of the parasitic effects, and an offset term that is purely due to parasitics. Thus, if the real part is plotted, there will be a peak exactly at the mechanical resonance point $\omega_{n} = \sqrt{k/m}$.

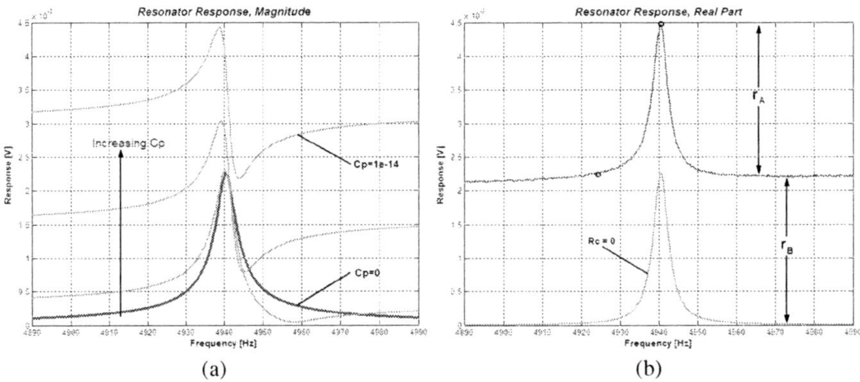

Fig. 5.22 (a) The deviation in the frequency response with increasing parasitic capacitance C_{p}. (b) The real part of the response is offset from zero, due to the parasitic resistance $R_{p} = 0$.

Starting with the experimentally acquired values of these terms over a frequency range that includes the resonance peak, every mechanical and electrical parameter of the dynamical system can be identified, including the parasitic effects. The following algorithm automatically extracts the parameters by examining the real and imaginary values of the response at special points (for simplicity, the response is divided by the amplifier gain K):

1- The exact mechanical resonance point is very easily identified from real part of the response. The frequency where the real part reaches its maximum is the resonant frequency $\omega_{n} = \sqrt{k/m}$. Starting with a given mass m of the resonator (mass is the parameter least affected by process variations, and easily estimated), the spring constant k is identified.

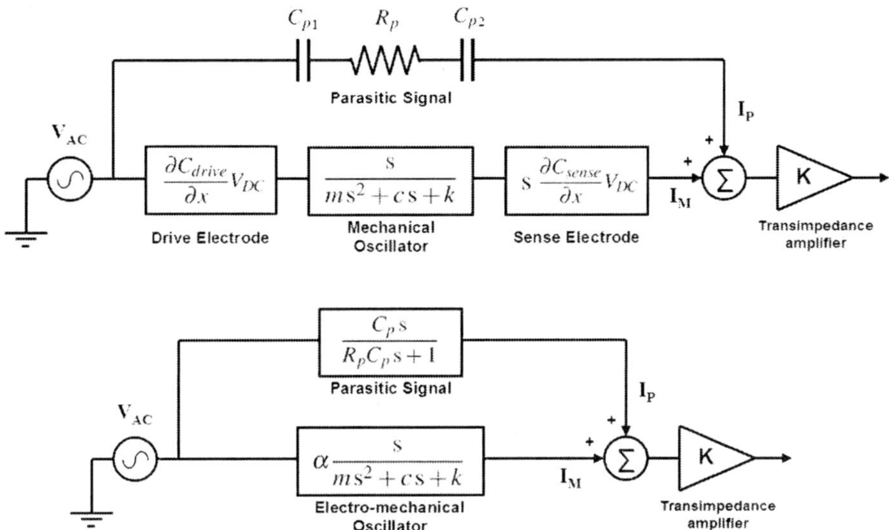

Fig. 5.23 The transfer function model of the overall system, including the lumped parasitic capacitance, and the substrate resistance.

2- The real part of the response includes only the parasitic effects at the frequencies away from the resonance point. This part evaluated at one frequency ω_l provides one non-linear equation with two unknowns R_p and C_p, and is equal to the offset in the real part (r_B in Figure 5.22b):

$$Re_l = \frac{R_p C_p^2 \omega_l^2}{(R_p C_p \omega_l)^2 + 1} = r_B. \tag{5.30}$$

3- At the resonance point, the imaginary part of an ideal system's response is zero, by the definition of resonance. Thus, the imaginary response at resonance is equal to the parasitic term, providing the second non-linear equation with two unknowns R_p and C_p:

$$Im_n = \frac{C_p \omega_n}{(R_p C_p \omega_n)^2 + 1} \tag{5.31}$$

These two equations are solved simultaneously to get the values of R_p and C_p.

4- At the resonance point, the real part of the response reduces to a very simple expression, plus the offset due to the parasitic effects:

$$Re_n = \frac{\alpha c \omega_n^2}{(k - m\omega_n^2)^2 + c^2 \omega_n^2} + r_B. \qquad \Rightarrow \qquad \frac{\alpha}{c} = r_A. \tag{5.32}$$

5- Evaluating the imaginary part of the response slightly away from resonance yields a second equation for the unknowns α and c. Since this point is used to evaluate damping, a point close to the half-power bandwidth points assures better estimation. The point ω_m used in the analysis is shown on Figure 5.24a.

$$Im_m - \frac{C_p\omega_m}{(R_pC_p\omega_m)^2 + 1} = \frac{\alpha(k - m\omega_m^2)\omega}{(k - m\omega_m^2)^2 + c^2\omega_m^2}. \tag{5.33}$$

Solving these two equations simultaneously, the values of the electrical gain α and the damping coefficient c are identified. Figure 5.24 shows the experimentally acquired response, and the simulated response with the identified parameters; verifying the estimation accuracy of the system parameters and parasitics.

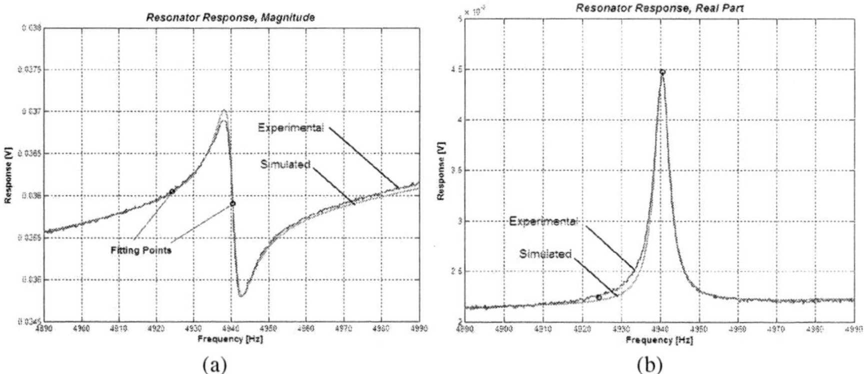

Fig. 5.24 The algorithm automatically extracts the system parameters from the real and imaginary parts of the frequency response. The experimentally acquired response, and the simulated response with the identified parameters are shown in (a) the magnitude, and (b) the real part of the response.

Having identified the parasitic terms in the real and imaginary parts of the response, these terms can be numerically filtered from the measured signal to reflect the actual mechanical dynamics, by subtracting the evaluated parasitic term at each frequency from the acquired frequency response trace. This system identification algorithm and the numerical filtering approach effectively provides accurate characterization results without the need for complicated detection electronics. Alternatively, a synchronous demodulation method using a high-frequency carrier could be used to electrically separate the response from parasitics, as will be explained the following sections.

5.6.2 Capacitive Sense-Mode Detection Circuits

Detection of the Coriolis response in the sense-mode is challenging, since it requires measurements of picometer-scale oscillations in the sense-mode, while the proof-mass oscillates with tens of microns amplitude in the drive-mode.

The most basic detection approach is to directly amplify the motional current due to the sense-mode oscillation. By imposing a constant DC bias voltage V_{DC} over a sense electrode with the capacitance $C_s = C_{sn} + \Delta C_s \sin \omega_d t$, the motional current becomes

$$i_s = \frac{d}{dt}\left[V_{DC} C_s(t) \right] = V_{DC} \omega_d \Delta C_s \cos \omega_d t \qquad (5.34)$$

To illustrate the magnitude of Coriolis motional currents, let us investigate a mode-mismatched gyroscope system with $f = 10$ kHz operation frequency and $\Delta f = 100$ Hz mismatch, sense-mode quality factor $Q_s = 1000$, and 10 μm drive-mode amplitude. For a 1°/s angular rate input, the sense-mode amplitude is 0.16 nm. If we have 100 detection electrodes on each side of the differential bridge with $L = 250$ μm length, $t = 50$ μm thickness and $d = 5$ μm gap, the total capacitance change is

$$\Delta C_s \approx 2N \frac{\varepsilon_0 t L}{d^2} y = 0.142 \text{ aF } / \ (°/s) \qquad (5.35)$$

Since the capacitance change ΔC_s levels are usually on the order of atto-Farads, the resulting motional current is also extremely small. For example, in the same system with a DC bias of $V_{DC} = 10$V, the resulting motional current is 8.9×10^{-12}A. Furthermore, the sense signal is at the same frequency as the parasitic feedthrough due to the drive voltage, which could be orders of magnitude larger.

The *synchronous demodulation* technique is commonly used to boost the amplitude the sense signals and separate them from the frequency band of the parasitic feedthrough signals. The method is based on imposing a high-frequency carrier signal on the structure, which is the common-mode of the differential capacitive bridge in the sense-mode (Figure 5.25). The current output from each sensing capacitor due to the carrier signal is converted into a voltage signal and amplified. The difference of the outputs is amplitude demodulated at the carrier signal frequency, yielding the Coriolis response signal at the driving frequency.

This scheme can be modeled as a sinusoidal carrier signal V_c imposed over the differential sensing capacitor C_{s+} and C_{s-}, and the sense current is amplitude modulated by the change in capacitance. The carrier signal and the sensing capacitance will then be of the form

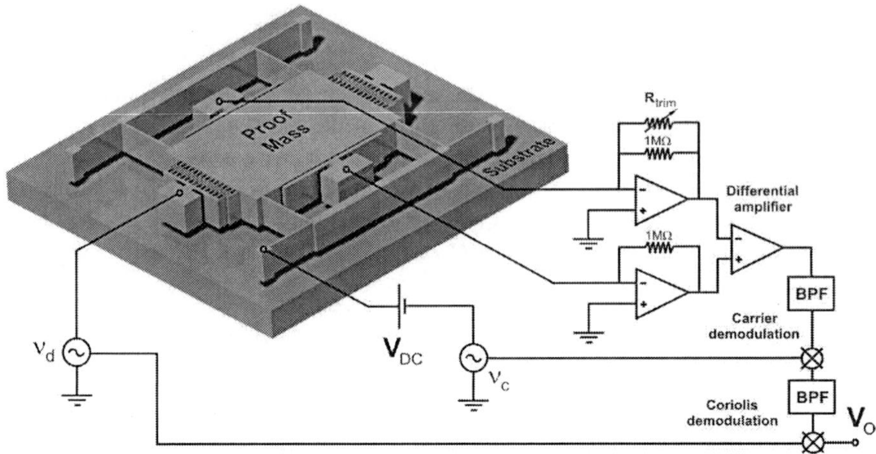

Fig. 5.25 Implementation of synchronous demodulation technique for detection of the Coriolis response, using a high-frequency carrier signal.

$$V_c = v_c \sin \omega_c t \tag{5.36}$$

$$C_{s+} = C_{sn+} + \Delta C_s \sin \omega_d t \tag{5.37}$$

$$C_{s-} = C_{sn-} - \Delta C_s \sin \omega_d t \tag{5.38}$$

where C_{sn+} and C_{sn-} are the nominal sensing capacitances, ΔC_s is the amplitude of capacitance change due to the Coriolis response, ω_d is the driving frequency, and ω_c is the carrier signal frequency. The sense currents from the sense capacitors are

$$i_{s+} = \frac{d}{dt}\left[V_c(t)C_{s+}(t) \right] = \frac{d}{dt}\left[C_{sn+}v_c \sin \omega_c t + \Delta C_s v_c \sin \omega_c t \sin \omega_d t \right]$$

$$i_{s-} = \frac{d}{dt}\left[V_c(t)C_{s-}(t) \right] = \frac{d}{dt}\left[C_{sn-}v_c \sin \omega_c t - \Delta C_s v_c \sin \omega_c t \sin \omega_d t \right] \tag{5.39}$$

The sense current from each capacitor is amplified by a trans-impedance amplifier with a gain of K, and converted into the voltage signals V_{s+} and V_{s-}

$$V_{s+} = K\omega_c C_{sn+}v_c \cos \omega_c t + \frac{K\Delta C_s v_c}{2}\left[(\omega_c + \omega_d)\sin(\omega_c + \omega_d)t - (\omega_c - \omega_d)\sin(\omega_c - \omega_d)t \right]$$

$$V_{s-} = K\omega_c C_{sn-}v_c \cos \omega_c t - \frac{K\Delta C_s v_c}{2}\left[(\omega_c + \omega_d)\sin(\omega_c + \omega_d)t - (\omega_c - \omega_d)\sin(\omega_c - \omega_d)t \right]$$

A differential amplifier takes the difference between the two voltage signals. When the nominal capacitances of the differential bridge are well matched, the carrier feedthrough terms $K\omega_c C_{sn} v_c \cos \omega_c t$ are canceled. The differential amplifier also cancels part of the drive feedthrough to the extent that the parasitic capacitances

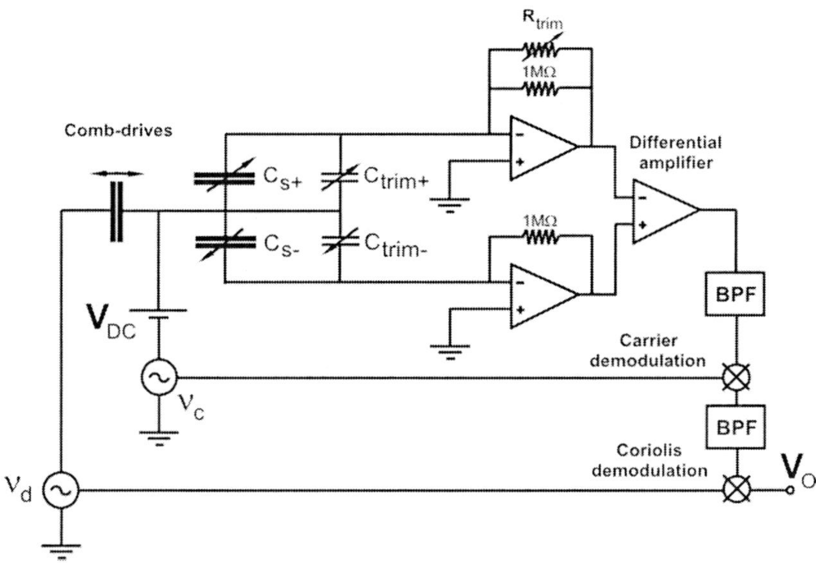

Fig. 5.26 The circuit-level model of the synchronous demodulation technique, including the trimming capacitors for capacitance matching.

are symmetric. After a band-pass filter around ω_c the resulting difference signal V_s becomes

$$V_s = V_{s+} - V_{s-} = K\Delta C_s \upsilon_c \left[(\omega_c + \omega_d)\sin(\omega_c + \omega_d)t - (\omega_c - \omega_d)\sin(\omega_c - \omega_d)t \right]$$

V_s is then amplitude demodulated at the carrier frequency ω_c by multiplying the sense signal V_s with a carrier reference signal

$$V_{cr} = \upsilon_{cr}\cos(\omega_c t + \theta) \tag{5.40}$$

After multiplication, the amplitude demodulated signal V_{sd} is obtained as

$$V_{sd} = +\tfrac{1}{2}K\Delta C_s \upsilon_c \upsilon_{cr}(\omega_c + \omega_d)\left[\cos(\omega_d t + \theta) - \cos\left((2\omega_c + \omega_d)t + \theta\right)\right]$$
$$-\tfrac{1}{2}K\Delta C_s \upsilon_c \upsilon_{cr}(\omega_c - \omega_d)\left[\cos(\omega_d t + \theta) - \cos\left((2\omega_c - \omega_d)t + \theta\right)\right]$$

The amplitude of the demodulated signal is maximized when $\theta = 90°$ for $\omega_c \gg \omega_d$ [133]. When the amplitude demodulated signal V_{sd} is band-pass filtered at the drive frequency ω_d, the high-frequency signals at $2\omega_c$, $(2\omega_c + \omega_d)$, and $(2\omega_c - \omega_d)$ are attenuated, leaving

$$V_{sd} = K\omega_c \Delta C_{so}\upsilon_c \upsilon_{cr}\sin\omega_d t \tag{5.41}$$

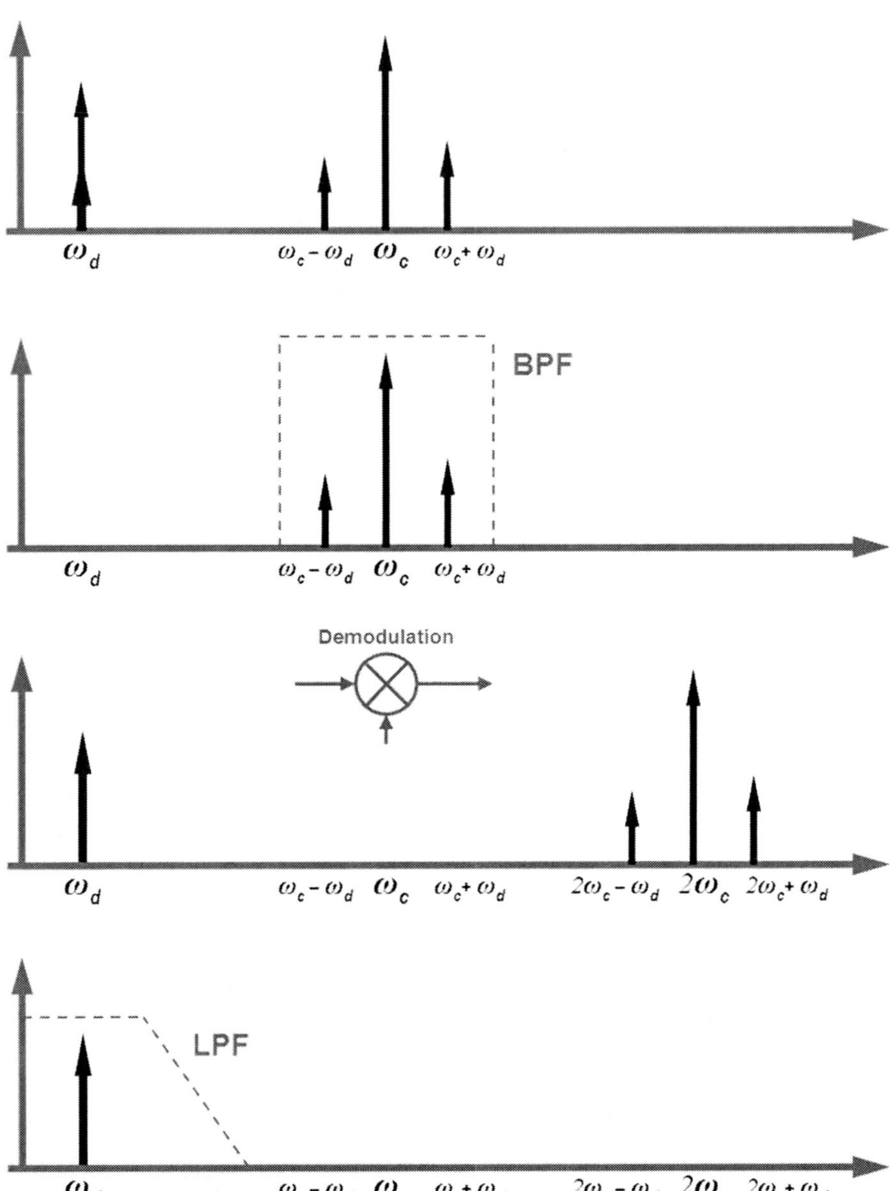

Fig. 5.27 Spectral density plots of the synchronous demodulation technique. First, the motion at ω_d modulates the carrier signal, resulting in the sidebands. Then the bandpass filter strips off the parasitic feedthrough at ω_d. After demodulation, the sidebands are down-converted to ω_d and lowpass filtered.

The signal V_{sd} is further amplitude demodulated at the driving frequency using a reference signal derived from the drive oscillator loop. If we consider the amplitude ratio of a synchronous demodulation output to basic motional current amplification, we have

$$\frac{|V_{sd}|}{|V_{motional}|} = \frac{K\omega_c \Delta C_{so} \upsilon_c \upsilon_{cr}}{2K\omega_d V_{DC}\Delta C_s} = \frac{\omega_c \upsilon_c \upsilon_{cr}}{2\omega_d} \tag{5.42}$$

This ratio illustrates that dramatically improved output signal amplitudes are achievable with the use of a high-frequency carrier signal, since $\omega_c \gg \omega_d$. In the practical implementation of the synchronous demodulation technique, the imbalance in the parasitic capacitances of the sense-mode differential capacitive bridge results in a large DC offset, corrupting the Coriolis signal. Utilizing trimming capacitors in parallel to the capacitors in the differential capacitive bridge, capacitance matching can be achieved, and the DC offset can be nulled.

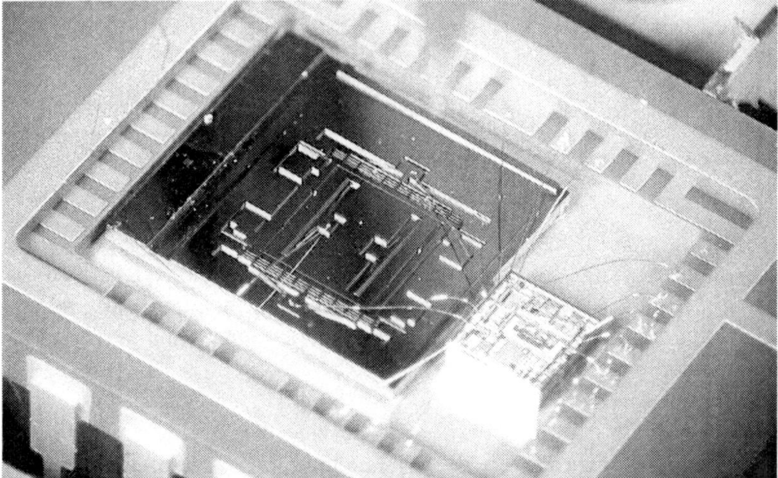

Fig. 5.28 Gyroscope packaged together with a capacitive readout chip by direct wire bonding.

Alternative detection techniques more suited for digital signal processing have also been implemented on micromachined gyroscopes. Switched capacitor techniques and Sigma-Delta ($\Sigma\Delta$) modulators [137–139] are among the most common methods.

5.6.3 Rate-Table Characterization

For characterizing the sensitivity of the gyroscopes to angular rate input, the gyroscopes are usually mounted on a rate table. Rate table equipment can provide very precise angular rates, while allowing electrical connections through slip-rings. An example multi-axis rate table by Ideal Aerosmith is shown in Figure 5.29. Rate tables are also available with climatic chambers to allow testing over temperature.

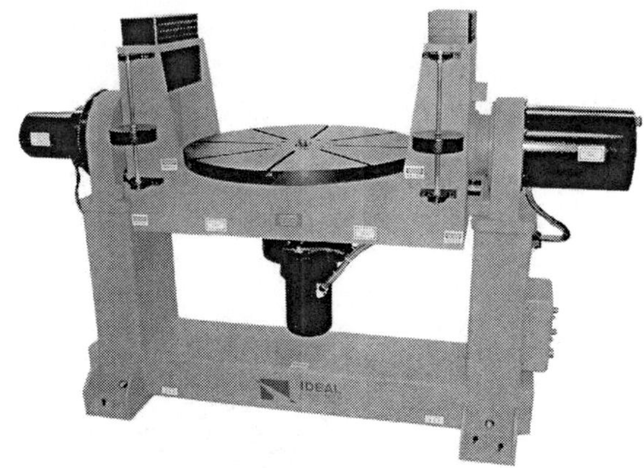

Fig. 5.29 A multi-axis rate table by Ideal Aerosmith.

In laboratory conditions, it is preferable to mount the gyroscope on a rate table together with its control and signal conditioning electronics, since it is not desirable to transfer low-level signals through slip-rings. Alternatively, the front-end amplifiers can be mounted together with the gyroscope, and remaining signal conditioning can be implemented externally.

For synchronous demodulation, lock-in amplifiers are usually very practical. Digital lock-in amplifiers can be used to provide the driving AC signal and carrier signal, and to synchronously demodulate the carrier readout and Coriolis signal with great precision. The output of the lock-in amplifier is further low-pass filtered to achieve the desired sensing bandwidth of the gyroscope. The experimental setup for implementing this characterization scheme is presented in Figure 5.30.

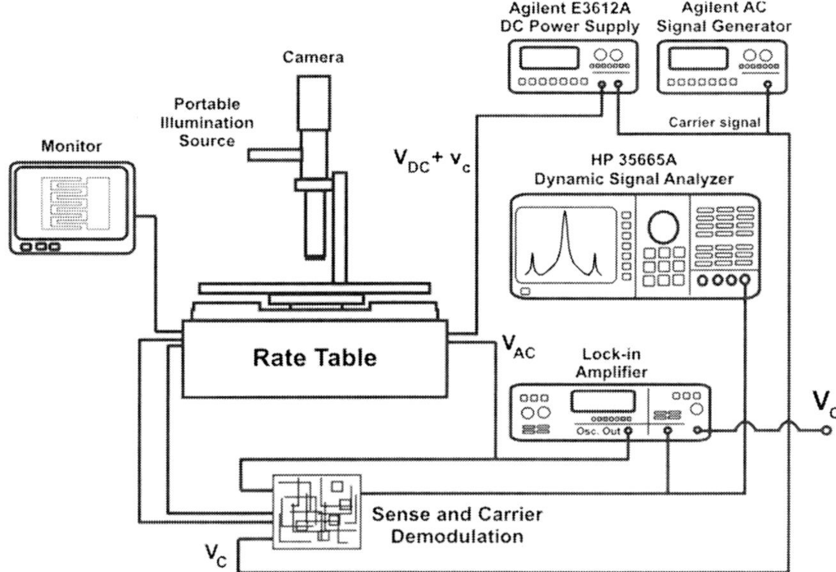

Fig. 5.30 Schematic of an experimental setup for rate-table evaluation of gyroscopes.

5.7 Summary

In this chapter, we presented an overview of electrostatic actuation and capacitive sensing in micromachined vibratory gyroscopes. Basic electrical design issues and system-level characterization of MEMS gyroscopes were discussed, with an introduction to optical and electrical methods. Mathematical theory and implementation of frequency response acquisition, system identification and parasitics filtering, and Coriolis signal detection using synchronous demodulation were covered.

Part II
Structural Approaches to Improve Robustness

Chapter 6
Linear Multi-DOF Architecture

This chapter introduces the Multi-DOF design concept that aims to eliminate the mode-matching challenge in conventional micromachined vibratory gyroscope systems. The chapter is organized as follows: first the 2-DOF sense-mode architecture, then the 2-DOF drive-mode architecture, and finally the 4-DOF system with 2-DOF drive and 2-DOF sense-modes are presented. The structure, operation principle, dynamics, example prototype design and characterization results of each architecture are covered.

6.1 Introduction

In previous chapters, it was illustrated that the performance of gyroscope systems with conventional 1-DOF drive and 1-DOF sense-mode oscillators is very sensitive to variations in system parameters that shift the drive or sense resonant frequencies. Especially under high quality factor conditions, even though high gain is achieved with close matching of the modes, extremely small frequency variations result in abrupt gain and phase changes. Furthermore, fluctuations in damping conditions directly affect the gain and the phases. In practical implementations, in order to achieve a stable response, the drive and sense mode resonant frequencies are intentionally separated by

$$\Delta f = f_s - f_d \tag{6.1}$$

Consequently, the conventional gyroscope system that relies on relative location of the drive and sense resonant peaks has strict mode-matching requirements, which renders the system response very sensitive to fabrication imperfections and fluctuations in operating conditions.

The Multi-DOF gyroscope design concepts introduced in this chapter provide drive and/or sense mode frequency responses with a flat operating frequency region where the response gain and phase are stable, in contrast to a conventional resonant

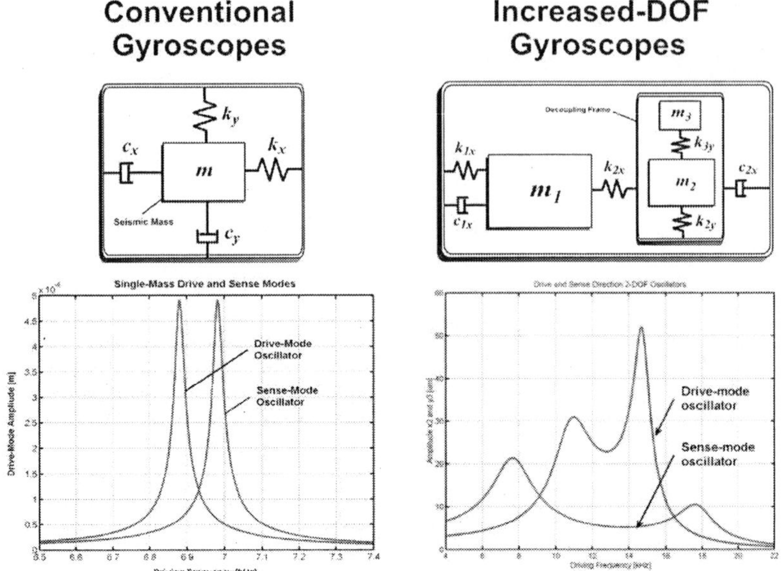

Fig. 6.1 Comparison of the single-mass conventional gyroscopes and the Multi-DOF design approach.

system (Figure 6.1). Thus, the sensing element dynamical system becomes inherently robust against structural and environmental parameter variations, and requires less demanding compensation schemes.

6.2 Fundamentals of 2-DOF Oscillators

Dynamical systems with two coupled proof-masses are known to have two resonance peaks: the in-phase mode and the anti-phase mode. At the in-phase mode frequency, the masses oscillate together in the same direction, while at the anti-phase mode frequency, the masses oscillate in the opposite directions.

In a generic 2-DOF oscillator system formed by the two masses m_1 and m_2 as in Figure 6.2, let us denote m_1 as the primary mass, and m_2 as the secondary mass. Assuming the excitation force F_d is applied on the primary mass m_1, the equations of motion can be expressed as

$$m_1\ddot{x}_1 + c_1\dot{x}_1 + k_1x_1 = k_2(x_2 - x_1) + F_d$$
$$m_2\ddot{x}_2 + c_2\dot{x}_2 + k_2x_2 = k_2x_1 \tag{6.2}$$

If the excitation force is a constant-amplitude sinusoidal drive force in the form $F_d = F_0\sin\omega_d t$, the steady-state response of the 2-DOF oscillator masses will be

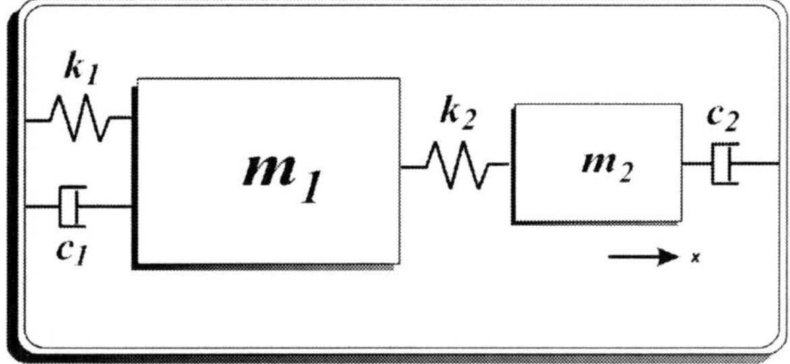

Fig. 6.2 A generic 2-DOF oscillator model.

$$X_1(j\omega) = \frac{F_0}{k_1} \frac{1 - (\frac{\omega}{\omega_2})^2 + j\omega\frac{c_2}{k_2}}{[1 + \frac{k_2}{k_1} - (\frac{\omega}{\omega_1})^2 + j\omega\frac{c_1}{k_1}][1 - (\frac{\omega}{\omega_2})^2 + j\omega\frac{c_2}{k_2}] - \frac{k_2}{k_1}}$$

$$X_2(j\omega) = \frac{F_0}{k_1} \frac{1}{[1 + \frac{k_2}{k_1} - (\frac{\omega}{\omega_1})^2 + j\omega\frac{c_1}{k_1}][1 - (\frac{\omega}{\omega_2})^2 + j\omega\frac{c_2}{k_2}] - \frac{k_2}{k_1}} \tag{6.3}$$

$$\omega_1 = \sqrt{\frac{k_1}{m_1}} \qquad \omega_2 = \sqrt{\frac{k_2}{m_2}} \tag{6.4}$$

where ω_1 and ω_2 are the resonant frequencies of the isolated primary and secondary mass-spring systems, respectively. The in-phase resonance frequency ω_{ip} and the anti-phase resonance frequency ω_{ap} of the oscillator can be found by equating the denominators of the undamped response to zero, yielding

$$\omega_{ip} = \sqrt{\frac{1}{2}\left(1 + \mu + \frac{1}{\gamma^2} - \sqrt{\left(1 + \mu + \frac{1}{\gamma^2}\right)^2 - \frac{4}{\gamma^2}}\right)} \, \omega_2$$

$$\omega_{ap} = \sqrt{\frac{1}{2}\left(1 + \mu + \frac{1}{\gamma^2} + \sqrt{\left(1 + \mu + \frac{1}{\gamma^2}\right)^2 - \frac{4}{\gamma^2}}\right)} \, \omega_2 \tag{6.5}$$

$$\mu = \frac{m_2}{m_1} \qquad \gamma = \frac{\omega_2}{\omega_1} = \sqrt{\frac{k_2 m_1}{k_1 m_2}} \tag{6.6}$$

where μ is the mass ratio, and γ is the ratio of the resonance frequencies of the isolated primary and secondary mass-spring systems [93].

When the oscillator is driven at the resonant frequency of the isolated secondary mass-spring system, i.e. $\omega_d = \omega_2$, the secondary mass moves to exactly cancel out the input force F_d applied on the active mass, similar to a vibration absorber, and the primary mass oscillation amplitude is minimized. This frequency, where maximum dynamic amplification is achieved, is called the *antiresonance frequency*.

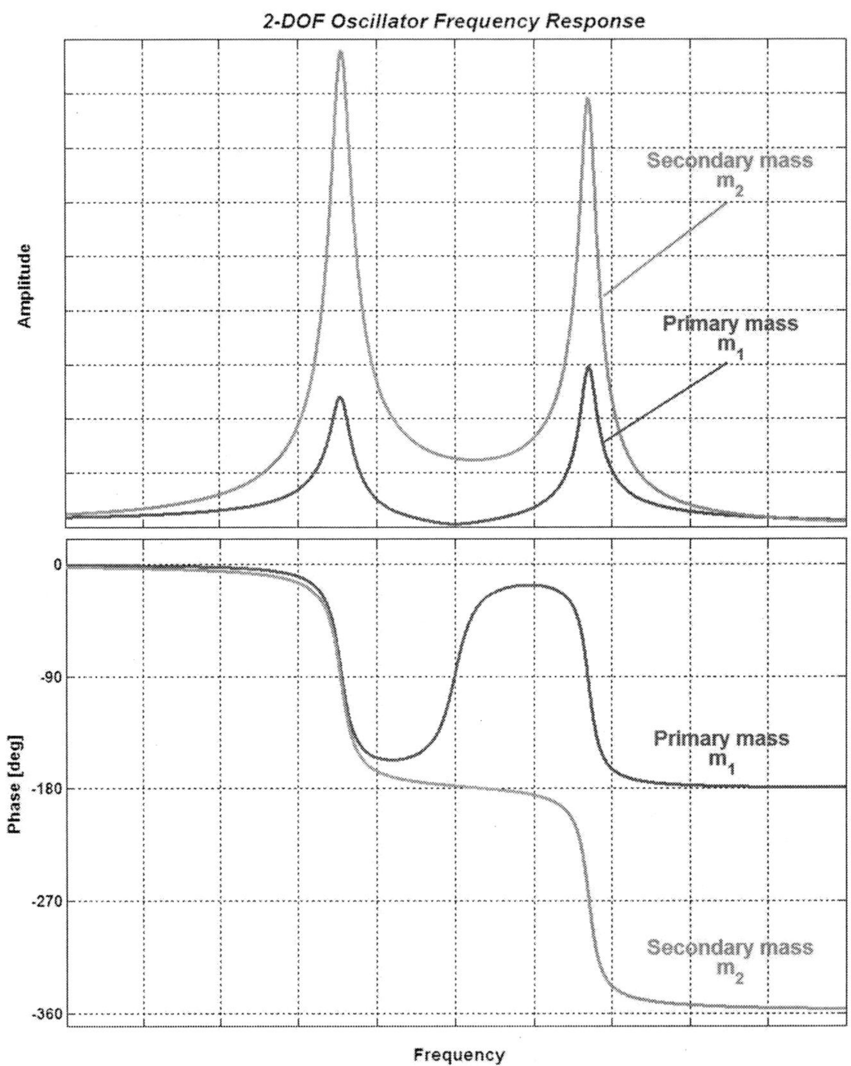

Fig. 6.3 The frequency response of a generic 2-DOF oscillator.

The frequency response amplitudes and phases of the primary and the secondary masses are presented in Figure 6.3. The following important conclusions can be drawn from the frequency response:

- There is a flat region in both the amplitude and phase of the secondary mass frequency response, between the two resonant frequencies.
- When operated within the flat region, variations in operation frequency have negligible effect on the oscillation amplitude and phase of the secondary mass.
- The oscillation amplitude of the secondary mass is considerably larger than the primary mass, which means that the secondary mass amplifies the oscillation of the primary mass.

The fact that variations in operation frequency results in minimal change in gain and phase within the flat region also means that parameter variations in the system which shift resonant frequencies have minimal effect. This point is illustrated in Figure 6.4, in which 5% variation in the k_2 spring is shown together with the nominal response. Within the flat region, both gain and phase are observed to remain stable.

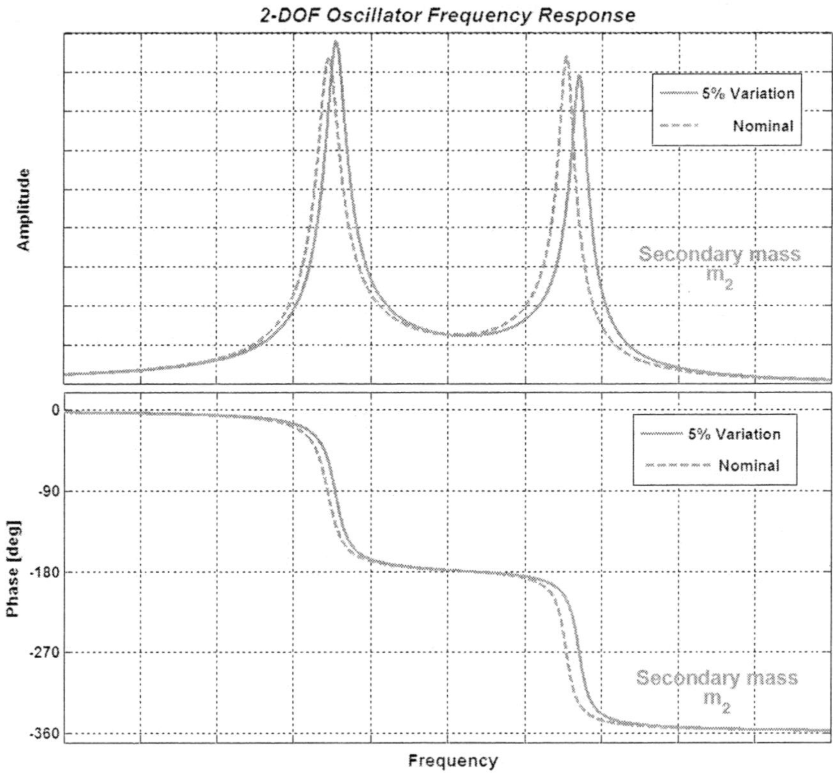

Fig. 6.4 The effect of a 5% variation in the k_2 spring on the frequency response of the 2-DOF oscillator. The gain and phase are affected minimally in the flat region.

Damping variations also have minimal effect on the secondary mass frequency response. Figure 6.2 shows a 50% variation in pressure together with the nominal system response. Within the flat region, gain and phase remain unaffected by damping changes, in contrast to the frequency regions near the resonance peaks.

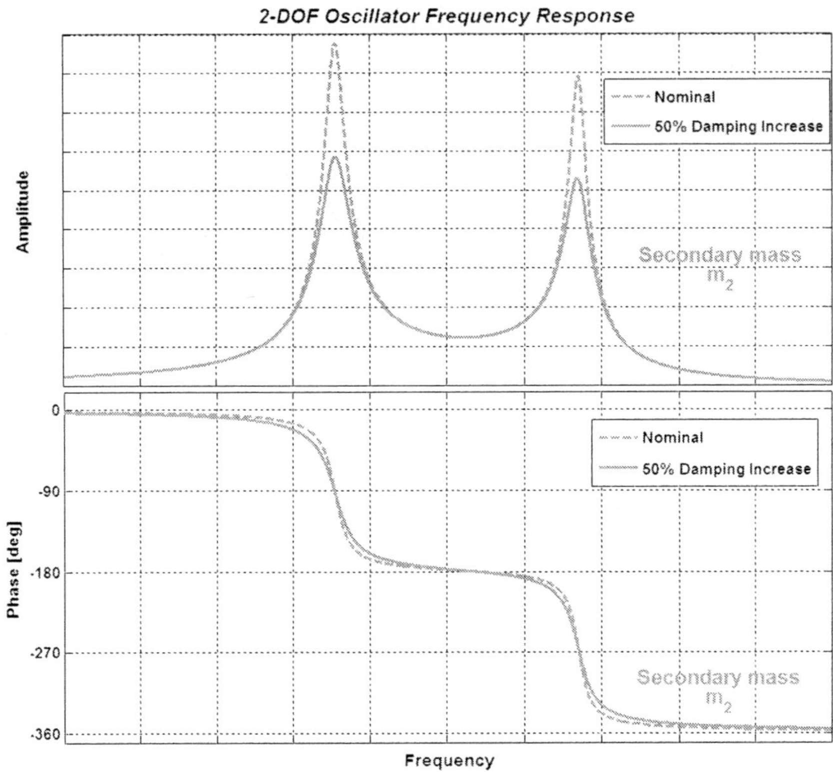

Fig. 6.5 The effect of a 50% increase in damping on the frequency response of the 2-DOF oscillator. Similarly, the gain and phase are affected minimally in the flat region.

These robust characteristics of the 2-DOF oscillators could provide many advantages if utilized in a micromachined gyroscope system. In the following sections, gyroscope system architectures that implement 2-DOF oscillators in the sense-mode, in the drive-mode and in both sense and drive modes are presented.

6.3 The 2-DOF Sense-Mode Architecture

In this architecture, the 2-DOF oscillator is utilized in the sense-mode, and the flat region is obtained in the sense-mode frequency response. The device is 1-DOF in the drive-mode, identical to a conventional gyroscope system, operated at resonance. This allows to use well-proven drive-mode control techniques, while providing robust gain and phase in the sense-mode.

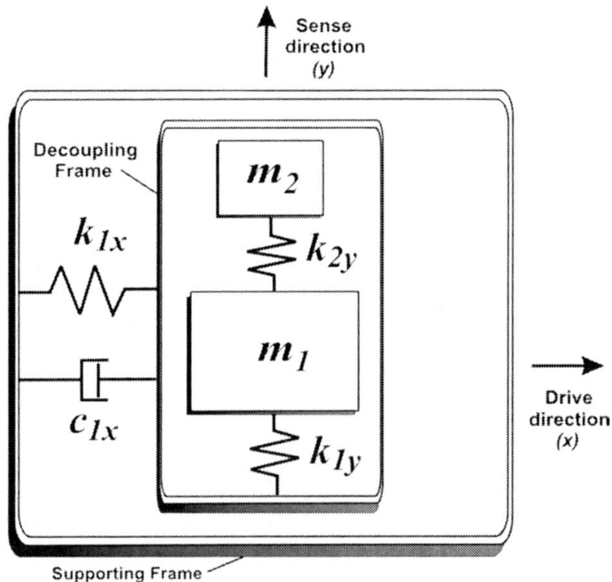

Fig. 6.6 Lumped mass-spring-damper model and the frequency response of the gyroscope system with 2-DOF sense-mode oscillator. The drive-mode resonant frequency is designed to be located at the center of the sense-mode flat region.

The overall gyroscope dynamical system is a total of 3-DOF, consisting of a 2-DOF sense-mode oscillator and a 1-DOF drive-mode oscillator, formed by two interconnected proof masses (Figure 6.6). The first mass, m_1, is free to oscillate both in the drive and sense directions, and is driven in the drive direction. The second mass, m_2, is constrained in the drive direction with respect to the first mass. Thus, m_2 forms the secondary mass of the 2-DOF sense-mode oscillator (Figure 6.6), and acts as the vibration absorber to dynamically amplify the sense mode oscillations of m_1. In the drive-direction, m_1 and m_2 oscillate together, and form a resonant 1-DOF oscillator.

Since the gyroscope structure oscillates as a 1-DOF resonator in the drive direction, the frequency response of the device has a single resonance peak in the drive-mode (Figure 6.7). The device is operated at resonance in the drive-mode using a self-oscillation loop and Automatic Gain Control (AGC) loop for amplitude con-

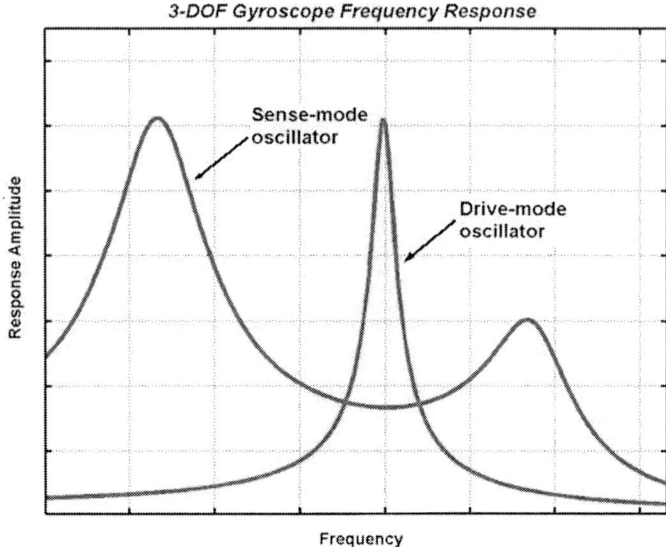

Fig. 6.7 The drive and sense frequency response of the gyroscope system with 2-DOF sense-mode oscillator. The drive-mode resonant frequency is designed to be located at the center of the sense-mode flat region.

trol, similar to a conventional system. The flat region of the sense-mode oscillator is designed to coincide with the drive-mode resonant frequency.

A bulk-micromachined prototype of the 2-DOF sense-mode gyroscope is shown in Figure 6.8. The secondary mass m_2 is suspended inside the primary mass m_1. The sense electrodes are attached to the secondary mass m_2, detecting the Coriolis response of the 2-DOF sense oscillator.

6.3.1 Gyroscope Dynamics

In the gyroscope structure design, a drive frame is implemented to mechanically decouple the drive and sense direction oscillations of m_1. The frame structure aims to minimize quadrature error and undesired electrostatic forces in the sense-mode due to drive-mode actuator imperfections. When m_1 is nested inside a drive-mode frame as in Figure 6.9, the sense-direction oscillations of the frame are constrained, and the drive-direction oscillations are well aligned with the designed drive direction.

With the drive-frame implementation, the following constraints define the dynamics of the 3-DOF system with 2-DOF sense-mode: the structure is stiff in the out-of-plane direction; the position vector of the decoupling frame is forced to lie along the drive-direction; m_1 oscillates purely in the sense-direction relative to the decoupling frame; m_1 and m_2 move together in the drive direction; and m_2 oscillates

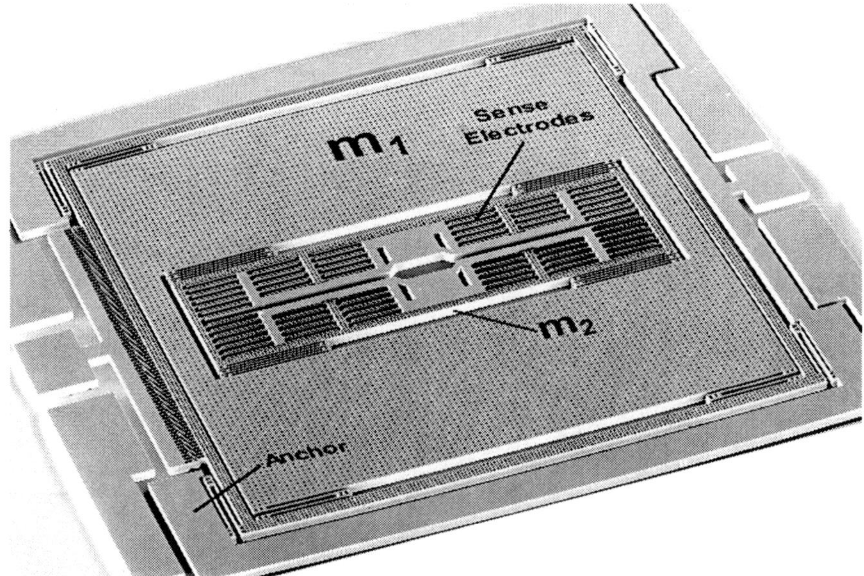

Fig. 6.8 Scanning electron micrograph of the prototype bulk-micromachined 3-DOF gyroscope with 2-DOF sense-mode.

purely in the sense-direction relative to m_1. Thus, the equations of motion of m_1 and m_2 decomposed into the drive and sense directions become

$$(m_1 + m_2 + m_f)\ddot{x}_1 + c_{1x}\dot{x}_1 + k_{1x}x_1 = (m_1 + m_2 + m_f)\Omega_z^2 x_1 + F_d(t)$$
$$m_1\ddot{y}_1 + c_{1y}\dot{y}_1 + k_{1y}y_1 = k_{2y}(y_2 - y_1) + m_1\Omega_z^2 y_1 - 2m_1\Omega_z\dot{x}_1 - m_1\dot{\Omega}_z x_1$$
$$m_2\ddot{y}_2 + c_{2y}\dot{y}_2 + k_{2y}y_2 = k_{2y}y_1 + m_2\Omega_z^2 y_2 - 2m_2\Omega_z\dot{x}_1 - m_2\dot{\Omega}_z x_1 \qquad (6.7)$$

where Ω_z is the z-axis angular rate, m_f is the mass of the decoupling frame, $F_d(t)$ is the driving electrostatic force applied to the active mass at the driving frequency ω_d. The Coriolis force that excites m_1 and m_2 in the sense direction is $2m_1\Omega_z\dot{x}_1$, and the Coriolis response of m_2 in the sense-direction (y_2) is detected for angular rate measurement.

6.3.2 Coriolis Response

The 2-DOF sense-mode dynamics can be further simplified with the assumptions that angular rate input is constant and limited to much lower rates than the operating frequency of the gyroscope, becoming

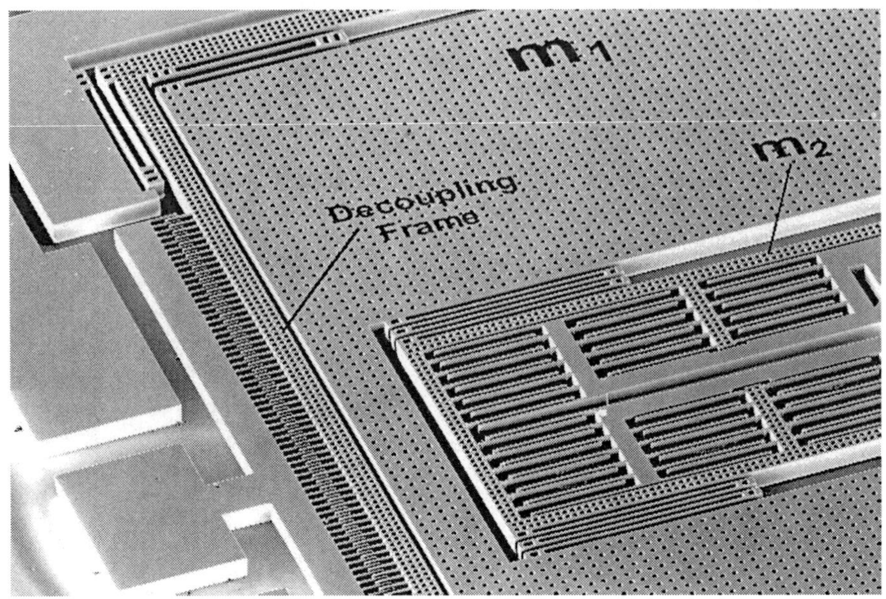

Fig. 6.9 The frame implementation for decoupling the drive and sense-mode oscillations of m_1.

$$m_1\ddot{y}_1 + c_{1y}\dot{y}_1 + (k_{1y} + k_{2y})y_1 = k_{2y}y_2 - 2m_1\Omega_z\dot{x}_1$$
$$m_2\ddot{y}_2 + c_{2y}\dot{y}_2 + k_{2y}y_2 = k_{2y}y_1 - 2m_2\Omega_z\dot{x}_1 \tag{6.8}$$

Based on the analysis by Adam Schofield, let us assume the operation frequency is at the mid-point between the two sense-mode peaks $\omega_{y_{ip}}$ and $\omega_{y_{ap}}$, such that

$$\omega_{y_{ip,ap}} = \omega_d \mp \frac{\Delta}{2}, \tag{6.9}$$

where ω_d is the operational frequency and Δ is the sense mode peak spacing, defined as $\Delta = \omega_{ap} - \omega_{ip}$. Thus, the three characteristic frequencies of the 2-DOF concept can be expressed using only the operational frequency, ω_d, and sense mode peak spacing, Δ.

With the definition of the following sense mode structural parameters

$$\omega_a^2 = \frac{k_{1y} + k_{2y}}{m_1}, \ \omega_b^2 = \frac{k_{2y}}{m_2}, \ \omega_c^2 = \frac{k_{2y}}{\sqrt{m_1 m_2}}, \ \mu^2 = \frac{m_2}{m_1} \tag{6.10}$$

the location of the in-phase and anti-phase sense-modes are obtained by solving the 2-DOF eigenvalue equation as

$$\omega_{y_{ip,ap}}^2 = \frac{1}{2}\left(\omega_a^2 + \omega_b^2 \mp \sqrt{\omega_a^4 + \omega_b^4 - 2\omega_a^2\omega_b^2 + 4\mu^2\omega_b^4}\right) \tag{6.11}$$

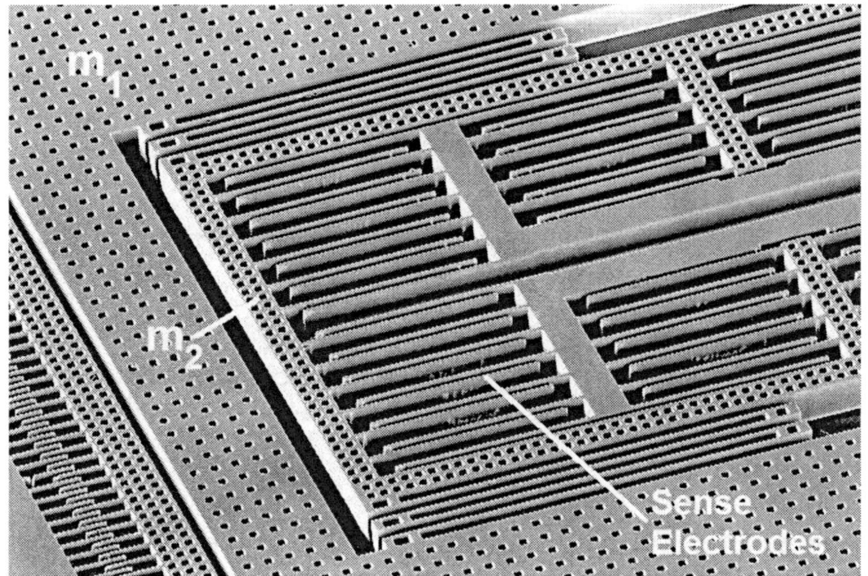

Fig. 6.10 Scanning electron micrograph of the sense-mode passive mass m_2 and the differential sensing electrodes.

The mechanical gain and bandwidth of a 2-DOF sense mode gyroscope are dependent on the region of constant amplitude between the peaks of the 2-DOF sense mode frequency response. Since this region is non-resonant, the upper bound of gain and lower bound of bandwidth can be determined by examining the system at zero damping. Starting from the sense mode equations of motion, the following transfer functions can be found

$$H_1(s) = \frac{s^2 + \left(1 + \mu^2\right)\omega_b^2}{s^4 + \left(\omega_b^2 + \omega_a^2\right)s^2 - \mu^2\omega_b^4 + \omega_a^2\omega_b^2} \tag{6.12}$$

$$H_2(s) = \frac{s^2 + \omega_a^2 + \omega_b^2}{s^4 + \left(\omega_b^2 + \omega_a^2\right)s^2 - \mu^2\omega_b^4 + \omega_a^2\omega_b^2} \tag{6.13}$$

where the input is Coriolis acceleration applied to both masses and the output is the displacement of sense mass m_1, (6.12), and sense mass m_2, (6.13), assuming zero damping.

The mechanical gain and bandwidth of the device, however, are only dependent on the magnitude of the smaller mass m_2, as this is the mass used for detection. Substituting $s = j\omega$ (where j is the imaginary unit) and taking the absolute value yields the gain of mass m_2 as

$$G_2(\omega) = \frac{\omega^2 - 2\omega_d^2 - \frac{\Delta^2}{2}}{\left(\omega^2 - \left(\omega_d - \frac{\Delta}{2}\right)^2\right)^2 \left(\omega^2 - \left(\omega_d + \frac{\Delta}{2}\right)^2\right)^2} \qquad (6.14)$$

The sense mode gain G_y, defined as the amplitude of mass m_2 at the operational frequency, $G_2(\omega_d)$, is of particular interest as this ultimately determines the scale factor of the device. Substituting $\omega = \omega_d$ into (6.14) yields

$$G_y(\omega_d) = \frac{\omega_d^2 + \frac{\Delta^2}{4}}{\left(\omega_d^2 - \frac{\Delta^2}{4}\right)\Delta^2} \qquad (6.15)$$

If we assume that the peak spacing is much smaller than the operational frequency (typical for most 2-DOF systems), $\omega_d \gg \Delta$, (6.15) further simplifies to

$$G_y(\omega_d) \approx \frac{1}{\Delta^2} \qquad (6.16)$$

Therefore, it can be seen from (6.16) that the major parameter affecting the sense mode gain of a multi-DOF sense mode gyroscope is the peak spacing, where smaller spacings ultimately result in larger gain values independent of operational frequency.

If we assume that the drive motion is a regulated constant amplitude sinusoid with an amplitude x_0 as in a typical gyroscope system, the sense-mode response amplitude for an angular rate input Ω_z becomes

$$y_0 \approx \frac{2x_0 \omega_d}{\Delta^2}\Omega_z \qquad (6.17)$$

Bandwidth

The mechanical bandwidth of a multi-DOF device is defined by the frequencies at which the amplitude of the sense mode frequency response increases by 3 dB versus the minimum value between the peaks. The frequency at which this minimum gain occurs can be found by solving for the zero derivative of (6.14), which gives

$$\omega_{min} = \sqrt{\omega_d^2 + \frac{3\Delta^2}{4}}, \qquad (6.18)$$

where ω_{min} is the minimum gain frequency. Substituting (6.18) into (6.14) gives

$$G_{\omega_{min}} = \frac{1}{\Delta^2}, \qquad (6.19)$$

which is the minimum gain value of the sense mode operational region.

The bandwidth can be determined using (6.19) by solving for the frequencies at which the amplitude increases by $\sqrt{2}$ from the minimum gain value which gives

$$\omega_{3dB1,2}^2 = \omega_d^2 + \left(1 + \sqrt{2}\right)\left(\frac{\Delta}{2}\right)^2$$

$$\mp \sqrt{\left(1 - \frac{\sqrt{2}}{2}\right)\omega_d^2 + 2\left(1 - \sqrt{2}\right)\left(\frac{\Delta}{2}\right)^4} \quad (6.20)$$

where ω_{3dB1} and ω_{3dB2} are the lower and higher 3 dB frequencies respectively. Together, the 3 dB points from (6.20) define the bandwidth of the 2-DOF sense system according to $BW_{3dB} = \omega_{3dB2} - \omega_{3dB1}$. If the peak spacing is again assumed to be much smaller than the operational frequency, $\omega_d \gg \Delta$, the bandwidth expression simplifies to

$$BW_{3dB} = \left(\sqrt{1 - \frac{\sqrt{2}}{2}}\right)\Delta$$

$$\approx 0.5412\Delta$$

which corresponds to a sense mode 3 dB bandwidth of roughly half of the peak spacing independent of operational frequency.

6.3.3 Illustrative Example

A bulk-micromachined prototype was designed as an illustrative example and fabricated in an SOI-based process discussed in Chapter 3. The dynamical system parameters of the prototype gyroscope with 2-DOF sense-mode are the following: the proof mass values are $m_1 = 2.46 \times 10^{-6}$kg, $m_2 = 1.54 \times 10^{-7}$kg, and the decoupling frame mass $m_f = 1.19 \times 10^{-7}$kg. The spring constants are $k_{1x} = 61.2$N/m, $k_{1y} = 78.4$N/m, and $k_{2y} = 3.36$N/m.

For the 1-DOF drive-mode oscillator, the effective proof-mass value becomes $m_{1x} = (m_1 + m_2 + m_f) = 2.74 \times 10^{-6}$kg. This yields a drive-mode resonant frequency of 752Hz.

In the sense-mode, the resonant frequencies of the isolated active and passive mass-spring systems are $\omega_{1y} = \sqrt{\frac{k_{1y}}{m_1}} = 897.7$ Hz and $\omega_{2y} = \sqrt{\frac{k_{2y}}{m_2}} = 732.2$ Hz, respectively; yielding a frequency ratio of $\gamma_y = \omega_{2y}/\omega_{1y} = 0.897$, and a mass ratio of $\mu_y = \frac{m_2}{m_1} = 0.0624$. With these parameters, the location of the expected in-phase and anti-phase resonance peaks in the sense-mode frequency response were calculated as $f_{y_{ip}} = 696.7$ Hz and $f_{y_{ap}} = 943.3$ Hz, based on the relation adapted from Equation (6.5)

$$f_{y_{ip,ap}} = \sqrt{\frac{1}{2}\left(1 + \mu_y + \frac{1}{\gamma_y^2} \pm \sqrt{\left(1 + \mu_y + \frac{1}{\gamma_y^2}\right)^2 - \frac{4}{\gamma_y^2}}\right)\frac{\omega_{2y}}{2\pi}} \qquad (6.21)$$

In the sense-mode frequency response, a flat region bandwidth of over 300Hz was experimentally demonstrated. The two resonance peaks in the sense-mode frequency response were observed as $f_{y_{ip}} = 693$ Hz and $f_{y_{ap}} = 940$ Hz. When the drive and sense-mode frequency responses of the prototype 3-DOF gyroscope are investigated together, the drive-mode resonant frequency is observed to be located inside the sense-mode flat region.

For rate-table characterization, synchronous demodulation technique was used. The carrier signal was imposed on the gyroscope structure, and the output from the differential sense-capacitors was amplified and synchronously amplitude demodulated at the carrier signal frequency using a lock-in amplifier. The Coriolis signal was finally demodulated at the driving frequency.

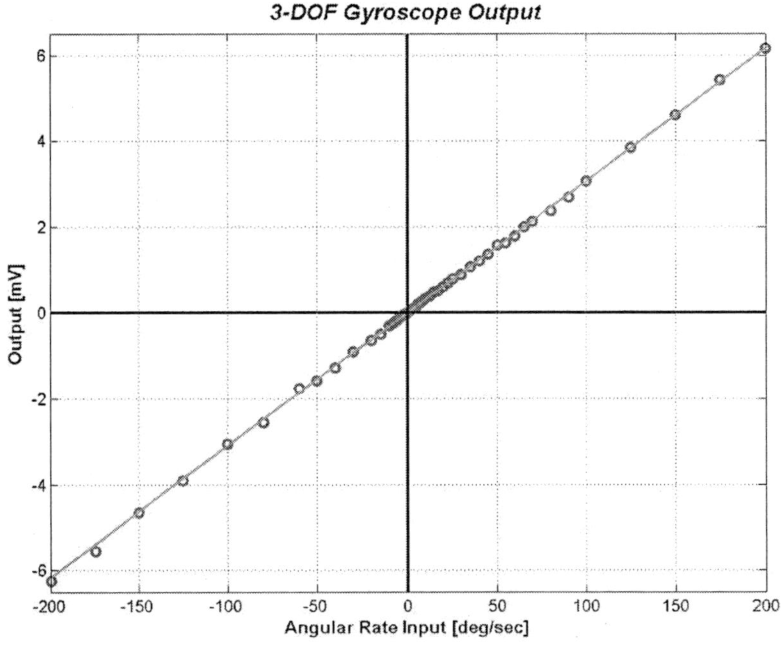

Fig. 6.11 The angular-rate input vs. voltage output plot obtained from the 3-DOF gyroscope with 2-DOF sense-mode, in the -200°/s to 200°/s input range.

With this technique, a sensitivity of 0.0308 mV/°/s was experimentally demonstrated while the device was operated in the flat-region of the sense-mode frequency response (Figure 6.11). The measured noise floor was 19.7 μV/$\sqrt{\text{Hz}}$ at 50Hz bandwidth, yielding a measured resolution of 0.64°/s/$\sqrt{\text{Hz}}$ at 50Hz bandwidth in atmospheric pressure. Details of the implementation and experimental results are discussed in [63, 77].

It should be noted that the primary limiting factors on the scale factor and noise performance are large capacitive gaps (over 10μm in this device) and detection electronics. The noise performance can be drastically improved with smaller capacitive gaps, close integration of detection electronics and higher performance synchronous demodulation circuits. The mechanical scale factor of the 2-DOF sense-mode system can actually be larger than a conventional gyroscope with mismatched modes, as will be illustrated on an example in Chapter 9.

6.3.4 Conclusions on the 2-DOF Sense-Mode Architecture

The most prominent benefit of the 3-DOF system with 2-DOF sense-mode is its compatibility with well-proven drive-mode control techniques, while the effect of parameter variations on the gain and phase of the sense-mode response is significantly suppressed. The concept and the operational principles were experimentally demonstrated. Bulk-micromachined prototypes were successfully operated in the flat region of the sense-mode, to measure angular rate with sufficient sensitivity and noise characteristics, while providing exceptional robustness.

6.4 The 2-DOF Drive-Mode Architecture

This system architecture is based on utilizing the 2-DOF oscillator in the drive-mode, so that the wide-band region is achieved in the drive-mode frequency response. By utilizing dynamic amplification in the drive-mode, large oscillation amplitudes of the sensing element is achieved with small actuation amplitudes, providing improved linearity and stability even with parallel-plate actuation.

The overall 3-DOF micromachined gyroscope is formed by a 2-DOF drive-mode oscillator and a 1-DOF sense-mode oscillator, composed of two interconnected proof masses (Figure 6.14). The first mass, m_1, is the only mass excited in the drive direction, and is constrained in the sense direction. The second mass, m_2, is free to oscillate both in the drive and sense directions. Thus, m_2 forms the secondary mass of the 2-DOF drive-mode oscillator (Figure 6.12).

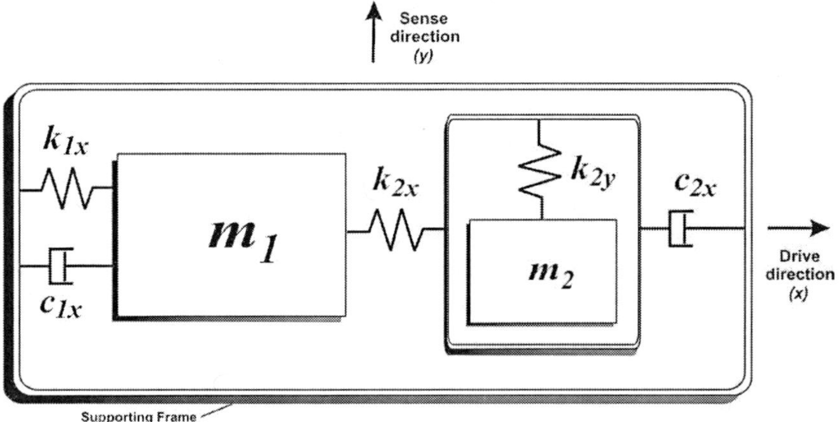

Fig. 6.12 Lumped mass-spring-damper model of the gyroscope system with 2-DOF drive-mode oscillator.

In the sense-direction, m_2 forms a resonant 1-DOF oscillator. Thus, in the presence of an input angular rate about the sensitive axis normal to the substrate (z-axis), the Coriolis force is induced on m_2, and only m_2 responds to the rotation-induced Coriolis force. The sense-mode response of m_2 is detected by the parallel-plate sensing electrodes to measure angular rate.

Since the dynamical system is a 1-DOF resonator in the sense direction, the frequency response of the device has a single resonance peak in the sense mode. The device is operated at the sense direction resonance frequency, which is designed to be located at the center of the flat region in the drive oscillator frequency response (Figure 6.13).

The drive-mode control system locks the operation frequency to the sense-mode resonant frequency. This could be achieved by tracking the amplitude or the phase of the quadrature signal. Since the operation frequency is within the drive-mode flat

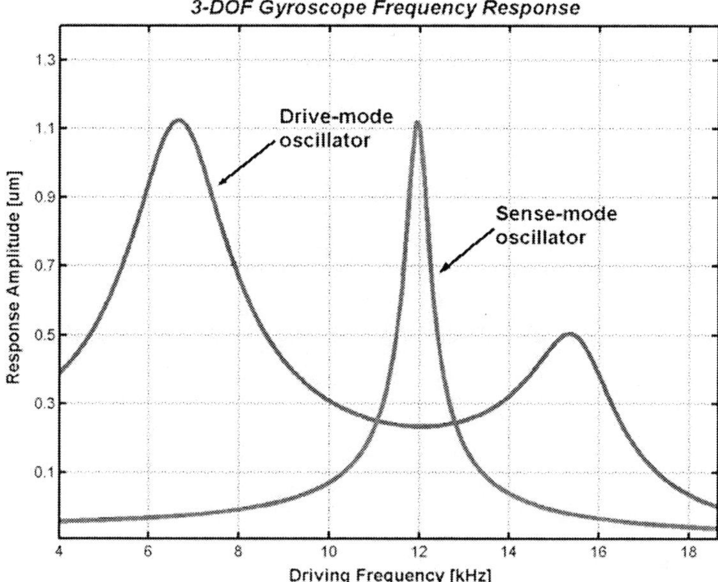

Fig. 6.13 The drive and sense frequency response of the gyroscope system with 2-DOF drive-mode oscillator. The device is operated at the sense direction resonance frequency, which is located within the flat region of the drive oscillator frequency response.

region, the drive amplitude and phase remain constant when the operation frequency shifts with the sense-mode frequency.

This concept allows to operate exactly at the sense-mode resonant peak to achieve excellent rate sensitivity. However, the major limitation is that it does not allow the use of a conventional self-oscillation and AGC loop in the drive-mode.

6.4.1 Gyroscope Dynamics

In this architecture, the driven mass m_1 oscillates only in the drive direction, and acts as a drive-frame that suppresses anisoelasticities due to fabrication imperfections. In order to minimize dynamical coupling between the drive and sense modes of m_2, a decoupling frame structure can be implemented inside m_1 as well. By nesting m_2 inside a drive-mode frame, the sense-direction oscillations of the frame are constrained, and the drive-direction oscillations are automatically forced to be in the designed drive direction. Then, m_2 is free to oscillate only in the sense-direction with respect to the frame, and the sense-mode response of m_2 will be perfectly orthogonal to the drive-direction. A bulk-micromachined prototype gyroscope implementing this architecture is shown in Figure 6.14.

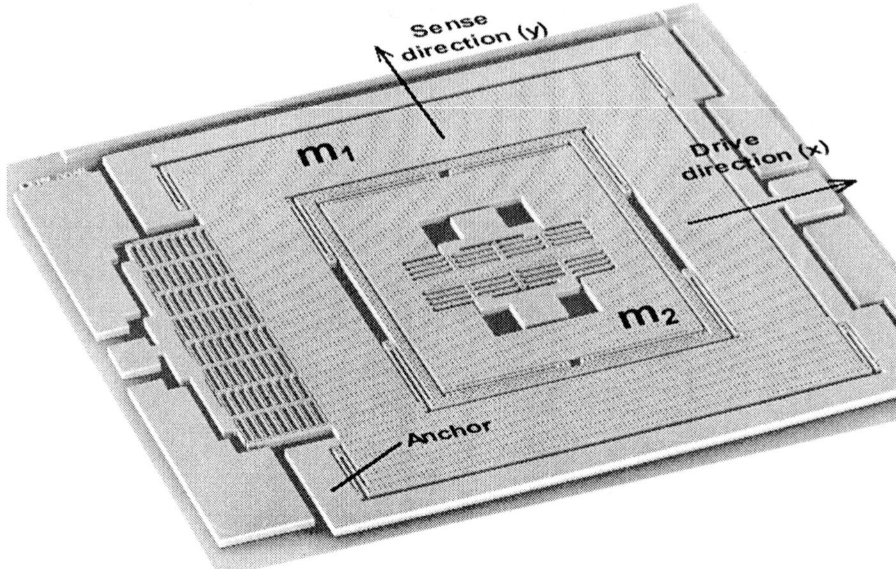

Fig. 6.14 Scanning electron micrograph of the prototype bulk-micromachined 3-DOF gyroscope with 2-DOF drive-mode.

For this dynamical system structure, the gyroscope dynamics can be analyzed in the non-inertial coordinate frame associated with the gyroscope. Each of the interconnected proof masses are assumed to be a rigid body with a position vector $\mathbf{r}$ attached to a rotating gyroscope reference frame B with an angular velocity of Ω, resulting in an absolute acceleration in the inertial frame A

$$\mathbf{a_A} = \mathbf{a_B} + \dot{\Omega} \times \mathbf{r_B} + \Omega \times (\Omega \times \mathbf{r_B}) + 2\Omega \times \mathbf{v_B} \tag{6.22}$$

where $\mathbf{v_B}$ and $\mathbf{a_B}$ are the velocity and acceleration vectors with respect to the reference frame B, respectively. Thus, the equations of motion of m_1 and m_2 can be expressed in the inertial frame as

$$m_1\mathbf{a_1} = \mathbf{F_1} + \mathbf{F_d} + \mathbf{F_r} - 2m_1\Omega \times \mathbf{v_1} - m_1\Omega \times (\Omega \times \mathbf{r_1}) - m_1\dot{\Omega} \times \mathbf{r_1}$$
$$m_2\mathbf{a_2} = \mathbf{F_2} - \mathbf{F_r} - 2m_2\Omega \times \mathbf{v_2} - m_2\Omega \times (\Omega \times \mathbf{r_2}) - m_2\dot{\Omega} \times \mathbf{r_2} \tag{6.23}$$

where $\mathbf{F_d}$ is the driving force applied on m_1, $\mathbf{F_1}$ is the net external force applied to m_1 including elastic and damping forces from the substrate, $\mathbf{F_2}$ is the net external force applied to m_2 including the damping force from the substrate, and $\mathbf{F_r}$ is the elastic reaction force between m_1 and m_2. In the gyroscope frame, $\mathbf{r_1}$ and $\mathbf{r_2}$ are the position vectors, and $\mathbf{v_1}$ and $\mathbf{v_2}$ are the velocity vectors of m_1 and m_2, respectively.

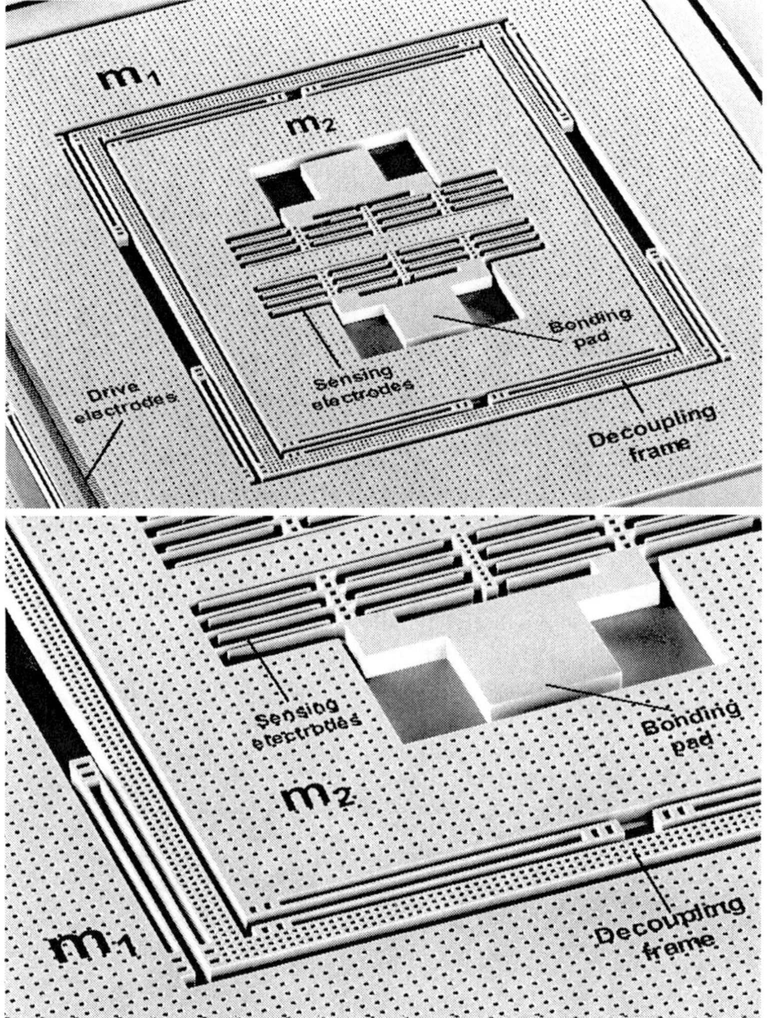

Fig. 6.15 The frame implementation for decoupling the drive and sense direction oscillations of m_2.

With the following constraints on the dynamical system, the equations of motion of m_1 and m_2 can be further simplified and decomposed into the drive and sense directions: the structure is stiff in the out-of-plane direction; the position vector of m_1 and the decoupling frame are forced to lie along the drive-direction, i.e. $y_1(t) = 0$; the decoupling frame and m_2 move together in the drive direction; and m_2 oscillates purely in the sense-direction relative to the decoupling frame. Thus, the equations of motion of m_1 and m_2 when subjected to an angular rate of Ω_z about the z-axis become

$$m_1\ddot{x}_1 + c_{1x}\dot{x}_1 + k_{1x}x_1 = k_{2x}(x_2 - x_1) + m_1\Omega_z^2 x_1 + F_d(t)$$

$$(m_2 + m_f)\ddot{x}_2 + c_{2x}\dot{x}_2 + k_{2x}(x_2 - x_1) = (m_2 + m_f)\Omega_z^2 x_2$$

$$m_2\ddot{y}_2 + c_{2y}\dot{y}_2 + k_{2y}y_2 = m_2\Omega_z^2 y_2 - 2m_2\Omega_z\dot{x}_2 - m_2\dot{\Omega}_z x_2 \qquad (6.24)$$

m_f is the mass of the decoupling frame, $F_d(t)$ is the driving electrostatic force applied to the active mass at the driving frequency ω_d, and Ω_z is the angular velocity applied on the gyroscope about the z-axis. The Coriolis force that excites m_2 in the sense direction is $2m_2\Omega_z\dot{x}_2$, and the Coriolis response of m_2 in the sense-direction (y_2) is detected for angular rate measurement.

6.4.2 Dynamical Amplification in the Drive-Mode

In a conventional implementation of a micromachined gyroscope, generally inter-digitated comb-drives are employed to achieve large drive-mode oscillation amplitudes. However, parallel-plate actuation provides much larger forces per area compared to comb-drives, with the expense of limited stable actuation range, and non-linear actuation forces. Even though parallel-plates allow much lower actuation voltages, the highly non-linear and unstable nature of parallel-plate actuation limits the actuation amplitude of the gyroscope.

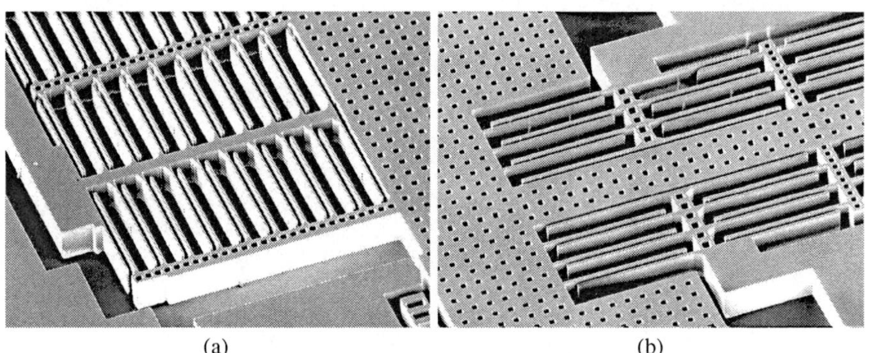

(a)   (b)

Fig. 6.16 (a) The parallel-plate actuator attached to m_1 in the drive-mode. (b) The differential capacitors for detecting sense-mode oscillations of m_2.

In the 3-DOF micromachined gyroscope system, m_2 forms the passive mass of the 2-DOF drive-mode oscillator, and acts as the vibration absorber of the driven mass m_1. While absorbing the oscillations of the active mass, m_2 itself achieves much larger drive-mode amplitudes than m_1. Thus, the actuation range of the parallel-plate actuators attached to the active mass m_1 is narrow (Figure 6.16a), while the sensing element m_2 oscillates with large drive-mode amplitudes, and generates larger Coriolis force. Thus, the nonlinear force profile and instability parallel-plate actuation is minimized.

6.4.3 Illustrative Example

A prototype 3-DOF gyroscope was designed for experimental demonstration of the design concept with the following dynamical system parameters: The proof mass values are $m_1 = 2.20 \times 10^{-6}$kg, $m_2 = 8.22 \times 10^{-7}$kg, and the decoupling frame mass $m_f = 1.73 \times 10^{-7}$kg. The spring constants are $k_{1x} = 305$N/m, $k_{2x} = 64.2$N/m, and $k_{2y} = 46.3$N/m. Thus, for the drive mode-oscillator, the active and passive proof-mass values become $m_{1x} = 2.20 \times 10^{-6}$kg and $m_{2x} = (m_2 + m_f) = 9.95 \times 10^{-7}$kg. Scanning electron micrograph of the bulk-micromachined prototype gyroscope is shown in Figure 6.17.

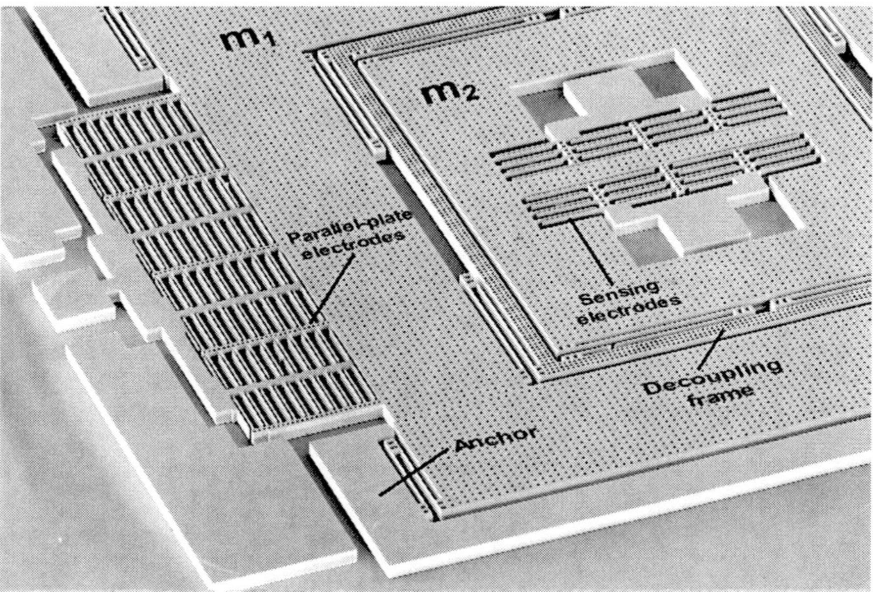

Fig. 6.17 Scanning electron micrograph of the prototype bulk-micromachined 3-DOF gyroscope.

In the drive-mode, the resonant frequencies of the isolated active and passive mass-spring systems are $\omega_{1x} = \sqrt{\frac{k_{1x}}{m_{1x}}} = 1.591$ kHz and $\omega_{2x} = \sqrt{\frac{k_{2x}}{(m_2+m_f)}} = 1.067$ kHz, respectively; yielding a frequency ratio of $\gamma_x = \omega_{2x}/\omega_{1x} = 0.682$, and a mass ratio of $\mu_x = \frac{(m_2+m_3)}{m_1} = 0.451$. With these parameters, the location of the two expected resonance peaks in the drive-mode frequency response were calculated as

$f_{x_{ip}} = 1.184$ kHz and $f_{x_{ap}} = 2.362$ kHz, based on the relation adapted from Equation (6.5)

$$f_{x_{ip,ap}} = \sqrt{\frac{1}{2}\left(1 + \mu_x + \frac{1}{\gamma_x^2} \pm \sqrt{\left(1 + \mu_x + \frac{1}{\gamma_x^2}\right)^2 - \frac{4}{\gamma_x^2}}\right)} \frac{\omega_{2x}}{2\pi} \qquad (6.25)$$

In the drive-mode frequency response, a flat region bandwidth of over 400Hz was experimentally demonstrated. The two resonance peaks in the drive-mode frequency response were observed as $f_{x_{ip}} = 583$ Hz and $f_{x_{ap}} = 1.058$ kHz, in close agreement with the theoretically estimated values. When the drive and sense-mode frequency responses of the prototype 3-DOF gyroscope are investigated together, the sense-mode resonant frequency is observed to be located inside the drive-mode flat region, defining the operation frequency.

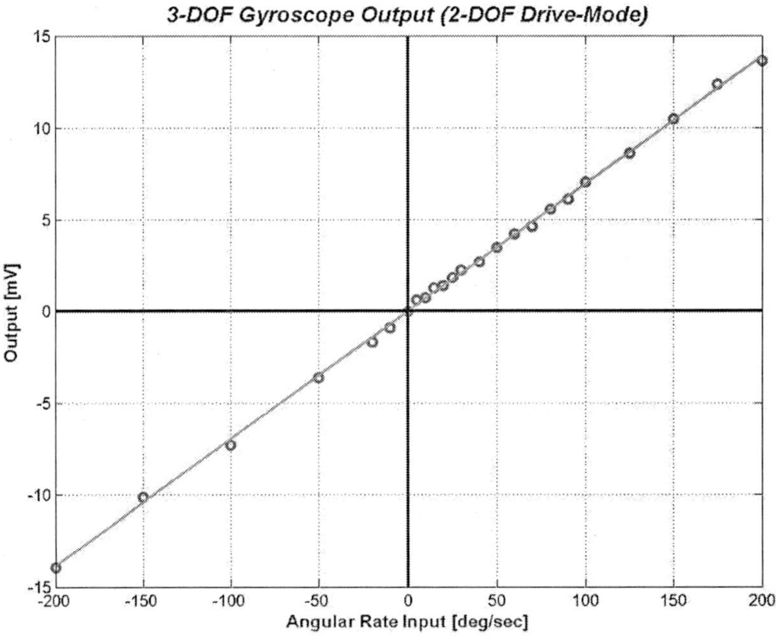

Fig. 6.18 The angular-rate input vs. voltage output plot obtained from the 3-DOF gyroscope with 2-DOF drive-mode. The device was operated at the in-phase drive resonant frequency.

To measure the angular rate response of the 3-DOF system, the synchronous demodulation technique was used. Due to the wide separation of the two drive-mode resonant peaks, the response amplitude in the drive-mode flat region was observed to be insufficient to generate a detectable Coriolis signal. Thus, the device was operated on the lower drive-mode resonance peak, at 916.1 Hz. At this frequency, the drive-mode passive-mass amplitude reached 20μm, while the driven mass amplitude was observed to be less than 2μm; providing a stable large-amplitude oscillation with parallel-plates. With this method, a sensitivity of 0.0694 mV/°/s, and a noise floor of 0.211 mV/$\sqrt{\text{Hz}}$ at 50Hz bandwidth was measured. This yields a measured resolution of 3.05°/s/$\sqrt{\text{Hz}}$ at 50Hz bandwidth.

6.4.4 Conclusions on the 2-DOF Drive-Mode Architecture

The major advantage of the 3-DOF system with 2-DOF drive-mode is the ability to achieve large drive-mode oscillation amplitude at the drive-mode secondary-mass, while the oscillation amplitude of the primary mass is drastically suppressed. This allows utilizing parallel-plate actuators with small gaps to produce large drive-mode amplitudes with low actuation voltages. This operation principle was experimentally demonstrated on bulk-micromachined prototypes. Even though this architecture does not allow the use of a conventional self-oscillation and AGC loop in the drive-mode, it can allow the use of simpler drive control schemes.

It should be noticed that, in order to achieve a sufficient drive-mode amplitude in the flat region of the frequency response, the two drive-mode resonant peaks have to be designed with close spacing. This requires a larger mass ratio m_1/m_2 in the 2-DOF drive-mode oscillator. Since m_1 does not contribute to the Coriolis force that excites the sense oscillator, the increase in die size due to m_1 does not benefit the sense-mode sensitivity. Thus, this concept is more suitable for applications that prioritize advantages of parallel-plate actuation and dynamical amplification of drive motion over die size.

6.5 The 4-DOF System Architecture

The 4-DOF system approach is based on combining the characteristics of the two previously described concepts, by utilizing a 2-DOF drive-direction oscillator and a 2-DOF sense-direction oscillator, which form a 4-DOF overall dynamical system. The frequency responses of both the drive and sense direction oscillators have two resonant peaks and a flat region between the peaks. The device is nominally operated in the flat regions of the response curves belonging to the drive and sense direction oscillators, where the gain is less sensitive to frequency fluctuations. The overall 4-DOF dynamical system is implemented with three interconnected proof masses, which provides the flexibility in defining the drive and sense mode dynamical parameters independently.

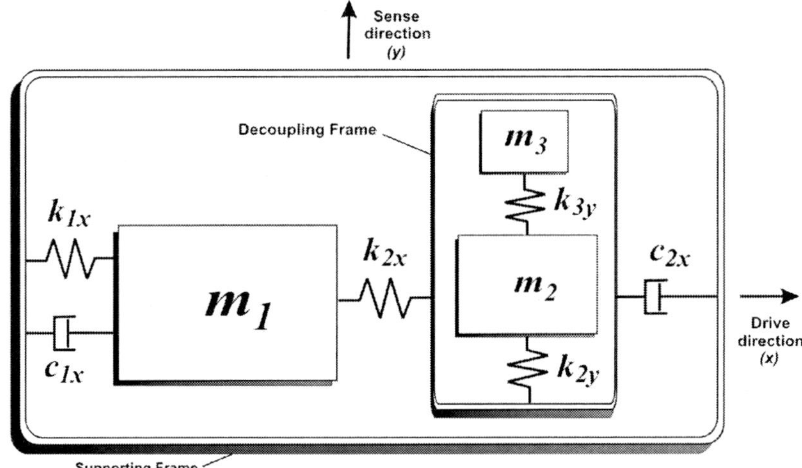

Fig. 6.19 Lumped mass-spring-damper model of the 4-DOF gyroscope dynamical system.

The first mass m_1, which is the only mass excited in the drive direction, is constrained in the sense direction, and is free to oscillate only in the drive direction. The second mass m_2 and third mass m_3 are constrained with respect to each other in the drive direction, thus oscillating as one combined mass in the drive direction. Thus, the first mass m_1 forms the primary mass and the combination of the second and third masses $(m_2 + m_3)$ form the secondary mass of the 2-DOF drive-direction oscillator, where m_1 is the driven mass (Figure 6.19).

In the sense direction, m_2 and m_3 are free to oscillate independently forming the 2-DOF sense-direction oscillator. The larger mass m_2 is the primary mass, which generates the Coriolis force that excites the sense oscillator. The small mass m_3 is the secondary mass, which amplifies the oscillations of m_2, and carries the sense electrodes used for detection of angular rate.

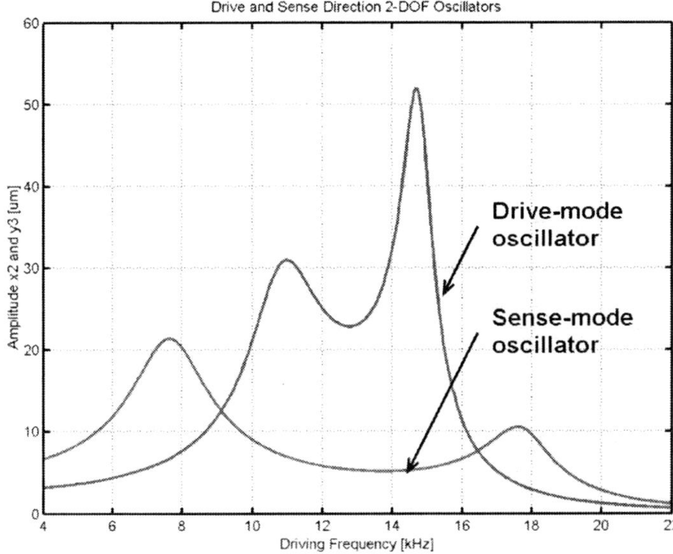

Fig. 6.20 The frequency responses of the 2-DOF drive and sense-mode oscillators, with the overlapped flat regions.

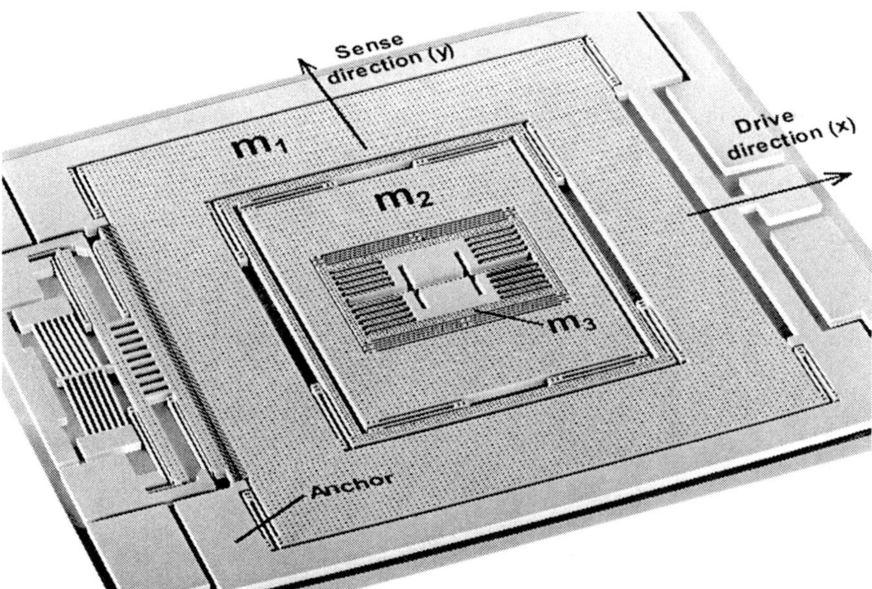

Fig. 6.21 The 4-DOF gyroscope concept is based on utilizing a 2-DOF drive-direction oscillator and a 2-DOF sense-direction oscillator, formed by three interconnected masses.

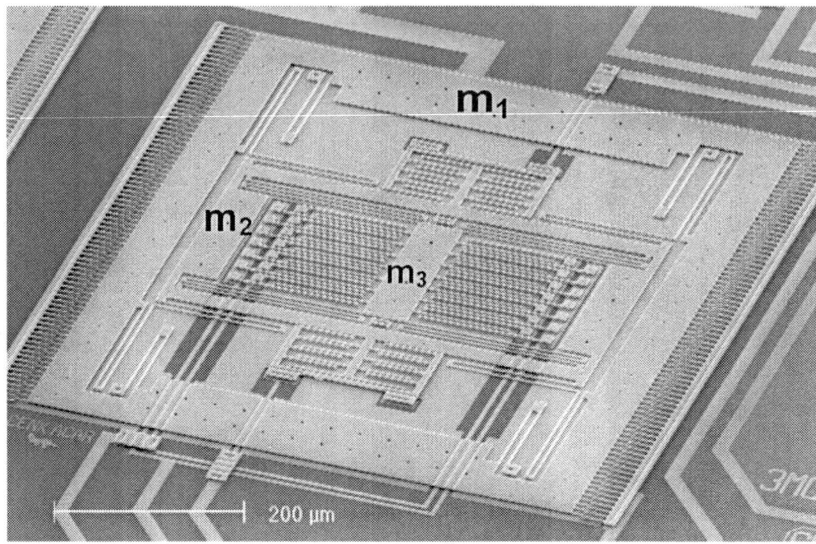

Fig. 6.22 Scanning electron micrograph of a surface-micromachined implementation of the 4-DOF gyroscope.

Mode-Decoupling in the 4-DOF System

In order to minimize quadrature error and bias due to dynamical coupling between the drive and sense modes, the drive and sense direction oscillators are mechanically decoupled. The driven mass m_1 oscillates only in the drive direction, and possible anisoelasticities due to fabrication imperfections are suppressed by the suspension fixed in the sense direction. The second mass m_2 oscillates in both drive and sense directions, and generates the rotation-induced Coriolis force that excites the 2-DOF sense-direction oscillator. The sense direction response of the third mass m_3, which comprises the vibration absorber of the 2-DOF sense-direction oscillator, is detected for measuring the input angular rate. Since the springs that couple the sense element m_3 to m_2 deform only for relative sense direction oscillations, mechanical coupling of the drive-mode oscillations into the sense-mode of m_3 is minimized, significantly enhancing gyroscopic performance.

The dynamic coupling between the drive and sense modes can be further suppressed by nesting m_2 inside a drive-mode frame as in Figure 6.23. The sense-direction oscillations of the frame are constrained by the suspension system, and the drive-direction oscillations of m_2 are automatically forced to be aligned with the designed drive direction. Thus, the drive and sense direction oscillators of m_2 are mechanically decoupled with the frame structure, leading to significantly suppressed quadrature error and zero-rate output.

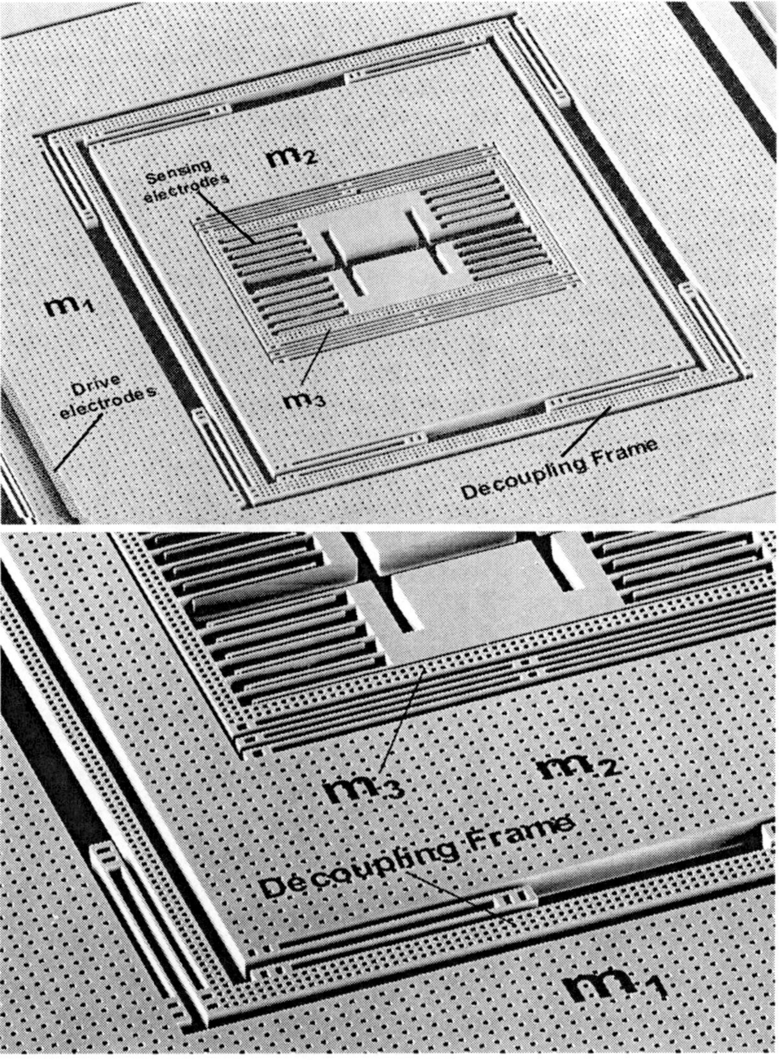

Fig. 6.23 The frame implementation for decoupling the drive and sense direction oscillations of m_2.

6.5.1 The Coriolis Response

The frequency responses of the 2-DOF drive direction oscillator and the 2-DOF sense direction oscillator have two resonant peaks and a flat region between the peaks. The device is nominally operated in the flat regions of the drive and sense direction oscillators, where the response amplitudes of the oscillators are less sensitive to parameter variations. In order to operate both of the drive and sense direction oscillators in their flat-region frequency bands, the flat regions of the oscillators have to be designed to overlap (Figure 6.20), by matching the drive and sense direction anti-resonance frequencies, as will be explained in the following sections. In contrast to the conventional gyroscopes, the flat regions with significantly wider bandwidths can be overlapped with sufficient precision in spite of fabrication imperfections and operation condition variations.

The response of the combined 4-DOF dynamical system to the rotation-induced Coriolis force will have a flat region in the frequency band coinciding to the flat regions of the independent drive and sense-mode oscillators (Figure 6.24). When the device is operated in this flat region, the oscillation amplitudes in both drive and sense directions are relatively insensitive to variations in system parameters and damping. Thus, by utilizing dynamical amplification in the 2-DOF oscillators instead of resonance, increased bandwidth and reduced sensitivity to structural and thermal parameter fluctuations and damping changes are achieved.

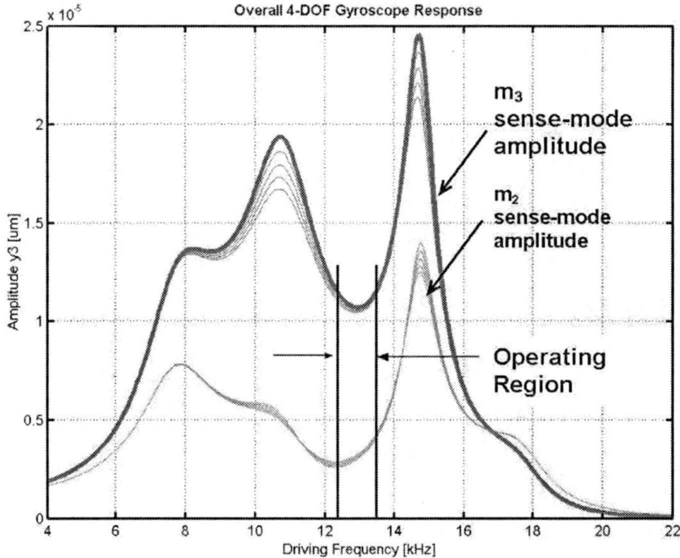

Fig. 6.24 The response of the overall 4-DOF gyroscope system. The oscillation amplitude is relatively insensitive to parameter variations and damping fluctuations in the flat operating region.

6.5.2 Dynamics of the 4-DOF Gyroscope

The dynamics of the 4-DOF gyroscope system is described in the non-inertial co-ordinate frame associated with the gyroscope. The 4-DOF system consists of three interconnected proof masses where each mass can be assumed to be a rigid body with a position vector **r** attached to a rotating reference frame B, resulting in an absolute acceleration in the inertial frame A

$$\mathbf{a_A} = \mathbf{a_B} + \dot{\Omega} \times \mathbf{r_B} + \Omega \times (\Omega \times \mathbf{r_B}) + 2\Omega \times \mathbf{v_B} \qquad (6.26)$$

where the subscript A denotes "relative to inertial frame A", B denotes "relative to rotating gyroscope frame B", $\mathbf{v_B}$ and $\mathbf{a_B}$ are the velocity and acceleration vectors with respect to the reference frame B, respectively, and Ω is the angular velocity of the gyroscope frame B relative to the inertial frame A. The term $2\Omega \times \mathbf{v_B}$ is the Coriolis acceleration, and excites the system in the sense direction. Thus, when a mass oscillating in the drive direction (x-axis) is subject to an angular rotation rate of Ω_z about the z-axis, the Coriolis acceleration induced in the sense direction (y-axis) is $a_y = 2\Omega_z \dot{x}(t)$.

Similarly, the equations of motion for the three proof masses observed in the non-inertial rotating frame can be expressed in the inertial frame as

$$m_1\mathbf{a_1} = \mathbf{F_1} + \mathbf{F_d} - 2m_1\Omega \times \mathbf{v_1} - m_1\Omega \times (\Omega \times \mathbf{r_1}) - m_1\dot{\Omega} \times \mathbf{r_1}$$
$$m_2\mathbf{a_2} = \mathbf{F_2} - 2m_2\Omega \times \mathbf{v_2} - m_2\Omega \times (\Omega \times \mathbf{r_2}) - m_2\dot{\Omega} \times \mathbf{r_2}$$
$$m_3\mathbf{a_3} = \mathbf{F_3} - 2m_3\Omega \times \mathbf{v_3} - m_3\Omega \times (\Omega \times \mathbf{r_3}) - m_3\dot{\Omega} \times \mathbf{r_3} \qquad (6.27)$$

where $\mathbf{F_1}$ is the net external force applied to m_1 including elastic and damping forces from the substrate and elastic interaction force from m_2; $\mathbf{F_2}$ is the net external force applied to m_2 including the damping force from the substrate and elastic interaction force from m_1 and m_3; $\mathbf{F_3}$ is the net external force applied to m_3 including the damping force from the substrate and the elastic interaction force from m_2; and $\mathbf{F_d}$ is the driving force applied to m_1. In the gyroscope frame, $\mathbf{r_1}$, $\mathbf{r_2}$, and $\mathbf{r_3}$ are the position vectors; and $\mathbf{v_1}$, $\mathbf{v_2}$, and $\mathbf{v_3}$ are the velocity vectors of m_1, m_2 and m_3, respectively. Since the first mass is fixed in the sense direction, i.e. $y_1(t) = 0$, and m_2 and m_3 move together in the drive direction, i.e. $x_2(t) = x_3(t)$; the 4-DOF equations of motion (along the x-axis and y-axis) of the three mass system subjected to an angular rate of Ω_z about the axis normal to the plane of motion (z-axis) become

$$m_1\ddot{x}_1 + c_{1x}\dot{x}_1 + k_{1x}x_1 = k_{2x}(x_2 - x_1) + m_1\Omega_z^2 x_1 + F_d(t)$$
$$(m_2 + m_3)\ddot{x}_2 + (c_{2x} + c_{3x})\dot{x}_2 + k_{2x}(x_2 - x_1) =$$
$$(m_2 + m_3)\Omega_z^2 x_2 + 2m_2\Omega_z\dot{y}_2 + 2m_3\Omega_z\dot{y}_3 + m_2\dot{\Omega}_z y_2 + m_3\dot{\Omega}_z y_3$$
$$m_2\ddot{y}_2 + c_{2y}\dot{y}_2 + k_{2y}y_2 = k_{3y}(y_3 - y_2) + m_2\Omega_z^2 y_2 - 2m_2\Omega_z\dot{x}_2 - m_2\dot{\Omega}_z x_2$$
$$m_3\ddot{y}_3 + c_{3y}\dot{y}_3 + k_{3y}(y_3 - y_2) = m_3\Omega_z^2 y_3 - 2m_3\Omega_z\dot{x}_3 - m_3\dot{\Omega}_z x_3. \qquad (6.28)$$

where $F_d(t)$ is the driving electrostatic force applied to the active mass at the driving frequency ω_d, and Ω_z is the angular velocity applied to the gyroscope about the z-axis. The terms $2m_2\Omega_z\dot{x}_2$ and $2m_3\Omega_z\dot{x}_3$ are the Coriolis forces that excite the system in the sense direction, and the Coriolis response of m_3 in the sense-direction (y_3) is detected for angular rate measurement.

6.5.3 Parameter Optimization

Since the foremost mechanical factor determining the performance of the gyroscope is the sense direction deflection of the sensing element m_3 due to the input rotation, the parameters of the dynamical system should be optimized to maximize the oscillation amplitude of m_3 in the sense direction.

However, the optimal compromise between amplitude of the response and bandwidth should be obtained to maintain robustness against parameters variations, while the response amplitude is sufficient for required sensitivity. The trade-offs between gain of the response (for higher sensitivity) and the system bandwidth (for increased robustness) will be typically guided by application requirements.

For the purpose of optimizing each parameter in the dynamical system, the overall 4-DOF gyroscope system can be decomposed into the 2-DOF drive direction oscillator (Figure 6.25a) and the 2-DOF sense direction oscillator (Figure 6.25b), to be analyzed separately in the next two sections.

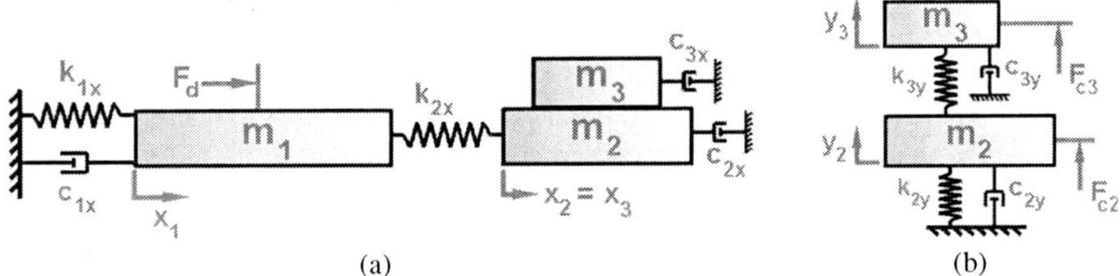

(a)   (b)

Fig. 6.25 (a) Lumped mass-spring-damper model for the 2-DOF drive-mode oscillator of the 4-DOF gyroscope. (b) Lumped mass-spring-damper model for the 2-DOF sense-mode oscillator.

6.5.3.1 Drive Mode Parameters

The main objective of parameter optimization in the drive mode is to maximize the rotation-induced Coriolis force generated by the second mass m_2. This force $F_{c2} = 2m_2\Omega_z\dot{x}_2$ is the dominant force exciting the 2-DOF sense-direction oscillator, and is proportional to the sensor sensitivity.

In the drive mode, the gyroscope is simply a 2-DOF system. The sinusoidal drive force is applied on the primary mass m_1 by the drive electrodes. The combination of the second and the third masses ($m_2 + m_3$) comprise the vibration absorber (secondary mass) of the 2-DOF oscillator, which mechanically amplifies the oscillations

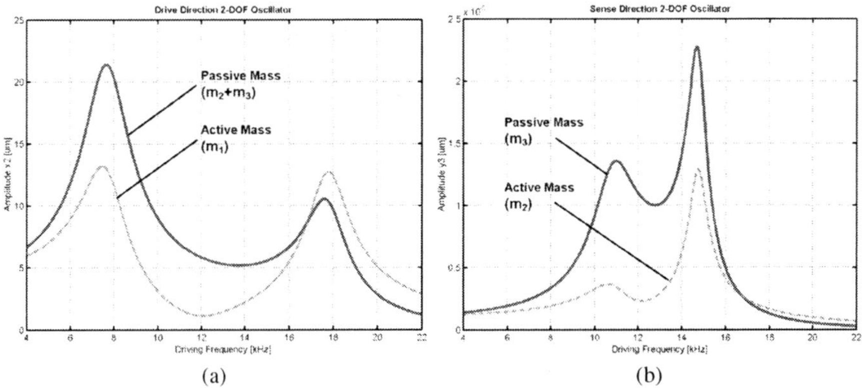

(a)   (b)

Fig. 6.26 (a) Frequency response of the 2-DOF drive-mode oscillator. (b) Frequency response of the 2-DOF sense-mode oscillator. The Coriolis force is acting on the active mass m_2, and the response is amplified by the passive mass m_3.

of m_1. Approximating the gyroscope by a lumped mass-spring-damper model (Figure 6.25a), the equations of motion in the drive direction can be expressed as

$$m_1\ddot{x}_1 + c_{1x}\dot{x}_1 + k_{1x}x_1 = k_{2x}(x_2 - x_1) + F_d$$
$$(m_2 + m_3)\ddot{x}_2 + c_{2x}\dot{x}_2 + k_{2x}x_2 = k_{2x}x_1. \tag{6.29}$$

When a constant-amplitude sinusoidal force $F_d = F_0 sin(\omega t)$ is applied on the active mass m_1 by the drive electrodes, the steady-state response of the 2-DOF system (Figure 6.26a) will be

$$X_1 = \frac{F_0}{k_{1x}} \frac{1 - (\frac{\omega}{\omega_{2x}})^2 + j\omega\frac{c_{2x}}{k_{2x}}}{[1 + \frac{k_{2x}}{k_{1x}} - (\frac{\omega}{\omega_{1x}})^2 + j\omega\frac{c_{1x}}{k_{1x}}][1 - (\frac{\omega}{\omega_{2x}})^2 + j\omega\frac{c_{2x}}{k_{2x}}] - \frac{k_{2x}}{k_{1x}}}$$

$$X_2 = \frac{F_0}{k_1} \frac{1}{[1 + \frac{k_{2x}}{k_{1x}} - (\frac{\omega}{\omega_{1x}})^2 + j\omega\frac{c_{1x}}{k_{1x}}][1 - (\frac{\omega}{\omega_{2x}})^2 + j\omega\frac{c_{2x}}{k_{2x}}] - \frac{k_{2x}}{k_{1x}}} \tag{6.30}$$

where $\omega_{1x} = \sqrt{\frac{k_{1x}}{m_1}}$ and $\omega_{2x} = \sqrt{\frac{k_{2x}}{(m_2+m_3)}}$ are the resonant frequencies of the isolated primary and secondary mass-spring systems, respectively. The two resonance peaks in the frequency response of the drive-mode oscillator can be found by equating the denominators of the undamped response to zero, yielding

$$\omega_{x-n1,2} = \sqrt{\frac{1}{2}\left(1 + \mu_x + \frac{1}{\gamma_x^2} \pm \sqrt{\left(1 + \mu_x + \frac{1}{\gamma_x^2}\right)^2 - \frac{4}{\gamma_x^2}}\right)} \, \omega_{2x} \tag{6.31}$$

where $\mu_x = \frac{(m_2+m_3)}{m_1}$ is the drive direction mass ratio, and $\gamma_x = \omega_{2x}/\omega_{1x} = \sqrt{\frac{k_{2x}m_1}{k_{1x}(m_2+m_3)}}$ is the ratio of the resonance frequencies of the isolated primary and secondary mass-spring systems in the drive-direction.

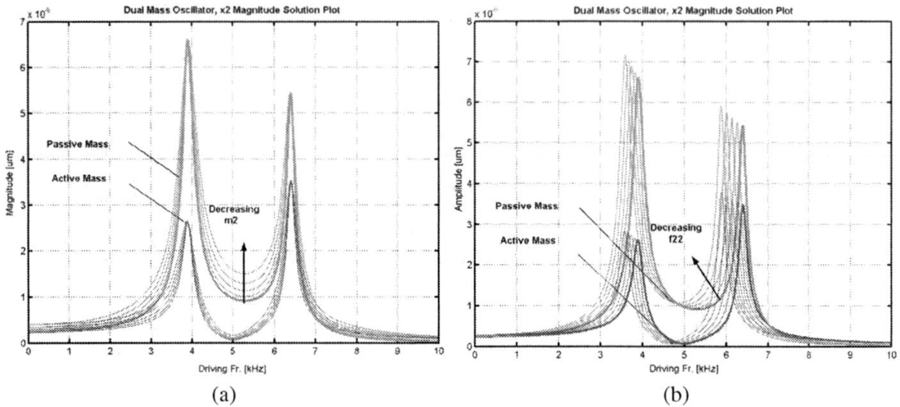

Fig. 6.27 Effect of (a) passive mass (m_2+m_3) variation, and (b) antiresonant frequency ω_{2x} variation on drive direction response.

Maximizing the Coriolis Force $F_{c2} = 2m_2\Omega_z\dot{x}_2$ generated by m_2 requires a large proof mass m_2, and large drive direction amplitude x_2. However, if the response of the passive mass in the drive direction is observed for varying m_2 values with m_1 being fixed, it is seen that for high oscillation amplitudes of passive mass, (m_2+m_3) should be minimized (Figure 6.27a). The anti-resonant frequency ω_{2x} of the isolated passive mass-spring system is determined according to gyroscope operating frequency specifications, noting that larger Coriolis forces are induced at higher frequencies, but the oscillation amplitudes become larger at lower frequencies (Figure 6.27b). Once ω_{2x} is fixed, the drive direction spring constant k_{2x} is obtained from ω_{2x} and (m_2+m_3).

The optimal drive direction mass ratio $\mu_x = \frac{(m_2+m_3)}{m_1}$ determining the mass of the active mass is dictated by low sensitivity to damping, response bandwidth and oscillation amplitude [93]. In order to achieve insensitivity to damping, the resonance peaks of the 2-DOF system response have to be separated far enough, which imposes a minimum value of μ_x. For a wide bandwidth, a large μ_x is required for large enough separation of the peaks; however, to prevent gain drop, the peak separation should be minimized (Figure 6.28a).

The degree of mechanical amplification depends on the ratio of the resonance frequencies of the isolated active and passive mass-spring systems, namely $\gamma_x = \omega_{2x}/\omega_{1x} = \sqrt{\frac{k_{2x}m_1}{k_{1x}(m_2+m_3)}}$. The optimal frequency ratio γ_x has to be determined such that γ_x is high enough for high mechanical amplification, and high oscillation amplitudes of passive mass (Figure 6.28b). From the optimal values of ω_{1x} and μ_x, the drive direction spring constant k_{1x} of the active mass is obtained. Finally, the damp-

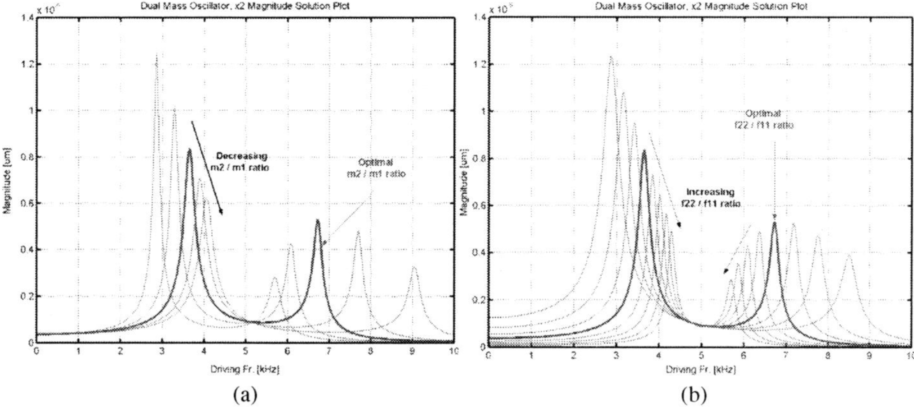

Fig. 6.28 Effect of (a) mass ratio $\mu_x = \frac{(m_2+m_3)}{m_1}$ variation, and (b) frequency ratio $\gamma_x = \omega_{2x}/\omega_{1x}$ variation on drive direction response.

ing conditions of the overall device have to be checked to verify that damping values are in the region where the response gain in the antiresonance region is insensitive to damping variations (Figure 6.29).

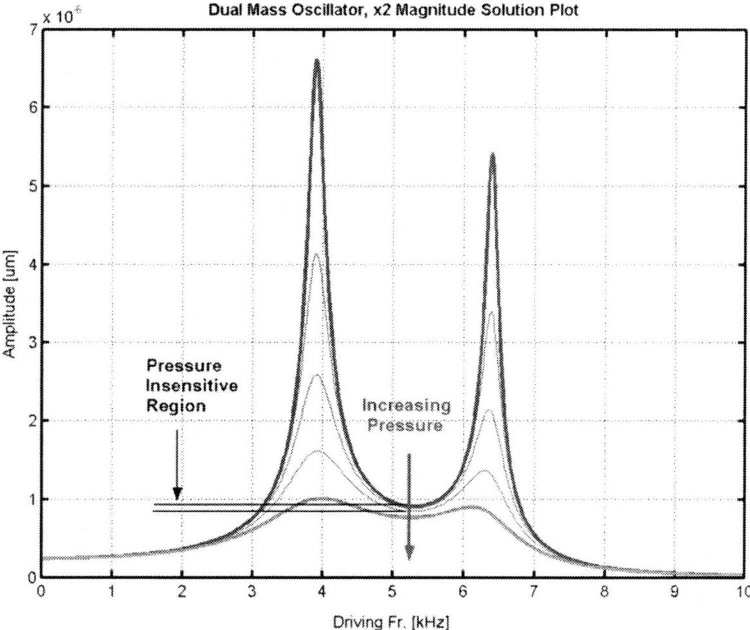

Fig. 6.29 Effect of damping on drive direction response. When the damping is under a critical value, the response in the flat region is insensitive to pressure changes.

6.5.3.2 Sense Mode Parameters

The 2-DOF sense-direction oscillator is formed by m_2 and m_3, where m_3 acts as the vibration absorber to achieve large sense direction oscillation amplitudes due to mechanical amplification (Figure 6.25b) . The objective of parameter optimization in the sense mode is to maximize y_3, which is the sense direction oscillation amplitude of the sensing element m_3.

The system is driven by the rotation-induced Coriolis forces $F_{c2} = 2m_2\Omega_z\dot{x}_2$ and $F_{c3} = 2m_3\Omega_z\dot{x}_2$ generated by m_2 and m_3, respectively. The dominant force exciting the 2-DOF sense-direction oscillator is F_{c2}, since the mass of the active mass m_2 is significantly larger than the mass of the passive mass m_3. The equations of motion of the lumped mass-spring-damper model of the sense direction oscillator become

$$m_2\ddot{y}_2 + c_{2y}\dot{y}_2 + k_{2y}y_2 = k_{3y}(y_3 - y_2) + 2m_2\Omega_z\dot{x}_2$$
$$m_3\ddot{y}_3 + c_{3y}\dot{y}_3 + k_{3y}y_3 = k_{3y}y_2 + 2m_3\Omega_z\dot{x}_2. \tag{6.32}$$

The response of the system to a constant-amplitude sinusoidal force is similar to that of the drive direction oscillator (Figure 6.26b), with the resonant frequencies of the isolated primary and secondary mass-spring systems of $\omega_{2y} = \sqrt{\frac{k_{2y}}{m_2}}$ and $\omega_{3y} = \sqrt{\frac{k_{3y}}{m_3}}$, respectively. When the frequency of the sinusoidal Coriolis force is matched with the resonant frequency of the isolated secondary mass-spring system, the secondary mass m_3 achieves maximum dynamic amplification. The two resonance peaks in the sense-mode frequency response are

$$\omega_{y-n1,2} = \sqrt{\frac{1}{2}\left(1 + \mu_y + \frac{1}{\gamma_y^2} \pm \sqrt{\left(1 + \mu_y + \frac{1}{\gamma_y^2}\right)^2 - \frac{4}{\gamma_y^2}}\right)}\,\omega_{2y} \tag{6.33}$$

where $\mu_y = \frac{m_3}{m_2}$ is the sense direction mass ratio, and $\gamma_y = \omega_{3y}/\omega_{2y} = \sqrt{\frac{k_{3y}m_2}{k_{2y}m_3}}$ is the sense-direction isolated system resonance frequency ratio.

The most important advantage of decoupling the 2-DOF drive and sense direction oscillators is that the Coriolis force that excites the sensing element is not generated by the sensing element. Instead, $F_{c2} = 2m_2\Omega_z\dot{x}_2$ generated by m_2 excites the active mass. The dynamics of the 2-DOF oscillator dictates that the mass of the passive mass m_3 has to be minimized in order to maximize its oscillation amplitude. Since the Coriolis Force $F_{c3} = 2m_3\Omega_z\dot{x}_2$ generated by m_3 is not required to be large, the sensing element m_3 can be designed to be as small as the mechanical design requirements and fabrication parameters allow.

Similarly, the optimal mass ratio $\mu_y = \frac{m_3}{m_2}$ in the sense direction determining the mass of the active mass m_2 is selected to achieve insensitivity to damping variation, a wide response bandwidth and a large oscillation amplitude [93]. The optimal ratio of the resonance frequencies of the isolated active and passive mass-spring systems $\gamma_y = \omega_{3y}/\omega_{2y} = \sqrt{\frac{k_{3y}m_2}{k_{2y}m_3}}$ is also selected to maximize oscillation amplitudes of passive mass.

6.5.3.3 Overall 4-DOF System Parameters

The frequency response of both of the 2-DOF drive and sense direction oscillators have two resonant peaks and a flat region between the peaks. To achieve maximum robustness against fluctuations in the system parameters, both of the 2-DOF oscillators have to be operated in the flat region of their response curves. Since the Coriolis forces that drive the sense-direction oscillator are at the same frequency as the electrostatic forces exciting the drive-direction oscillator, the flat-region frequency band of the oscillators have to be overlapped, by designing the drive and sense anti-resonance frequencies to match. Thus, the requirement $\omega_{3y} = \omega_{2x}$, i.e. $\sqrt{\frac{k_{3y}}{m_3}} = \sqrt{\frac{k_{2x}}{(m_2+m_3)}}$, determines the optimal system parameters, together with the optimized ratios $\mu_x = \frac{(m_2+m_3)}{m_1}$, $\gamma_x = \frac{\omega_{2x}}{\omega_{1x}}$, $\mu_y = \frac{m_3}{m_2}$, and $\gamma_y = \frac{\omega_{3y}}{\omega_{2y}}$. Since the flat regions have significantly wider bandwidths, they can be overlapped with sufficient precision without feedback control in the presence of imperfections, in contrast to the conventional gyroscopes.

Exciting the 2-DOF drive-direction oscillator at its anti-resonance frequency results in minimal oscillation amplitudes of the electrostatically driven mass (Figure 6.26a) . Thus, by minimizing the travel distance of the actuators, higher actuation stability and linearity is achieved by means of mechanical amplification. Also, since the 2-DOF sense-direction oscillator is excited at its anti-resonance frequency, the sense-direction oscillation amplitude of m_2 is minimized (Figure 6.26b) . This results in a minimal coupling between the oscillation modes, leading to reduced zero-rate drift of the gyroscope.

6.5.4 Illustrative Example

For experimental demonstration of the 3-mass gyroscope design concept, a prototype gyroscope was designed specifically for the one-mask SOI-based bulk-micromachining process. The dynamical system parameters of the device were selected as follows: For the drive mode-oscillator, the proof mass values are $m_1 = 2.18 \times 10^{-6}$kg and $(m_2 + m_3) = 9.76 \times 10^{-7}$kg, and the spring constants are $k_{1x} = 315$N/m and $k_{2x} = 168$N/m; for the sense mode-oscillator, the proof mass values are $m_2 = 7.67 \times 10^{-7}$kg and $m_3 = 7.29 \times 10^{-8}$kg, and the spring constants are $k_{2y} = 144$N/m and $k_{3y} = 13.6$N/m.

In order to estimate the actual stiffness values in the fabricated gyroscopes, test structures with the exact same suspension members and known proof-masses were

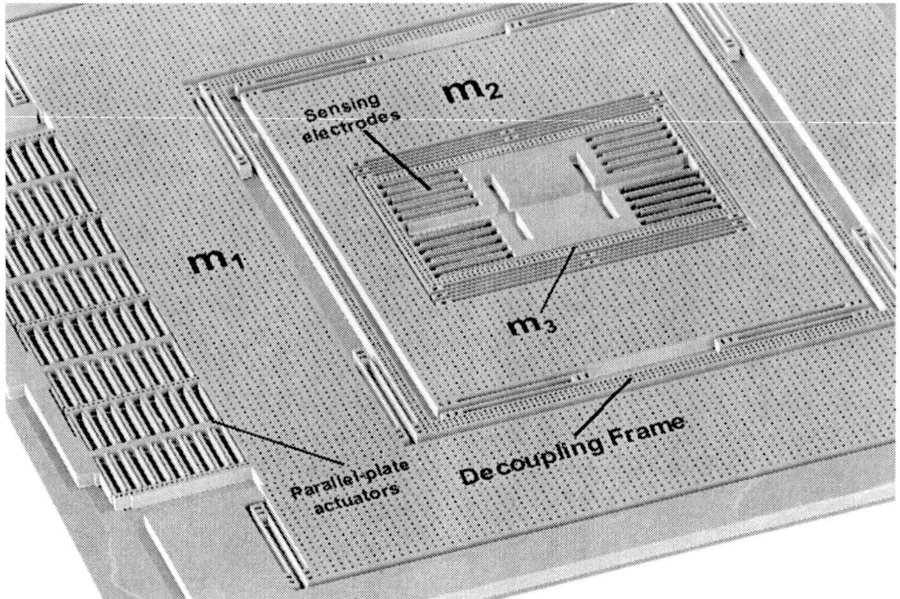

Fig. 6.30 Scanning electron micrograph of the parallel-plate actuated prototype 4-DOF gyroscope.

fabricated on the same wafer. The resonant frequencies of the test structures were measured, and the stiffness values of each suspension member in the gyroscope system were calculated.

In the drive-mode, the resonant frequencies of the isolated active and passive mass-spring systems are $\omega_{1x} = \sqrt{\frac{k_{1x}}{m_1}} = 1.601$ kHz and $\omega_{2x} = \sqrt{\frac{k_{2x}}{(m_2+m_3)}} = 1.747$ kHz, respectively; yielding a frequency ratio of $\gamma_x = \omega_{2x}/\omega_{1x} = 1.091$, and a mass ratio of $\mu_x = \frac{(m_2+m_3)}{m_1} = 0.448$. With these parameters, the location of the two expected resonance peaks in the drive-mode frequency response were calculated as $f_{x-n1} = 1.184$ kHz and $f_{x-n2} = 2.362$ kHz.

In the sense-mode, the resonant frequencies of the isolated active and passive mass-spring systems are $\omega_{2y} = \sqrt{\frac{k_{2y}}{m_2}} = 1.789$ kHz and $\omega_{2x} = \sqrt{\frac{k_{3y}}{m_3}} = 1.783$ kHz, respectively; yielding a frequency ratio of $\gamma_y = \omega_{3y}/\omega_{2y} = \sqrt{\frac{k_{3y}m_2}{k_{2y}m_3}} = 0.9968$, and a mass ratio of $\mu_y = \frac{m_3}{m_2} = 0.0950$. The resulting estimated values of the two resonance peaks in the sense-mode frequency response are $f_{y-n1} = 1.532$ kHz and $f_{y-n2} = 2.082$ kHz.

In the drive-mode frequency response, a flat region of over 800Hz was experimentally demonstrated. The two resonance peaks in the drive-mode frequency response were observed as $f_{x-n1} = 1.20$ kHz and $f_{x-n2} = 2.34$ kHz, very close to the theoretically estimated values.

The location of the two resonance peaks in the sense-mode frequency response were measured as $f_{y-n1} = 1.52$kHz and $f_{y-n2} = 2.07$kHz, in very close agreement

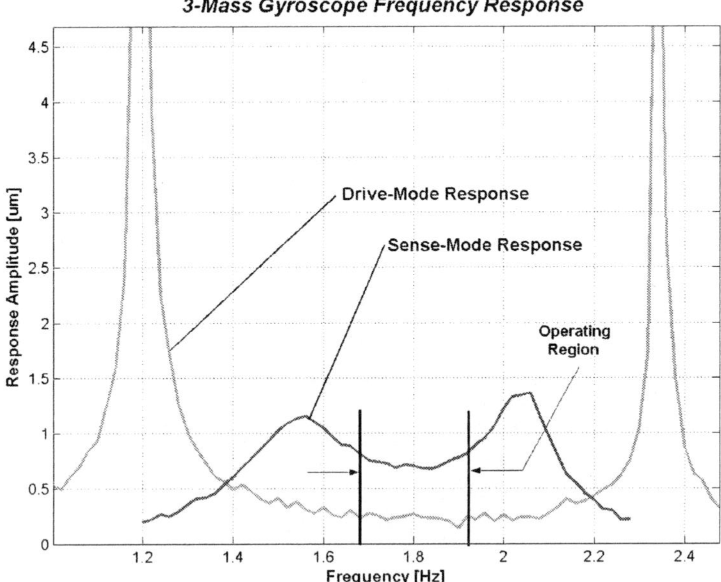

Fig. 6.31 Experimental demonstration of the overlapping flat regions in the the drive and sense-mode frequency responses, illustrating the operation frequency region.

with the theoretically estimated values. At the first resonance peak, the passive mass m_3 was observed to achieve 4.62 times dynamic amplification of the active mass vibration. The flat operation region in the sense-mode was experimentally demonstrated as almost 300Hz, overlapping with the drive-mode flat region (Figure 6.31).

6.5.5 Conclusions on the 4-DOF System Architecture

The 4-DOF system architecture provides robustness both in the drive and sense-modes, and utilizes dynamical amplification in the drive and sense directions to achieve large oscillation amplitudes. Employing three proof masses to form the decoupled oscillators allows the Coriolis force that excites the sensing element to be generated by a larger intermediate proof mass, resulting in larger Coriolis forces for increased sensor sensitivity. Also, by utilizing a 2-DOF drive-mode large drive-mode oscillation amplitudes in the drive-mode passive-mass are achieved, while the deflection of the parallel-plate actuators is significantly suppressed. The major drawback of the approach is the necessity to employ a large primary proof-mass in the drive-mode, which is not utilized in generating Coriolis force, similar to the 2-DOF drive mode architecture. Similarly, this concept also does not allow the use of a conventional self-oscillation and AGC loop in the drive-mode, and is not compatible with well-proven drive control techniques.

6.6 Demonstration of 2-DOF Oscillator Robustness

In order to characterize parametric sensitivity of the 2-DOF oscillator system frequency response, a prototype gyroscope structure with actuation electrodes attached to the primary mass was designed and fabricated. The primary mass actuation electrodes accurately emulate the Coriolis force generated by the primary mass.

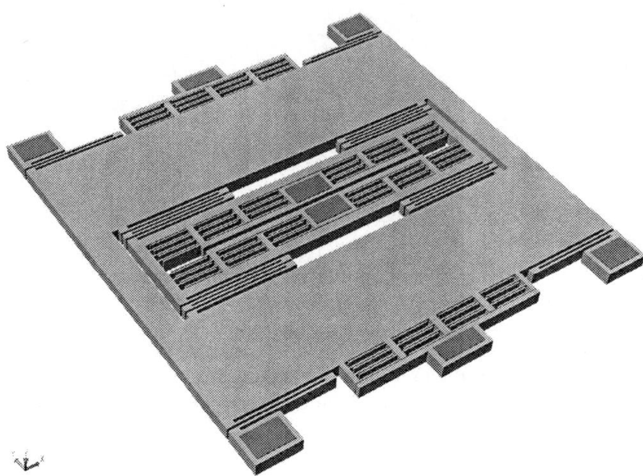

Fig. 6.32 The 2-DOF oscillator test structure, with actuation electrodes attached to the primary mass.

The frequency response of the 2-DOF oscillator was electrostatically detected under varying pressure and temperature conditions in an MMR vacuum probe station. The response signal was acquired using off-chip transimpedance amplifiers connected to an HP Signal Analyzer in sine-sweep mode. Two-port actuation and detection was utilized, where one probe was used to impose the DC bias voltage on the gyroscope structure through the anchor, one probe was used to apply the AC drive voltage on the actuation port attached to the primary mass, and the detection port on the secondary mass was directly connected to the transimpedance amplifier.

6.6.0.1 Pressure Variations

Figure 6.33 presents the experimentally measured amplitude and phase responses of the sense-mode passive mass m_3 at 5, 15, and 30 Torr acquired in an MMR vacuum probe station, after numerical parasitic filtering. The oscillation amplitude in the two resonance peaks were observed to increase with decreasing pressures. However, the change in the response amplitude in the flat operating region is insignificant, as anticipated by the theoretical analysis. This experimentally demonstrates the damping insensitivity of the sense-mode response in the flat operating region.

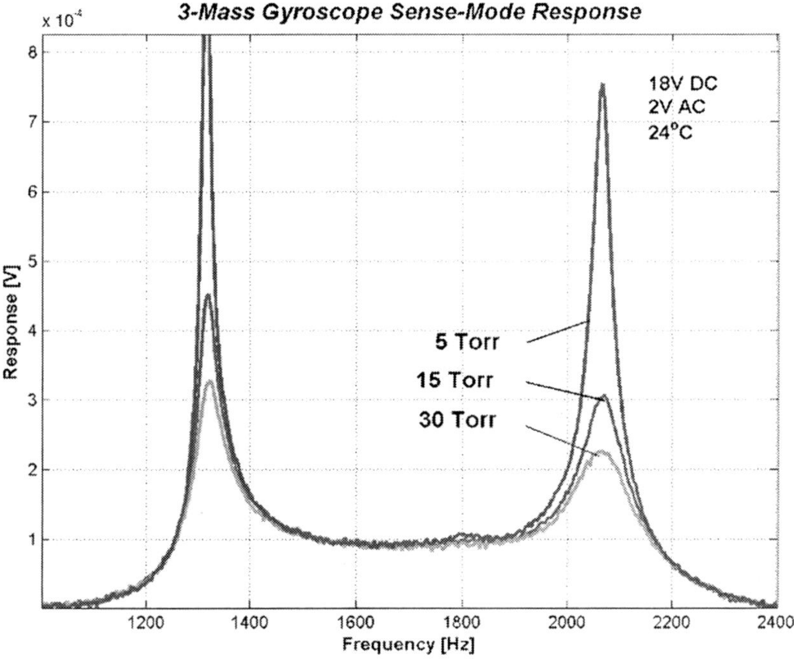

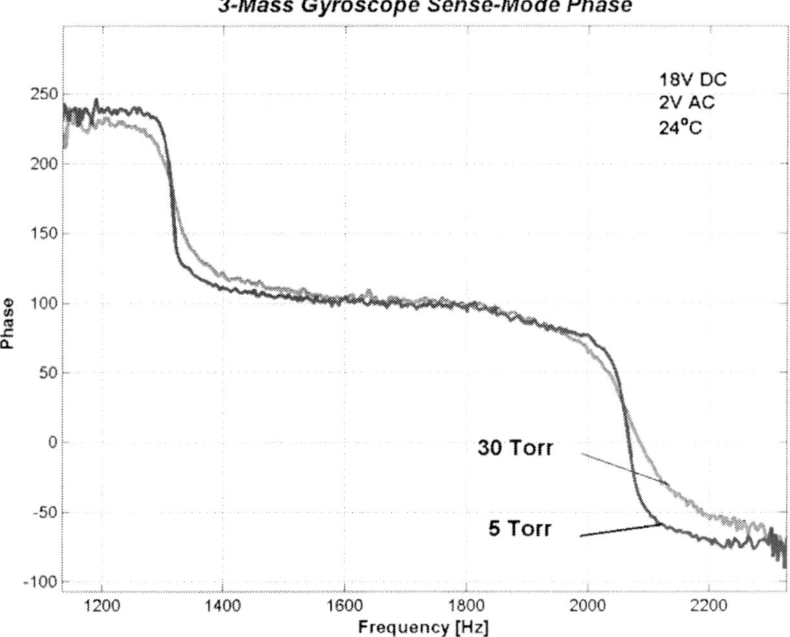

Fig. 6.33 Electrostatically acquired amplitude and phase response of m_3, with changing pressure values, after the parasitics are numerically filtered.

Furthermore, the phase of the sensing m_3 was observed to stay constant in the operating frequency band, while the phase changes were observed at the two resonance peaks as expected (Figure 6.33). Thus, it is experimentally verified that a constant-phase response is achieved in the operating region, in contrast to the abrupt phase changes at the resonance peak of the conventional gyroscopes.

6.6.0.2 Temperature Variations

The sensitivity of the prototype 4-DOF gyroscopes to temperature variations was characterized by heating the vacuum chamber of the MMR probe station, and continuously monitoring the temperature of the sample using a solid-state temperature sensor attached to the stage carrying the sample.

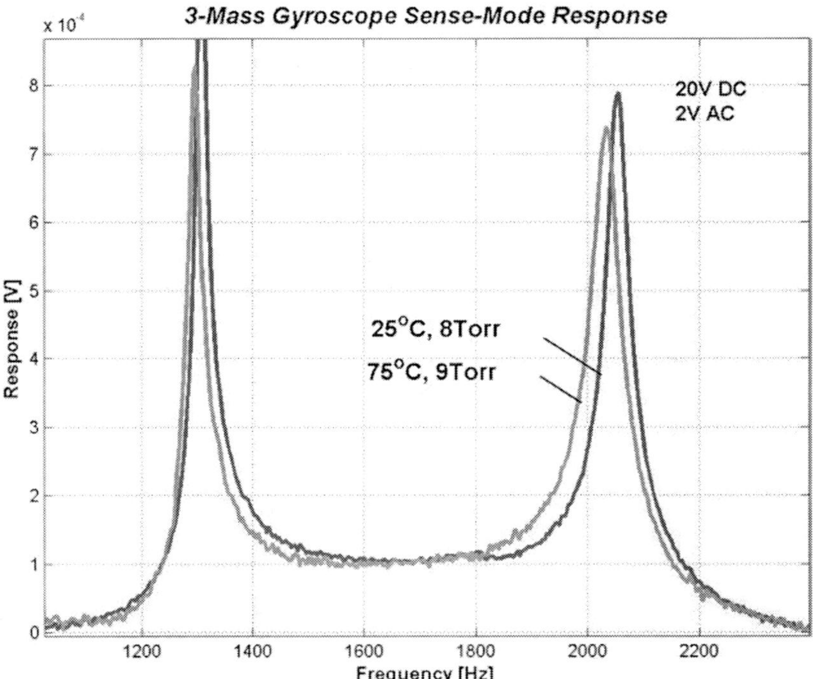

Fig. 6.34 The frequency response of the sense-mode passive mass m_3, at 25°C and 75°C. The response gain at the operating region is observed to stay constant.

Figure 6.34 presents the capacitively acquired frequency response of the sensing element m_3 at the temperatures 25°C and 75°C. The response amplitude in the flat operating region was observed to be less than 2% for the 50°C variation in temperature, experimentally demonstrating the improved robustness against temperature variations.

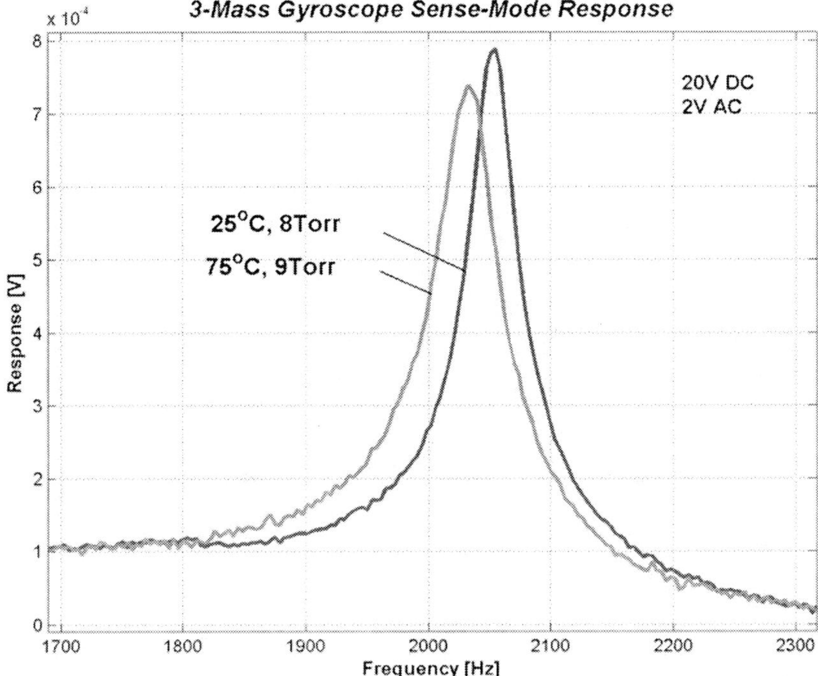

Fig. 6.35 Close-up of the frequency response of the sense-mode passive mass m_3, at 25°C and 75°C, showing the frequency shift at the resonance peak, and the constant response at the operating region.

When the change in the response gain at the resonance peaks are considered (Figure 6.35), it is observed that the frequency shift due to the temperature change results in a maximum of over 40% drop in the gain. The response amplitude in the flat operating region is observed to remain unchanged also in Figure 6.35.

6.6.0.3 Electrostatic Tuning

In order to observe the effects of larger stiffness variations on the system response, the frequency response of the sense-mode passive mass m_3 was acquired with different DC bias voltages. Figure 6.36 presents the experimental frequency response measurements for 18V to 21V DC bias at 4 Torr pressure. The electrostatic negative spring effect was observed to result in 30Hz shift in the first resonance peak and 45Hz shift in the second resonance peak, however, the response amplitude in the flat operating region was observed to change insignificantly.

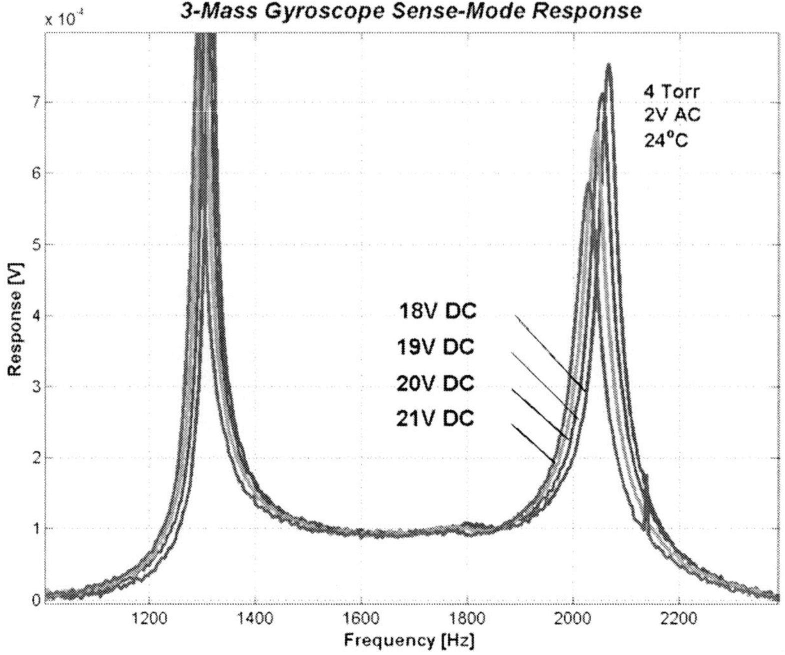

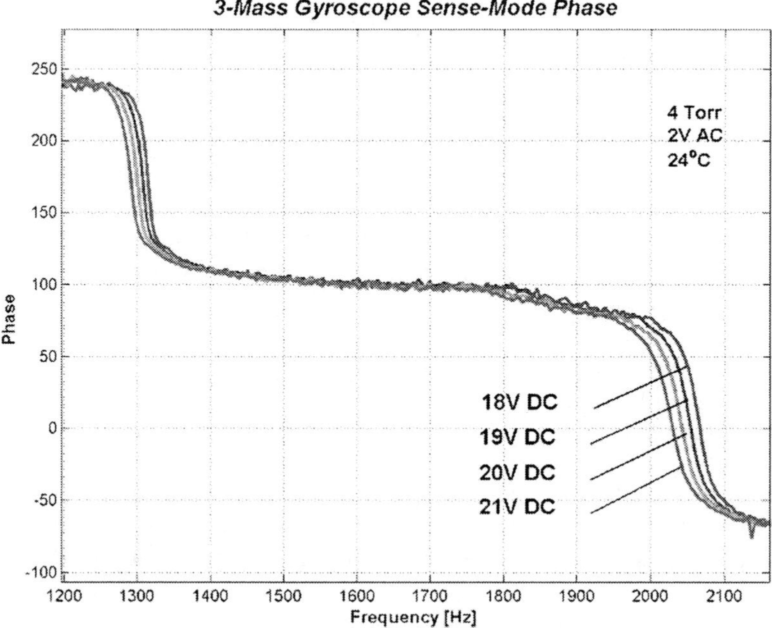

Fig. 6.36 Electrostatically detected amplitude and phase response of the sense-mode passive mass m_3, with changing DC bias.

6.7 Summary

In this chapter, we introduced design concepts based on utilizing Multi-DOF oscillators to eliminate the mode-matching problem in conventional micromachined vibratory gyroscope systems. The 2-DOF sense-mode architecture, the 2-DOF drive-mode architecture, and the 4-DOF system with 2-DOF drive and 2-DOF sense-modes were analyzed.

The 2-DOF sense-mode architecture was demonstrated to be the most suitable concept for low-cost commercial applications due to its compatibility with well-proven drive-mode control techniques. The flat region in the sense-mode frequency response significantly suppresses the effects of parameter variations on the gain and phase of the sense-mode response, while the drive-mode is operated at resonance with a self-oscillator/AGC loop identical to conventional systems.

The 2-DOF drive-mode architecture allows to achieve large drive-mode oscillation amplitudes by taking advantage of dynamical amplification. However, the large primary drive mass required for the drive oscillator does not contribute to the Coriolis force, increasing the die size without improving sense-mode sensitivity. Furthermore, the drive controller design is challenging since well-proven drive control techniques cannot be used. The 4-DOF system that combines the 2-DOF sense-mode and the 2-DOF drive-mode architectures also comes with the disadvantages associated with the 2-DOF drive-mode.

Chapter 7
Torsional Multi-DOF Architecture

This chapter presents a torsional implementation of the 3-DOF system with 2-DOF drive-mode, which aims to achieve large actuation and detection capacitances in surface micromachining. The design concept is based on employing a 2-DOF drive-mode oscillator comprised of a sensing plate suspended inside two gimbals. By utilizing dynamic amplification of torsional oscillations in the drive-mode instead of resonance, large oscillation amplitudes of the sensing element is achieved with small actuation amplitudes, providing improved linearity and stability despite parallel-plate actuation. The device operates at resonance in the sense direction for improved sensitivity, while the drive direction amplitude is inherently constant within the same frequency band.

In this chapter, the structure, operation principle, and a surface micromachined implementation of the design concept are presented. Detailed analysis of the mechanics and dynamics of the torsional system are covered, and the preliminary experimental results verifying the basic operational principles of the design concept are reported.

7.1 Introduction

Batch-fabrication of micromachined gyroscopes in VLSI compatible surface micromachining technologies constitutes the key factor in low-cost production and commercialization. The first integrated commercial MEMS gyroscopes produced by Analog Devices have been fabricated utilizing surface micromachining technology [114]. However, the limited thickness of structural layers attained in current surface-micromachining processes results in very small sensing capacitances and higher actuation voltages, restricting the performance of the gyroscope. Various devices have been proposed in the literature that employ out-of-plane actuation and detection, with large capacitive electrode plates [115,116]. However, highly non-linear and unstable nature of parallel-plate actuation limits the actuation amplitude of the gyroscope. In this chapter, we present a surface-micromachined torsional gyroscope design utilizing dynamical amplification of rotational oscillations to achieve large oscillation amplitudes about the drive axis without resonance (Figures 7.2 and 7.3); thus addressing the issues of electrostatic instability while providing large sense capacitance. The approach suggests to employ a three-mass structure with two gimbals and a sensing plate. Large oscillation amplitudes in the passive gimbal, which contains the sensing plate, are achieved by amplifying the small oscillation amplitude of the driven gimbal (active gimbal). Thus, the actuation range of the parallel-plate actuators attached to the active gimbal is narrow, minimizing the nonlinear force profile and instability. The design concept is expected to overcome the small actuation and sensing capacitance limitation of surface-micromachined gyroscopes, while achieving improved excitation stability and robustness against fabrication imperfections and fluctuations in operation conditions.

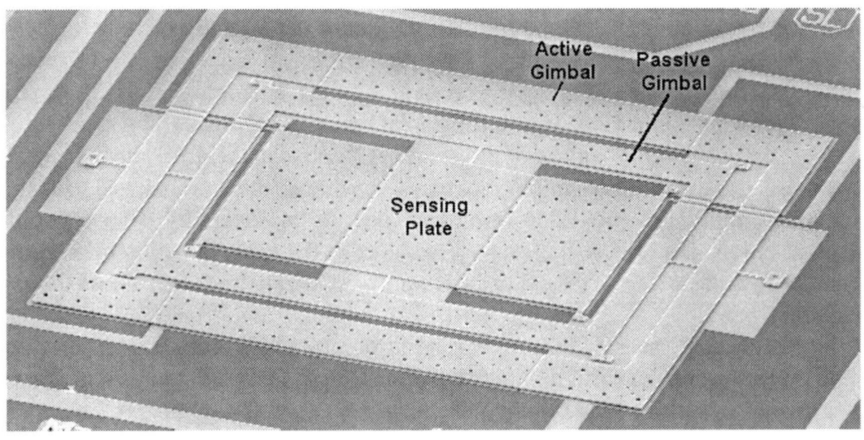

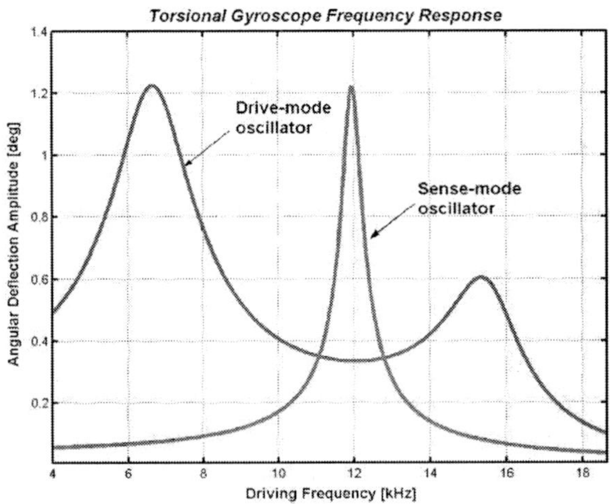

Fig. 7.1 The torsional 3-DOF gyroscope overcomes the small actuation and sensing capacitance limitation of surface-micromachining, while achieving improved drive stability and robustness.

7.2 Torsional 3-DOF Gyroscope Structure and Theory of Operation

The overall torsional gyroscope system is composed of three interconnected rotary masses: the active gimbal, the passive gimbal, and the sensing plate (Figure 7.2). The active gimbal and the passive gimbal are free to oscillate only about the drive axis x. The sensing plate oscillates together with the passive gimbal about the drive axis, but is free to oscillate independently about the sense axis y, which is the axis of response when a rotation along z-axis is applied.

The active gimbal is driven about the x-axis by parallel-plate actuators formed by the electrode plates underneath. The combination of the passive gimbal and the sensing plate comprises the vibration absorber of the driven gimbal. Thus, a torsional 2-DOF oscillator is formed in the drive direction. The frequency response of the 2-DOF drive oscillator has two resonant peaks and a flat region between the peaks, where the response amplitude is less sensitive to parameter variations (Figure 7.4a). The sensing plate, which is the only mass free to oscillate about the sense axis, forms the 1-DOF torsional resonator in the sense direction.

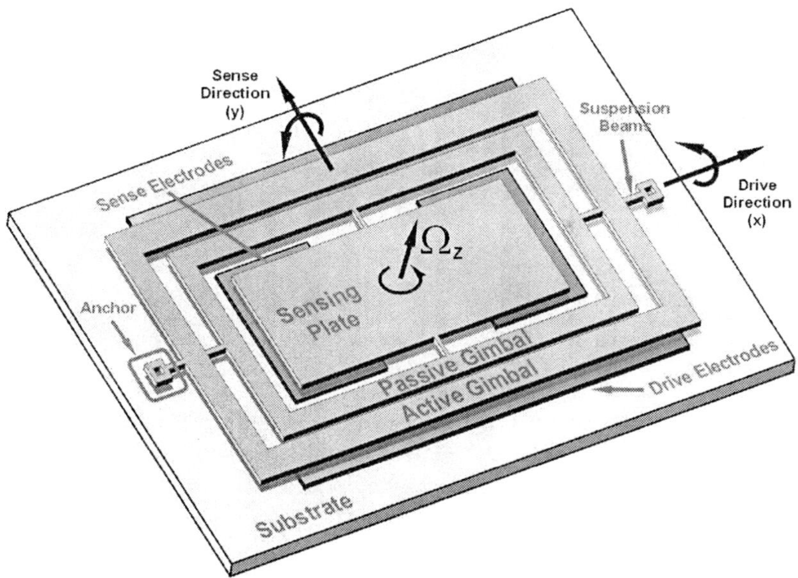

Fig. 7.2 Conceptual schematic of the torsional micromachined gyroscope with non-resonant drive.

In the presence of an input angular rate about the sensitive axis normal to the substrate (z-axis), only the sensing plate responds to the rotation-induced Coriolis torque. The oscillations of the sense plate about the sense axis are detected by the electrodes placed underneath the plate. Since the dynamical system is a 1-DOF resonator in the sense direction, the frequency response of the device has a single resonance peak in the sense mode. To define the operation frequency band of the system, sense direction resonance frequency of the sensing plate is designed to coincide with the flat region of the drive oscillator (Figure 7.4a). This allows operation at resonance in the sense direction for improved sensitivity, while the drive direction amplitude is inherently constant in the same frequency band, in spite of parameter variations or perturbations. Thus, the proposed design eliminates the mode-matching requirement by utilizing dynamic amplification of rotational oscillations instead of resonance in drive direction, leading to reduced sensitivity to structural and thermal parameter fluctuations and damping variations, while attaining sufficient performance with resonance in the sense-mode.

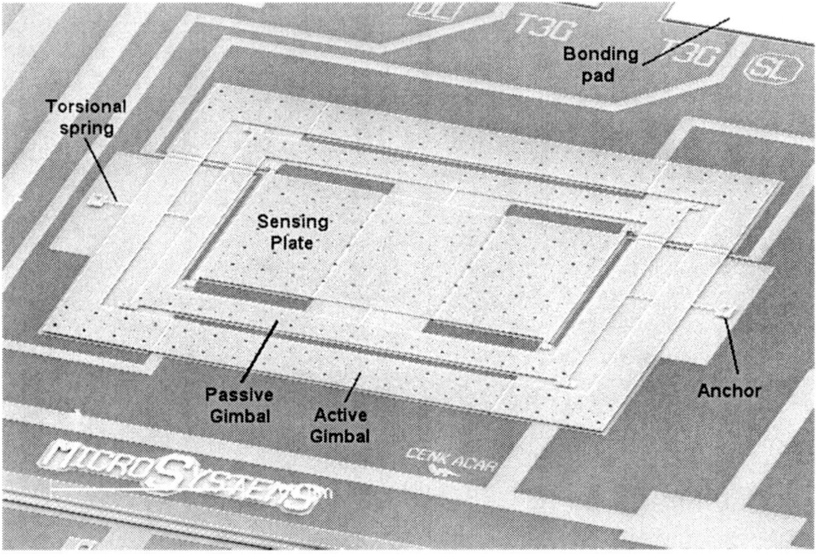

Fig. 7.3 SEM micrograph of the fabricated prototype torsional micromachined gyroscopes.

A prototype of the torsional gyroscope has been fabricated in a standard surface micromachining process (Figure 7.3), and the design objectives have been verified experimentally. The basic operational principles of the design concept had been experimentally demonstrated on linear in-plane prototype gyroscopes in the previous chapter, including the flat driving frequency band within where the drive-mode amplitude varies insignificantly, and mechanical amplification of active mass oscillation by the sensing element.

7.2.1 The Coriolis Response

The design concept is based on operating at the sense-direction resonance frequency of the 1-DOF sensing plate, in order to attain the maximum possible oscillation amplitudes in response to the induced Coriolis torque. The frequency response of the 2-DOF drive direction oscillator has two resonant peaks and a flat region between the peaks (Figure 7.4a). When the active gimbal is excited in the flat frequency band, amplitudes of the drive-direction oscillations are insensitive to parameter variations due to any possible fluctuation in operation conditions of the device. Moreover, the maximum dynamic amplification of active gimbal oscillations by the passive gimbal occurs in this flat operation region, at the antiresonance frequency. Thus, in order to operate the sense-direction resonator at resonance, while the 2-DOF drive direction oscillators operates in the flat-region frequency bands, the flat region of the drive-oscillator have to be designed to overlap with the sense-direction resonance peak (Figure 7.4a). This can be achieved by matching the drive direction anti-resonance frequency with the sense direction resonance frequency. However, in contrast to the conventional gyroscopes, the flat region with significantly wider bandwidth can be easily overlapped with the resonance peak without feedback control with sufficient precision in spite of fabrication imperfections and variations in operation conditions (Figure 7.4b).

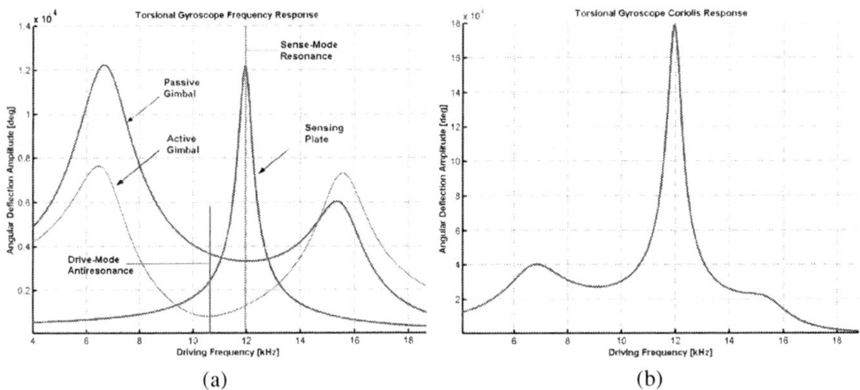

Fig. 7.4 (a) The frequency responses of the 2-DOF drive and 1-DOF sense-mode oscillators. (b) The response of the overall 3-DOF gyroscope system. The drive-direction oscillation amplitude is insensitive to parameter variations and damping fluctuations in the flat operating region, eliminating mode-matching requirement.

The drive-mode control loop sets the operation system locks the operation frequency to the sense-mode resonant frequency. Similar to the linear implementation, this could be achieved by tracking the amplitude or the phase of the quadrature signal. Thus, the device is operated exactly at the sense-mode resonant peak to achieve excellent rate sensitivity. However, the fact that a conventional self-oscillation and AGC loop cannot be used in the drive-mode is a disadvantage.

By utilizing dynamical amplification in the 2-DOF drive-oscillator instead of resonance, increased bandwidth and reduced sensitivity to structural and thermal parameter fluctuations and damping changes can be achieved, while sense-direction resonance provides high sensitivity of the device. Consequently, the design concept allows to build z-axis gyroscopes utilizing surface-micromachining technology with large sense capacitances, while resulting in improved robustness and long-term stability over the operating time of the device.

7.2.2 Gyroscope Dynamics

The dynamics of each rotary proof-mass in the torsional gyroscope system is best understood by attaching non-inertial coordinate frames to the center-of-mass of each proof-mass and the substrate (Figure 7.5). The angular momentum equation for each mass will be expressed in the coordinate frame associated with that mass. This allows the inertia matrix of each mass to be expressed in a diagonal and time-invariant form. The absolute angular velocity of each mass in the coordinate frame of that mass will be obtained using the appropriate transformations. Thus, the dynamics of each mass reduces to

$$
\begin{aligned}
\mathbf{I}_s \dot{\omega}_s^s + \omega_s^s \times (\mathbf{I}_s \omega_s^s) &= \tau_{se} + \tau_{sd} \\
\mathbf{I}_p \dot{\omega}_p^p + \omega_p^p \times (\mathbf{I}_p \omega_p^p) &= \tau_{pe} + \tau_{pd} \\
\mathbf{I}_a \dot{\omega}_a^a + \omega_a^a \times (\mathbf{I}_a \omega_a^a) &= \tau_{ae} + \tau_{ad} + M_d
\end{aligned}
\tag{7.1}
$$

where $\mathbf{I}_s, \mathbf{I}_p$, and $\mathbf{I}_a$ denote the diagonal and time-invariant inertia matrices of the sensing plate, passive gimbal, and active gimbal, respectively, with respect to the associated body attached frames. Similarly, ω_s^s, ω_p^p, and ω_a^a denote the absolute angular velocity of the sensing plate, passive gimbal, and active gimbal, respectively, expressed in the associated body frames. The external torques $\tau_{se}, \tau_{pe}, \tau_{ae}$ and $\tau_{sd}, \tau_{pd}, \tau_{ad}$ are the elastic and damping torques acting on the associated mass, whereas M_d is the driving electrostatic torque applied to the active gimbal.

If we denote the drive direction deflection angle of the active gimbal with by θ_a, the drive direction deflection angle of the passive gimbal by θ_p, the sense direction deflection angle of the sensing plate by ϕ (with respect to the substrate), and the absolute angular velocity of the substrate about the z-axis by Ω_z (Figure 7.5); the homogeneous rotation matrices from the substrate to active gimbal ($R_{sub \to a}$), from active gimbal to passive gimbal ($R_{a \to p}$), and from passive gimbal to the sensing plate ($R_{p \to s}$), respectively, become

$$R_{sub \to a} = \begin{bmatrix} 1 & 0 & 0 \\ 0 & cos\theta_a & -sin\theta_a \\ 0 & sin\theta_a & cos\theta_a \end{bmatrix} \tag{7.2}$$

$$R_{a \to p} = \begin{bmatrix} 1 & 0 & 0 \\ 0 & cos(\theta_p - \theta_a) & -sin(\theta_p - \theta_a) \\ 0 & sin(\theta_p - \theta_a) & cos(\theta_p - \theta_a) \end{bmatrix} \tag{7.3}$$

$$R_{p \to s} = \begin{bmatrix} cos\phi & 0 & sin\phi \\ 0 & 1 & 0 \\ -sin\phi & 0 & cos\phi \end{bmatrix} \tag{7.4}$$

Using the obtained transformations, the total absolute angular velocity of the sensing plate can be expressed in the non-inertial sensing plate coordinate frame as

$$\omega_s^s = \begin{bmatrix} 0 \\ \dot{\phi} \\ 0 \end{bmatrix} + R_{p \to s} \begin{bmatrix} \dot{\theta}_p \\ 0 \\ 0 \end{bmatrix} + R_{p \to s} R_{a \to p} R_{sub \to a} \begin{bmatrix} 0 \\ 0 \\ \Omega_z \end{bmatrix} \tag{7.5}$$

The absolute angular velocities of the active and passive gimbals are obtained similarly, in the associated non-inertial body frame. Substitution of the angular velocity vectors into the derived angular momentum equations yields the dynamics of the sensing plate about the sense axis (y-axis), and the active and passive gimbal dynamics about the drive axis (x-axis)

$$I_y^s \ddot{\phi} + D_y^s \dot{\phi} + [K_y^s + (\Omega_z^2 - \dot{\theta}_p^2)(I_z^s - I_x^s)]\phi =$$
$$(I_z^s + I_y^s - I_x^s)\dot{\theta}_p \Omega_z + I_y^s \theta_p \dot{\Omega}_z + (I_z^s - I_x^s)\phi^2 \dot{\theta}_p \Omega_z$$
$$(I_x^p + I_x^s)\ddot{\theta}_p + (D_x^p + D_x^s)\dot{\theta}_p + [K_x^p + (I_y^p - I_z^p + I_y^s - I_z^s)\Omega_z^2]\theta_p =$$
$$K_x^p \theta_a - (I_z^s + I_x^s - I_y^s)\dot{\phi}\Omega_z - I_x^s \phi \dot{\Omega}_z$$
$$I_x^a \ddot{\theta}_a + D_x^a \dot{\theta}_a + K_x^a \theta_a = K_x^p (\theta_p - \theta_a) + M_d \tag{7.6}$$

where I_x^s, I_y^s, and I_z^s denote the moments of inertia of the sensing plate; I_x^p, I_y^p, and I_z^p are the moments of inertia of the passive gimbal; I_x^a, I_y^a, and I_z^a are the moments of inertia of the active gimbal; D_x^s, D_x^p, and D_x^a are the drive-direction damping ratios, and D_y^s is the sense-direction damping ratio of the sensing plate; K_y^s is the torsional stiffness of the suspension beam connecting the sensing plate to the passive gimbal, K_x^p is the torsional stiffness of the suspension beam connecting the passive gimbal to the active gimbal, and K_x^a is the torsional stiffness of the suspension beam connecting the active gimbal to the substrate.

With the assumptions that the angular rate input is constant, i.e. $\dot{\Omega}_z = 0$, and the oscillation angles are small, the rotational equations of motion can be further simplified, yielding:

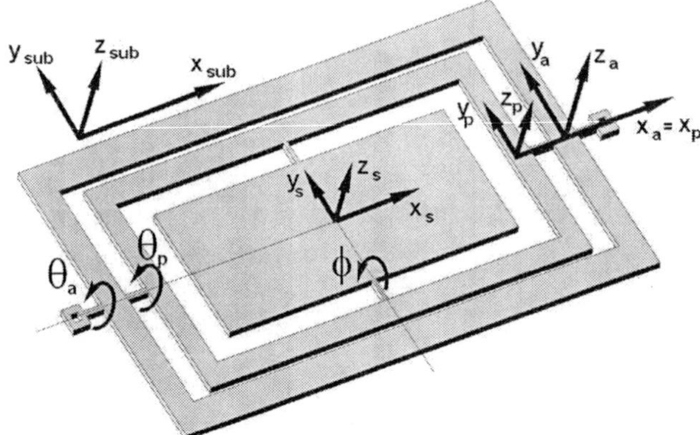

Fig. 7.5 Illustration of the non-inertial coordinate frames attached to the sensing plate, passive gimbal, active gimbal, and the substrate.

$$I_y^s \ddot{\phi} + D_y^s \dot{\phi} + K_y^s \phi = (I_z^s + I_y^s - I_x^s)\dot{\theta}_p \Omega_z$$
$$(I_x^p + I_x^s)\ddot{\theta}_p + (D_x^p + D_x^s)\dot{\theta}_p + K_x^p \theta_p = K_x^p \theta_a$$
$$I_x^a \ddot{\theta}_a + D_x^a \dot{\theta}_a + K_x^a \theta_a = K_x^p(\theta_p - \theta_a) + M_d \tag{7.7}$$

It should be noticed in the sense-direction dynamics that, the term $(I_z^s + I_y^s - I_x^s)\dot{\theta}_p(t)\Omega_z$ is the Coriolis torque that excites the sensing plate about the sense axis, with ϕ being the detected deflection angle about the sense axis for angular rate measurement.

7.2.3 Cross-Axis Sensitivity

The response of the sensing plate to the angular input rates (Ω_x and Ω_y) orthogonal to the sensitive axis (z-axis) can be modeled similarly, using the derived homogeneous transformation matrices $R_{p\to s}$, $R_{a\to p}$, and $R_{sub\to a}$, and expressing the total absolute angular velocity of the sensing plate as

$$\omega_{s,xy}^s = \begin{bmatrix} 0 \\ \dot{\phi} \\ 0 \end{bmatrix} + R_{p\to s}\begin{bmatrix} \dot{\theta}_p \\ 0 \\ 0 \end{bmatrix} + R_{p\to s}R_{a\to p}R_{sub\to a}\begin{bmatrix} \Omega_x \\ \Omega_y \\ 0 \end{bmatrix} \tag{7.8}$$

With the derived total absolute angular velocity in the presence of cross-axis inputs ($\omega_{s,xy}^s$), and the assumption that the input rates are constant ($\dot{\Omega}_x = \dot{\Omega}_y = 0$); the equation of motion of the sensing plate about the sense-axis becomes:

$$\mathbf{I}_s \dot{\omega}^s_{s,xy} + \omega^s_{s,xy} \times (\mathbf{I}_s \omega^s_{s,xy}) = \tau_{se} + \tau_{sd}$$

$$I^s_y \ddot{\phi} + D^s_y \dot{\phi} + [K^s_y + (\dot{\theta}^2_p + 2\dot{\theta}_p \Omega_x + \Omega^2_x)(I^s_x - I^s_z)]\phi = (I^s_x - I^s_z)(\theta_p \dot{\theta}_p \Omega_y + \theta_p \Omega_x \Omega_y)$$

For small oscillation angles and small magnitudes of the cross-axis inputs Ω_x and Ω_y, the equation of motion reduces to:

$$I^s_y \ddot{\phi} + D^s_y \dot{\phi} + K^s_y \phi = (I^s_z - I^s_x)(2\phi \dot{\theta}_p \Omega_x - \theta_p \dot{\theta}_p \Omega_y) \tag{7.9}$$

When the excitation terms on the right side of this equation are compared to the excitation component $(I^s_z + I^s_y - I^s_x)\dot{\theta}_p \Omega_z$ due to Ω_z, it is seen that the additional factors ϕ and θ_p in these terms make them orders of magnitude (over 10^{-5} times) smaller than the Coriolis excitation. Thus, the cross-axis sensitivity of the ideal system is negligible, provided that the sensor is aligned perfectly within the sensor package.

7.3 Illustration of a MEMS Implementation

This section describes the principle elements of a MEMS implementation of the conceptual design presented in Section 2. First, the suspension system design for the torsional system is investigated with the derivation of the stiffness values. The capacitive sensing and actuation details is followed by the discussion of achieving dynamic amplification in the drive mode, along with an approach for determining optimal system parameters to maximize sensor performance. Finally, the sensitivity and robustness analyses of the system are presented.

7.3.1 Suspension Design

The suspension system of the device that supports the gimbals and the sensing plate is composed of thin polysilicon beams with rectangular cross-section functioning as torsional bars. The active gimbal is supported by two torsional beams of length L^a_x anchored to the substrate, aligned with the drive axis, so that the gimbal oscillates only about the drive axis. The passive gimbal is also attached to the active gimbal with two torsional beams of length L^p_x aligned with the drive axis, forming the 2-DOF drive-direction oscillator. Finally, the sensing plate is connected to the passive gimbal using two torsional beams of length L^s_y lying along the sense axis, allowing it to oscillate about the sense axis independent from the gimbals (Figure 7.6).

Assuming each torsional beam is straight with a uniform cross-section, and the structural material is homogeneous and isotropic; the torsional stiffness of each beam with a length of L can be modeled as [106]

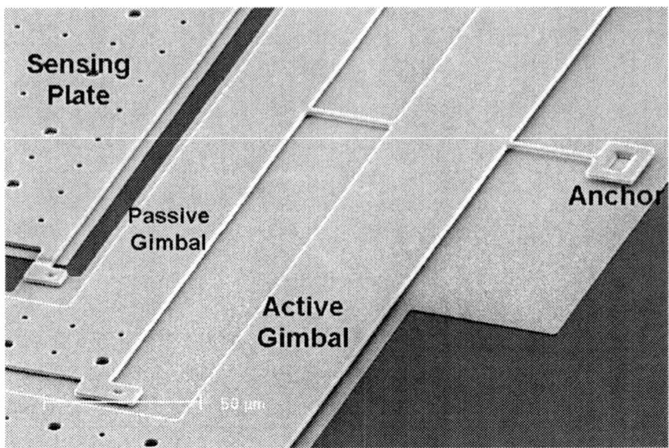

Fig. 7.6 SEM micrograph of the torsional suspension beams in the prototype gyroscopes.

$$K = \frac{SG + \sigma J}{L} \tag{7.10}$$

where $G = \frac{E}{2(1-v)}$ is the shear modulus with the elastic modulus E and Poisson's ratio v; σ is the residual stress; and $J = \frac{1}{12}(wt^3 + tw^3)$ is the polar moment of inertia of the rectangular beam cross-section with a thickness of t and a width of w. The cross-sectional coefficient S can be expressed for the same rectangular cross-section as [106]

$$S = \left(\frac{t}{2}\right)^3 \frac{w}{2} \left[\frac{16}{3} - 3.36\frac{t}{w}\left(1 - \frac{t^4}{12w^4}\right)\right] \tag{7.11}$$

Assuming the same thickness t and width w for each beam, the torsional stiffness values in the equations of motion of the ideal gyroscope dynamical system model can be calculated as follows:

$$K_x^a = 2\frac{SG + \sigma J}{L_x^a}, \; K_x^p = 2\frac{SG + \sigma J}{L_x^p}, \; K_y^s = 2\frac{SG + \sigma J}{L_y^s} \tag{7.12}$$

For the presented prototype design (Figure 7.3), the suspension beams lengths are $L_x^a = L_x^p = L_y^s = 30\mu$m, with the width of 2μm and a structural thickness of 2μm; resulting in the stiffness values of $K_x^a = K_x^p = K_y^s = 1.04 \times 10^{-18}$ kg m^2/s^2.

7.3.2 Finite Element Analysis

In order to verify the validity of the assumptions in the theoretical analysis, the operational modes and the other resonance modes of the system were simulated using Finite Element Analysis (FEA) package MSC Nastran/Patran.

The geometry of the device was optimized to match the drive-mode resonant frequency of the isolated passive mass-spring system $\omega_x^p = \sqrt{K_x^p/(I_x^a + I_x^s)}$ with the sense mode resonance frequency of the sensing plate $\omega_y = \sqrt{K_y^s/I_y^s}$. Theoretical analysis of the device geometry, which is presented in detail in Figure 7.3, and a structural thickness of 2μm yields $K_x^p = K_y^s = 1.04 \times 10^{-18}$ kg m^2/s^2, $(I_x^a + I_x^s) = 4.97 \times 10^{-18}$ kg m^2 and, $I_y^s = 4.94 \times 10^{-18}$kg m^2; resulting in $\omega_x^p = 7.285$kHz and $\omega_y = 7.263$kHz.

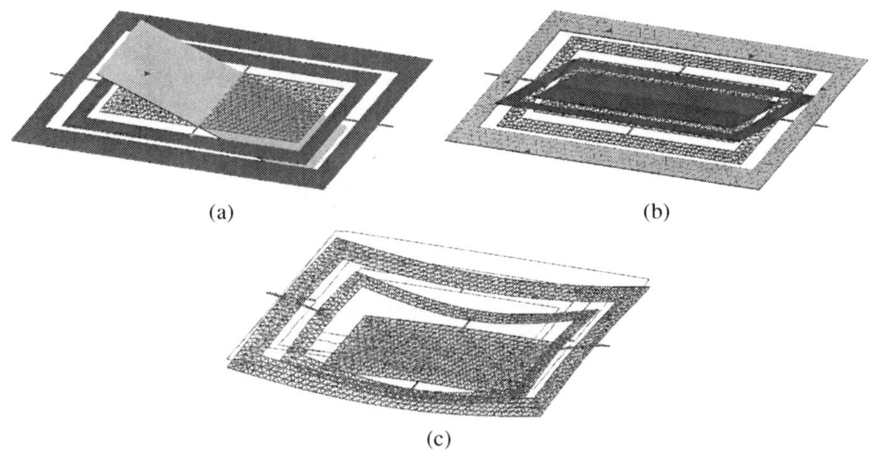

(a)   (b)

(c)

Fig. 7.7 Finite Element Analysis simulation of the torsional system: (a) Sense-mode resonance frequency of the sensing plate at ω_y =7.457kHz. (b) Drive-mode resonant frequency of the isolated passive mass-spring system at ω_{xp} =7.097kHz. (c) The undesired mode with the lowest frequency at 8.735kHz, being the linear out-of-plane mode.

Through FEA simulations, the sense mode resonance frequency of the sensing plate about the sense axis was obtained to be ω_y =7.457kHz (Figure 7.7a), with a 2.28% discrepancy from the theoretical calculations. The drive-mode resonant frequency of the isolated passive mass-spring system was obtained as the first mode (Figure 7.7b) at ω_x^p =7.097kHz. The 5.91% discrepancy from theoretical analysis is attributed in part to the reduced stiffness due to the compliance of the active and passive gimbal frame structures. More importantly, the torsional beams exhibit linear deflections as well, and do not undergo purely torsional deflections; which also provides considerable additional compliance.

The undesired resonance modes of the structure were also observed to be sufficiently separated from the operational modes. In the FEA simulations, the undesired

mode with the lowest frequency was observed to be the linear out-of-plane mode
(Figure 7.7c), at 8.735kHz.

7.3.3 Electrostatic Actuation

The active gimbal is excited about the drive axis by the electrodes symmetrically
placed underneath its edges. Applying a voltage of $V_1 = V_{DC} + v_{AC}sin(\omega_d t)$ to electrode 1 on one side of the gimbal, and $V_2 = V_{DC} - v_{AC}sin(\omega_d t)$ to electrode 2 on
the opposing side, a balanced actuation scheme is imposed.

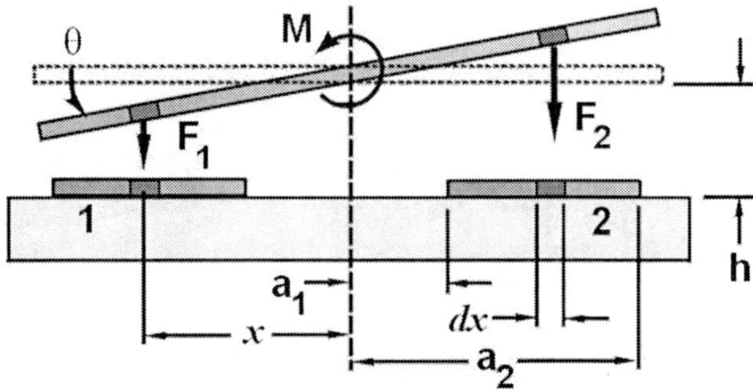

Fig. 7.8 Cross-section of the torsional electrostatic parallel-plate actuation electrodes attached to
the active gimbal.

The net moment M_d that drives the active gimbal is the sum the positive and negative resultant moments applied by electrode 1 and electrode 2, respectively. These
moments can be expressed by integrating the moments generated by the infinite
number of infinitesimally-small capacitors of width dx, located by a distance of x
from the center of rotation (Figure 7.8). Neglecting the fringing field effects, with
symmetric electrodes of width $(a_2 - a_1)$ and length L placed by a distance of a_1 from
the center-line, the net actuation moment M_d, which is a function of the deflection
angle θ_a, can be expressed as:

$$M_d = M_1 - M_2 = \int_{a1}^{a2} x\,dF_1 - \int_{a1}^{a2} x\,dF_2$$

$$= \int_{a1}^{a2} x\frac{\varepsilon_0 V_1^2 L dx}{2(h - xtan\theta_a)^2} - \int_{a1}^{a2} x\frac{\varepsilon_0 V_2^2 L dx}{2(h + xtan\theta_a)^2} \tag{7.13}$$

where h is the elevation of the structure from the substrate, and $\varepsilon_0 = 8.85 \times 10^{-12} F/m$ is the permittivity of air. Assuming small angles of actuation, which are achieved due to dynamic amplification of oscillations as will be explained in the next section, the net electrostatic moment reduces to the expression used in the simulations of the dynamic system:

$$M_d = \frac{\varepsilon_0 L}{2\theta_a^2} \left(\frac{1}{1 - \frac{a_2}{h}\theta_a} - \frac{1}{1 - \frac{a_1}{h}\theta_a} + \ln \frac{h - a_2\theta_a}{h - a_1\theta_a} \right) V_1^2$$

$$- \frac{\varepsilon_0 L}{2\theta_a^2} \left(\frac{1}{1 + \frac{a_2}{h}\theta_a} - \frac{1}{1 + \frac{a_1}{h}\theta_a} + \ln \frac{h + a_2\theta_a}{h + a_1\theta_a} \right) V_2^2 \qquad (7.14)$$

For the presented prototype design (Figure 7.3), the drive mode electrodes underneath the active gimbal are 380μm $\times 60\mu$m; resulting in a total of 0.252μN force per each electrode with a 1V actuation voltage. The total moment applied by each electrode at the deflection of $\theta_a = 0$ is 5.29×10^{-13}Nm.

7.3.4 Optimization of System Parameters

Since the foremost mechanical factor determining the performance of the gyroscope is the angular deflection ϕ of the sensing plate about the sense axis due to the input rotation, the parameters of the dynamical system should be optimized to maximize ϕ. However, the optimal compromise between amplitude of the response and bandwidth should be obtained to maintain robustness against parameters variations, while the response amplitude is sufficient for required sensitivity. The trade-offs between gain of the response (for higher sensitivity) and the system bandwidth (for increased robustness) will be typically guided by application requirements.

For a given input rotation rate Ω_z, in order to maximize the Coriolis torque $(I_z^s + I_y^s - I_x^s)\dot{\theta}_p(t)\Omega_z$ that excites the sensing plate about the sense axis, the oscillation amplitude of the passive gimbal about the drive axis should be maximized. In the drive mode, the gyroscope is simply a 2-DOF torsional system. The sinusoidal electrostatic drive moment M_d is applied to the active gimbal. The combination of the passive gimbal and the sensing plate comprise the vibration absorber of the 2-DOF oscillator, which mechanically amplifies the oscillations of the active gimbal. Approximating the 2-DOF oscillator by a lumped mass-spring-damper model, the equations of motion about the drive axis can be expressed as:

$$I_x^a \ddot{\theta}_a + D_x^a \dot{\theta}_a + K_x^a \theta_a = K_x^p (\theta_p - \theta_a) + M_d$$
$$(I_x^p + I_x^s)\ddot{\theta}_p + (D_x^p + D_x^s)\dot{\theta}_p + K_x^p \theta_p = K_x^p \theta_a \qquad (7.15)$$

where I_x^s, I_y^s, and I_z^s are the moments of inertia of the sensing plate; I_x^p, I_y^p, and I_z^p are the moments of inertia of the passive gimbal; I_x^a, I_y^a, and I_z^a are the moments of inertia of the active gimbal; D_x^s, D_x^p, and D_x^a are the drive-direction damping ratios; K_x^p is the torsional stiffness of the suspension beam connecting the passive gimbal to the active gimbal, and K_x^a is the torsional stiffness of the suspension beam connecting the active gimbal to the substrate.

When the driving frequency ω_{drive} is matched with the resonant frequency of the isolated passive mass-spring system (ω_x^p), i.e. $\omega_{drive} = \omega_x^p = \sqrt{\frac{K_x^p}{(I_x^a + I_x^s)}}$, maximum dynamic amplification is achieved. Thus, if the drive direction anti-resonance frequency ω_x^p and the sense direction resonance frequency $\omega_y = \sqrt{\frac{K_y^p}{I_y^s}}$ are designed to match, maximum dynamic amplification in drive mode is achieved, the Coriolis torque drives the sensing plate into resonance, and the drive-mode oscillator is excited in the flat frequency band. The optimal design condition can be summarized as follows:

$$\sqrt{\frac{K_x^p}{(I_x^a + I_x^s)}} = \sqrt{\frac{K_y^p}{I_y^s}} = \omega_{drive} \tag{7.16}$$

7.3.5 Sensitivity and Robustness Analyses

The response of the complete electro-mechanical system of the torsional gyroscope was simulated by incorporating the presented electro-mechanical modeling. With the sense-direction resonance frequency of 7.457 kHz as obtained from the finite element analysis simulations, the effective sense direction response amplitude of the sensing capacitors to a 1 °/s input angular rate was found to be $1.6 \times 10^{-5} \mu m$. It is assumed that the gyroscope is vacuum packaged so that the pressure within the encapsulated cavity is equal to 100 miliTorrs (13.3 Pa), and that the passive gimbal oscillates in the whole $2 \mu m$ gap. The response of a torsional gyroscope with a resonant drive-mode and the same geometry to the same input is $0.53 \times 10^{-5} \mu m$, since the stable drive-mode actuation range is limited to $0.66 \times 10^{-5} \mu m$. However, the required actuation voltage amplitude for the anti-resonant mode is 3.9 times larger than the resonant drive-mode approach. It should also be noticed that, in the presented prototype design (Figure 7.3) with the sensing electrode area of $200 \mu m \times 130 \mu m$ (nominal capacitance of 11.51pF); 1 °/s input angular rate results in a total capacitance change of 29.2fF, which is considerably larger compared to in-plane surface-micromachined gyroscope designs.

In the case of a potential shift in the sense-mode resonance frequency, e.g. due to temperature fluctuations, residual stresses, or fabrication variations, the response amplitude is sustained at a constant value to a great extent without active tuning of resonance frequencies. For example, a 5% mismatch in the sense-mode resonance frequency of the sensing plate (ω_y) and the drive-mode anti-resonance frequency (ω_x^p) results in only 2.5% error in the response amplitude (Figure 7.9). Without

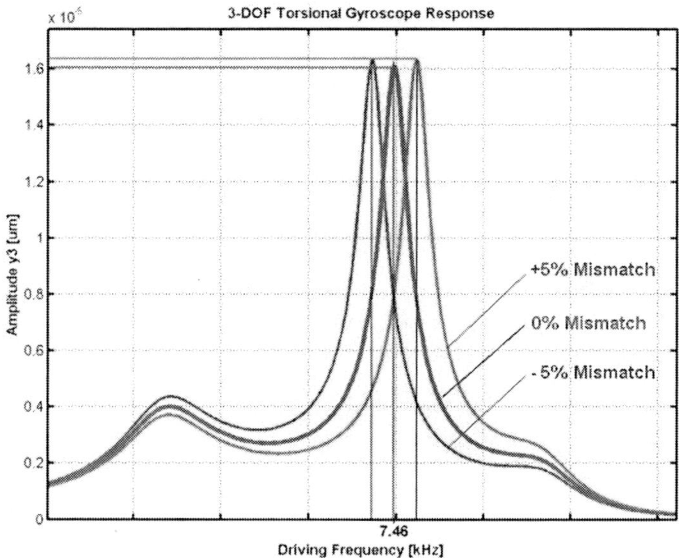

Fig. 7.9 The response of the complete torsional gyroscope system. 5% mismatch between the sense-mode resonance frequency (ω_y) and the drive-mode anti-resonance frequency (ω_x^p)results in only 2.5% error in the response amplitude.

active compensation, a conventional 2-DOF gyroscope can exhibit over 60% error for the same frequency mismatch under the same operation conditions.

7.4 Experimental Characterization

The response of the fabricated prototype gyroscopes have been characterized electrostatically under vacuum, and optically using a Sensofar PLμ Confocal Imaging Profiler and Polytec Scanning Laser Doppler Vibrometer under atmospheric pressure.

The sense-mode resonance frequency and the drive-mode antiresonance frequency of the prototype gyroscope were measured in a cryogenic MMR Vacuum Probe Station. The frequency response of the device was acquired using off-chip transimpedance amplifiers connected to an HP Signal Analyzer in sine-sweep mode. Due to the large actuation and sensing capacitances, actuation voltages as low as 0.7V to 1.8V DC bias, and 30mV AC were used under 40mTorr vacuum. For sense-mode resonance frequency detection, one-sided actuation was utilized, where one sensing electrode was used for driving, and the other for detection. For detecting the drive-mode antiresonant frequency, which is equal to the resonant frequency of the isolated passive mass-spring system (ω_x^p), a separate test structure that consists of

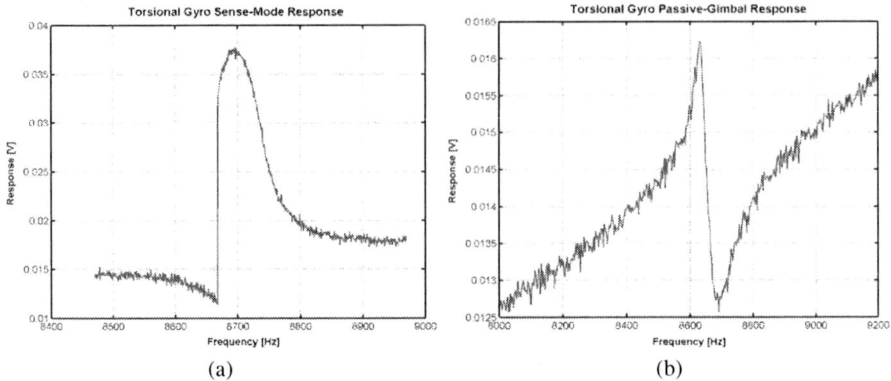

(a)   (b)

Fig. 7.10 (a) Sense-mode, and (b) drive-mode frequency response acquired using off-chip tran-simpedance amplifiers, yielding ω_y= 8.725 kHz, and ω_x^p=8.687 kHz for 0.7V DC bias.

the passive gimbal-sensing plate assembly was used, with the one-sided actuation scheme, and the same actuation voltages.

The resonance peaks in the sense and drive modes were observed very clearly for the 0.7V - 1.8V DC bias range. With a 0.7V DC bias, the sense-mode resonance frequency was measured to be 8.725 kHz (Figure 7.10a), and the drive-mode anti-resonance frequency was measured to be 8.687 kHz (Figure 7.10b). We were also able to electrostatically tune both the drive and sense mode resonance frequencies by several hundred Hertz with only 1V DC bias tuning range (Figure 7.11a). It was observed that ω_x^p and ω_y are exactly matched for $V_{dc} = 1.0$V. Furthermore, it was possible to observe both of the drive and sense resonance modes in the same sweep, by driving the structure with one drive electrode, and sensing the response with one sense electrode. Figure 7.11b shows the frequency response with 1.3V DC bias, where the primary excited drive-mode appears as the large resonance peak, and the sense-mode appears as the secondary small peak.

The experimentally measured resonance frequencies, and the estimated values via theoretical analysis and finite element analysis are summarized in Table 1. To investigate the origins of over 1kHz discrepancy between the measurements and FEA results, the prototype was analyzed using a PLμ Confocal Imaging Profiler, for obtaining the structural parameters, such as layer thickness, elevation, suspension beam geometry, and any possible curling in the structure (Figure 7.12a). The PLμ optical profiler acquires confocal images using a proprietary confocal arrangement, fast scanning devices and high contrast algorithms; and includes a motorized x-y stage and a PZT scanning device [117]. In the confocal microscope, all structures out of focus are suppressed by the detection pinhole, employing an arrangement of diaphragms. Thus, light rays from outside the focal plane are rejected, and three-dimensional data sets of the devices are acquired by out-of-plane scanning.

Using the PLμ profiler, the structural layer thickness was measured as 1.97 μm, elevated by 2.1 μm from the substrate, with extremely small curling (10 nm elevation difference between the middle section and the edges). However, due to the

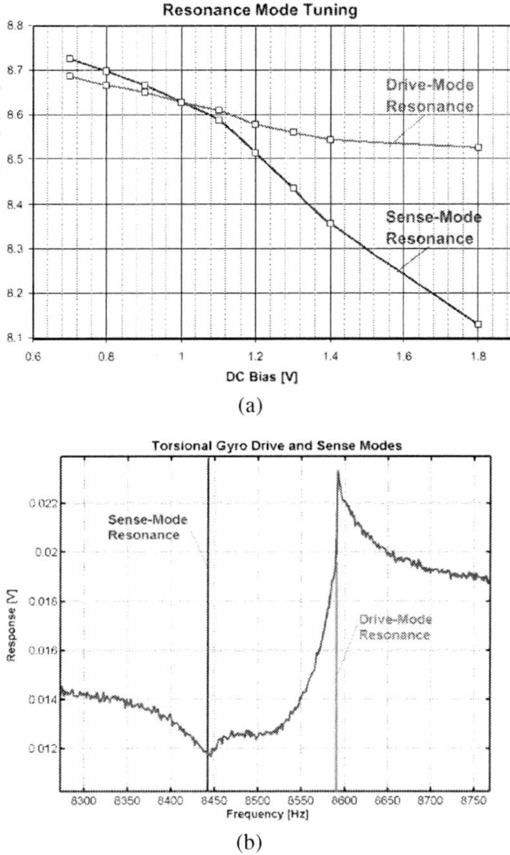

Fig. 7.11 (a) Electrostatic tuning of the drive and sense mode resonance frequencies by changing the DC bias. Notice that ω_x^p and ω_y are exactly matched for $V_{dc} = 1.0$V. (b) Drive and sense modes observed in the same sweep with 1.3V DC bias.

Table 7.1 Summary of the estimated and measured resonance frequencies.

	Theoretical Analysis	FEA Results	FEA Results with $l = 27.1\mu$m	Experimental $V_{dc} = 0.7$V	Measured Q [40 mTorr]
Drive Mode (ω_x)	7.28 kHz	7.097 kHz	8.364 kHz	8.687 kHz	50
Sense Mode (ω_y)	7.26 kHz	7.457 kHz	8.642 kHz	8.725 kHz	100

corner rounding effects in photo-lithography, the effective length of the torsional suspension beams were observed as 27.1μm (Figure 7.12b). When the FEA was repeated with the 27.1μm suspension length, the results agreed with experimental measurements to a great extent, with 3.6% discrepancy in the drive-mode, and 0.95% discrepancy in the sense-mode (Table 1).

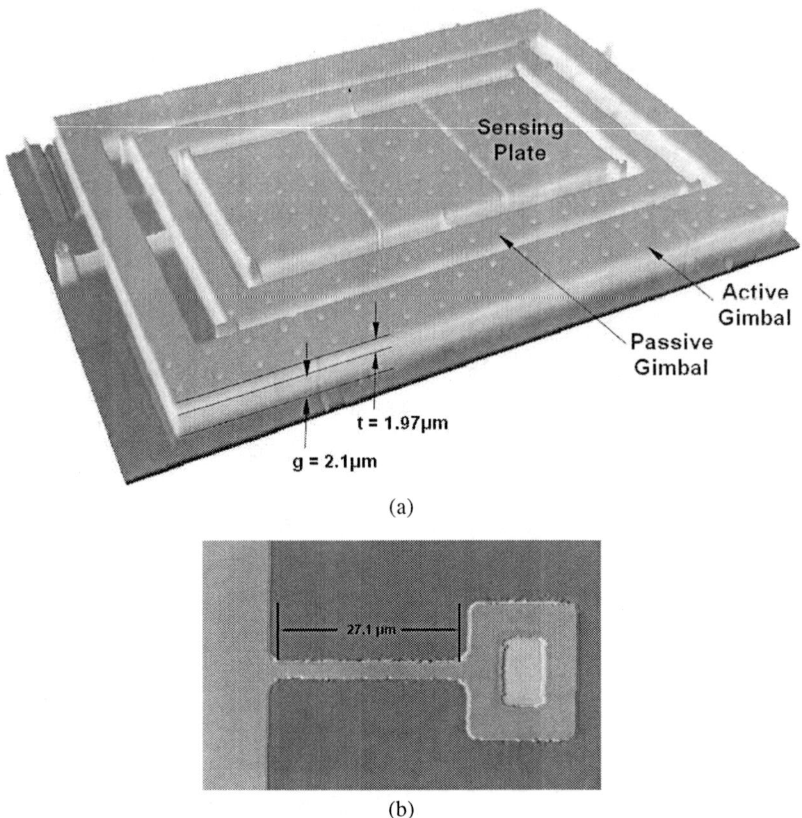

Fig. 7.12 (a) Confocal imaging profiler scan of the structure, for obtaining layer thickness, elevation, suspension beam geometry, and curling. (b) Effective length of the torsional suspension beams were observed as 27.1μm due to corner rounding effects.

In order the verify the mode-shapes of the structure at the measured frequencies, a Polytec Scanning Laser Doppler Vibrometer was used under atmospheric pressure for dynamic optical profiling. Laser Doppler vibrometry (LDV) is a non-contact vibration measurement technique using the Doppler effect, based on the principle of the detection of the Doppler shift of coherent laser light, that is scattered from a small area of the test object. Laser vibrometers are typically two-beam interferometric devices which detect the phase difference between an internal reference and the measurement beam, which is focused on the the target and scattered back to the interferometer [119].

Dynamic response of the device in the drive and sense modes was characterized using the Scanning Laser Doppler Vibrometer in scanning mode, which allows to measure the response of a dense array of points on the whole gyroscope structure. Dynamic excitation of the sensing plate about the sense-axis at the experimentally measured sense-mode resonance frequency ($\omega_y = 8.725$ kHz) revealed that only the

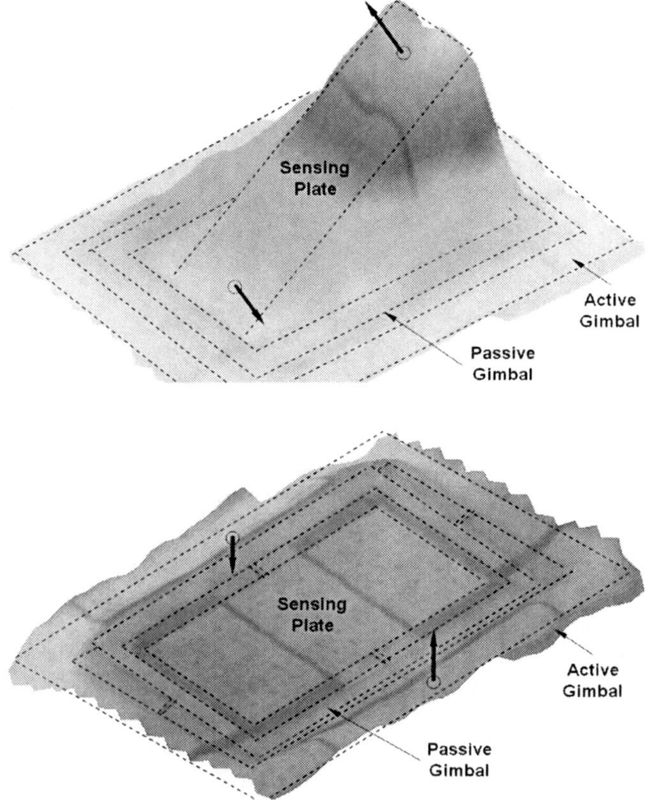

Fig. 7.13 The sense-mode and the drive-mode dynamic response measurements using the LDV in the scanning mode.

sensing plate responds in the sense mode, verifying that the 1-DOF resonator formed in the sense-mode is decoupled from the drive-mode (Figure 7.13a), in agreement with the intended design and finite element analysis simulations. Dynamic excitation of the active gimbal about the drive-axis at frequencies away from the anti-resonance frequencies verified that the active gimbal oscillates independent from the passive gimbal-sensing plate assembly (Figure 7.13b), constituting the active mass of the 2-DOF oscillator.

Most prominently, dynamic amplification of the active gimbal oscillations by the passive gimbal was successfully demonstrated. At the drive-mode anti-resonance frequency, which was measured to be 8.687 kHz, the passive gimbal was observed to achieve over 1.7 times larger oscillation amplitudes than the driven active gimbal (Figure 7.14). This translates into attaining over 2.4 times larger drive-mode deflection angles at the sensing plate than the active gimbal.

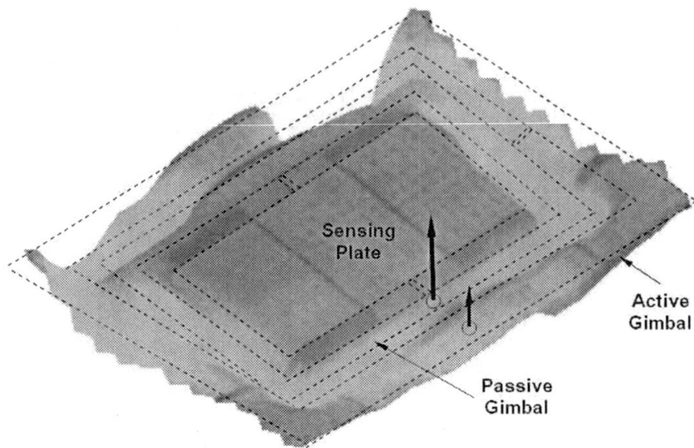

Fig. 7.14 Scanning mode LDV measurements at the anti-resonance frequency, demonstrating dynamic amplification of the active gimbal oscillations by the passive gimbal. The passive gimbal was observed to achieve over 1.7 times larger oscillation amplitudes than the driven active gimbal.

7.5 Summary

In this chapter, the design concept of a 3-DOF gimbal-type torsional micromachined gyroscope with 2-DOF drive mode was introduced. The analysis of the system dynamics and structural mechanics of the torsional system were presented, along with the experimental results verifying the design objectives.

The proposed approach is based on achieving large oscillation amplitudes in the passive gimbal by amplifying the small oscillation amplitude of the driven gimbal. This allows to minimize the nonlinear force profile and instability due to parallel-plate actuation of the active gimbal, and to achieve large drive-mode amplitudes in the sensing element with small actuation voltages.

Chapter 8
Distributed-Mass Architecture

This chapter presents an approach that is expected to yield robust vibratory MEMS gyroscopes with improved gain characteristics, while retaining the wide bandwidth. The approach is based on utilizing multiple drive-mode oscillators with incrementally spaced resonance frequencies to achieve wide-bandwidth response in the drive-mode. Enhanced mode-decoupling is achieved by distributing the linear drive-mode oscillators radially and symmetrically, to form a multi-directional linear drive-mode and a torsional sense-mode; minimizing quadrature error and zero-rate-output.

8.1 Introduction

Gyroscope systems that offer improved robustness by increasing the degree-of-freedom of the dynamical system have been introduced in the previous chapters. Even though increased-DOF gyroscope systems provide significantly increased bandwidth (over several hundred Hz), this is achieved with the expense of sacrificing response gain.

This chapter presents a novel approach that provides wider drive-mode bandwidth than conventional MEMS gyroscopes, with less sacrifice in response gain compared to previously reported wide-bandwidth devices. The concept is based on utilizing multiple drive-mode oscillators (Figure 8.1) with incrementally spaced resonance frequencies.

8.2 The Approach

Since the Coriolis force and the sense-mode response are directly proportional to the drive-mode oscillation amplitude, it is generally desired to enhance the drive-mode amplitude by increasing the Q factor with vacuum packaging and operating at the peak of the drive-mode resonance curve. However, large drive-mode amplitude and

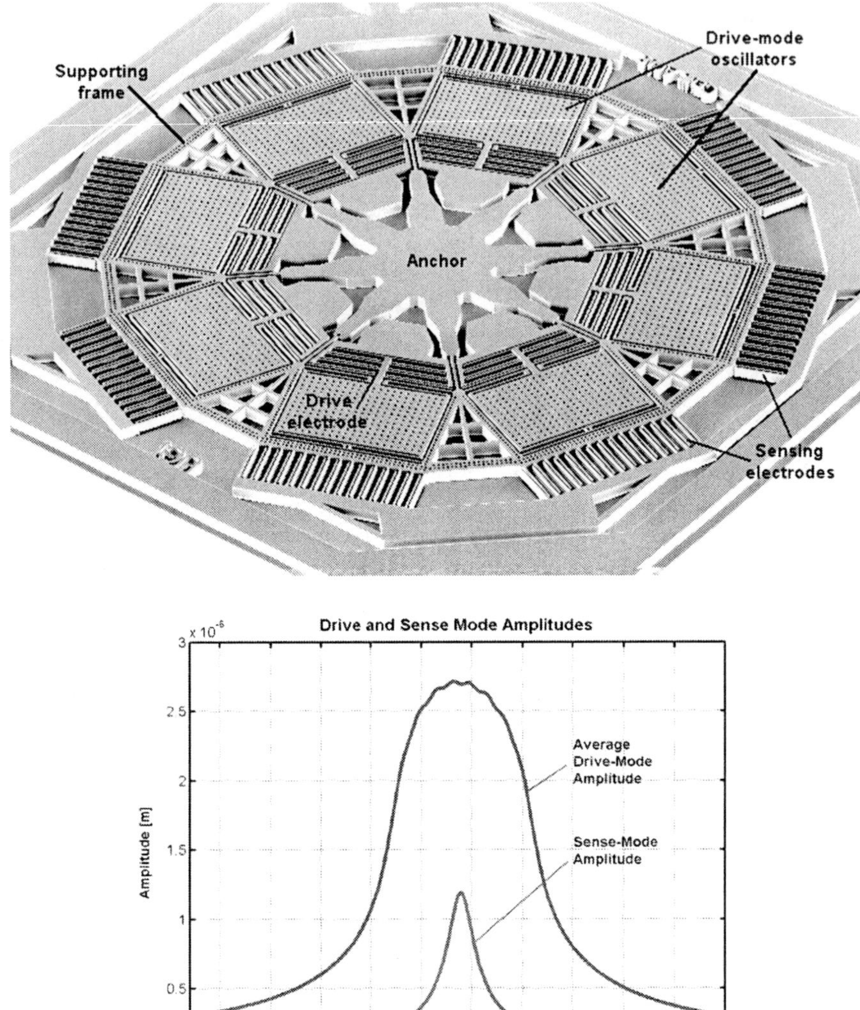

Fig. 8.1 The distributed-mass gyroscope concept achieves wide-bandwidth response in the drive-mode by utilizing multiple drive-mode oscillators with incrementally spaced resonance frequencies. Quadrature error is also minimized with enhanced mode-decoupling.

bandwidth can not be achieved at the same time, with a 1-DOF drive system. The proposed approach explores the possibility of increasing the drive-mode response bandwidth of micromachined gyroscopes, by utilizing multiple resonators with incrementally spaced resonant frequencies in the drive-mode. The drive and sense modes are effectively decoupled by forming a multi-directional linear drive-mode that transmits the Coriolis force into a torsional sense-mode.

The design concept is based on forming multiple drive-mode oscillators, distributed symmetrically around the center of a supporting frame. The distributed drive-mode oscillators are driven in-phase towards the center of symmetry, and are structurally constrained in the tangential direction with respect to the supporting frame. Each oscillator is driven at the same drive frequency. In the presence of an angular rotation rate about the z-axis, a sinusoidal Coriolis force at the drive frequency is induced on each proof mass in the direction orthogonal to each drive-mode oscillation directions (Figure 8.2). Thus, each of the induced Coriolis force vectors lie in the tangential direction, combining to generate a resultant torque on the supporting frame. The net Coriolis torque excites the supporting frame into torsional oscillations about the z-axis, which are detected by sense capacitors for angular rate measurement.

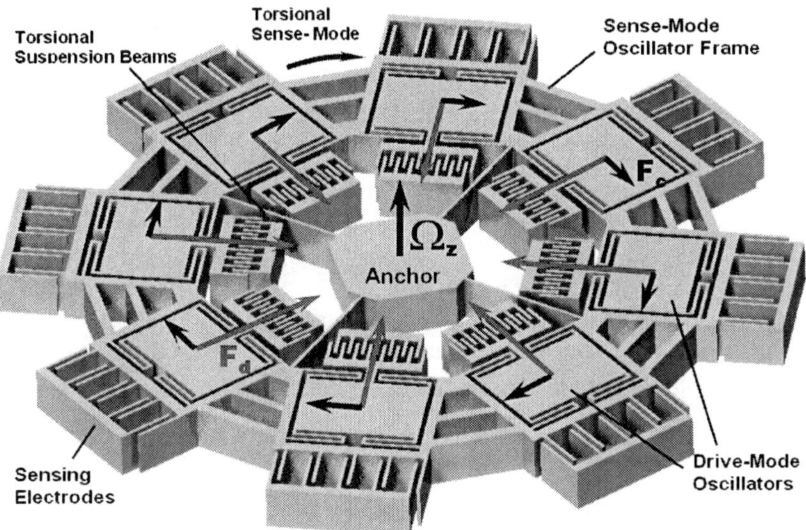

Fig. 8.2 Conceptual illustration of the Distributed-Mass Gyroscope with 8 symmetric drive-mode oscillators.

The multi-directional and axi-symmetric nature of the drive-mode oscillators offers several structural benefits over a conventional gyroscope design:

- Instability and drift due to mechanical coupling between the drive and sense modes is minimized, since the structure is designed to completely decouple the multi-directional linear drive-mode and the rotational sense-mode. Thus, zero-rate-output and quadrature error are significantly reduced in the presence of structural imperfections.
- The sensing electrodes are attached to the supporting frame, and do not respond to the drive-mode vibrations owing to the structural decoupling. This minimizes the noise in the response induced by the drive-mode oscillations.

- The torsional sense mode rejects external linear accelerations and vibrations.
- Since the drive forces applied to the drive-mode oscillators cancel out in all directions due to the radial symmetry, the net force on the structure is effectively suppressed. This results in near-zero reaction force induced on the anchor, thus minimizing energy emission to the substrate.
- The central single anchor structure minimizes the effects of packaging stresses and thermal gradients.
- The symmetry of the drive-mode oscillator structure about several axes also cancels the effects of directional residual stresses, and elastic anisotropy of the structural material.

8.2.1 The Coriolis Response

In the proposed approach, the distributed drive-mode oscillators are driven towards the center, and constrained in the tangential direction with respect to the supporting frame. The constrained dynamics of each proof-mass along the associated drive axis with respect to the supporting frame reduces to

$$m_i\ddot{x}_i + c_x\dot{x}_i + k_x x_i = F_{di} \tag{8.1}$$

where m_i is the ith proof-mass, and x_i is the drive-mode response of the ith mass. Thus, in the presence of an angular rotation rate about the z-axis, the Coriolis forces, which are proportional to drive direction oscillation amplitudes, induced on each proof mass are

$$F_{ci} = 2m_i\Omega_z\dot{x}_i \tag{8.2}$$

The rotation-induced Coriolis forces are orthogonal to each of the drive-mode oscillation directions. Thus, each of the induced Coriolis force vectors lie in the tangential direction, combining to form a resultant torque on the supporting frame. The net Coriolis torque generated as the combination of each Coriolis force becomes

$$\overrightarrow{M_c} = \sum_{i=1}^{n} \overrightarrow{r}_c \times \overrightarrow{F}_{ci} = \sum_{i=1}^{n} 2r_c m_i\Omega_z\dot{x}_i \cdot \hat{k} \tag{8.3}$$

where $\overrightarrow{r}_c$ is the position vector of the oscillator center-of-mass, and $\hat{k}$ is the unit vector in the z-direction. The Coriolis torque $\overrightarrow{M_c}$ excites the supporting frame into torsional oscillations about the z-axis, which is detected by the sense capacitors, providing measurement of angular rate. Assuming the rate input is constant and smaller compared to the driving frequency, the simplified equation of motion of the supporting frame in the sense-direction is

$$I_z\ddot{\phi} + D_z\dot{\phi} + K_z\phi = M_c, \tag{8.4}$$

where ϕ is the torsional deflection of the supporting frame, I_z denotes the moment of inertia of the supporting frame combined with the proof masses, D_z is the sense-mode torsional damping ratio, and K_z is the torsional stiffness of the suspension structure.

8.2.2 Wide-Bandwidth Operation for Improving Robustness

In the presented design concept, a wide-bandwidth operation region is achieved in the drive-mode frequency response, by designing or actively tuning the resonance frequency of each drive-mode oscillator to be incrementally spaced (Figure 8.3a). Since the tangential Coriolis forces induced on each proof mass jointly generate a resultant torque on the supporting frame, a "levelled" total Coriolis torque is achieved over a wide range of driving frequencies (Figure 8.3b). The device is nominally operated in this levelled region of the Coriolis torque frequency response, so that fluctuations in system parameters that shift oscillator resonance frequencies will not result in a significant change in the total Coriolis torque. If the sense-mode resonance frequency is designed to be accommodated in the same frequency band (Figure 8.3b), the requirement on the degree of mode-matching is relaxed, and robustness against structural and thermal parameter fluctuations is achieved.

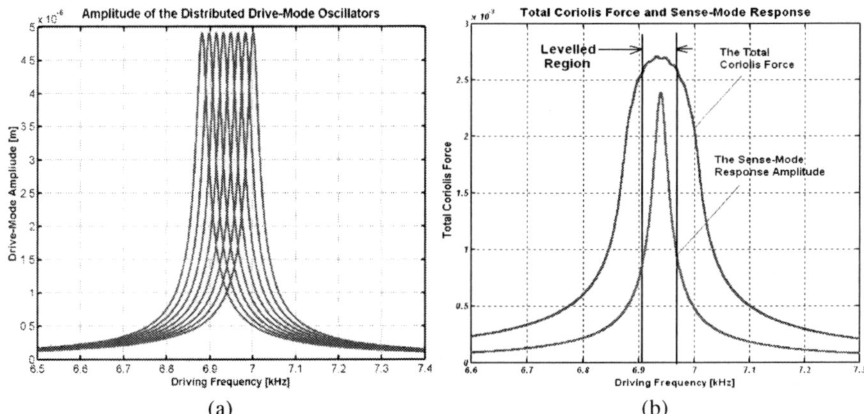

Fig. 8.3 (a) The frequency responses of the distributed drive-mode oscillators; (b) The frequency spectrum of the total Coriolis torque generated by the distributed drive-mode oscillators.

8.2.2.1 Driving Scheme

The drive-mode oscillators are driven at the same frequency within the levelled frequency region. This assures that the sinusoidal Coriolis forces induced on each drive-mode oscillator are at the same frequency. Thus, the sinusoidal Coriolis forces are superposed, and generate a resultant moment that excites the torsional sense-mode at the driving frequency.

The forced oscillation amplitude of each oscillator will be different depending on the location of the drive frequency within the operation region, but the total drive-mode response will be constant at a known value. This requires the quality factor and resonance amplitude of all resonators to be equal as in an ideal system. To compensate for quality factor variations among the resonators, the control system has to identify the drive-mode parameters of each oscillator during calibration and start-up, and apply the appropriate drive signal to each oscillator so that the resonance amplitude of each is equal to a pre-set value.

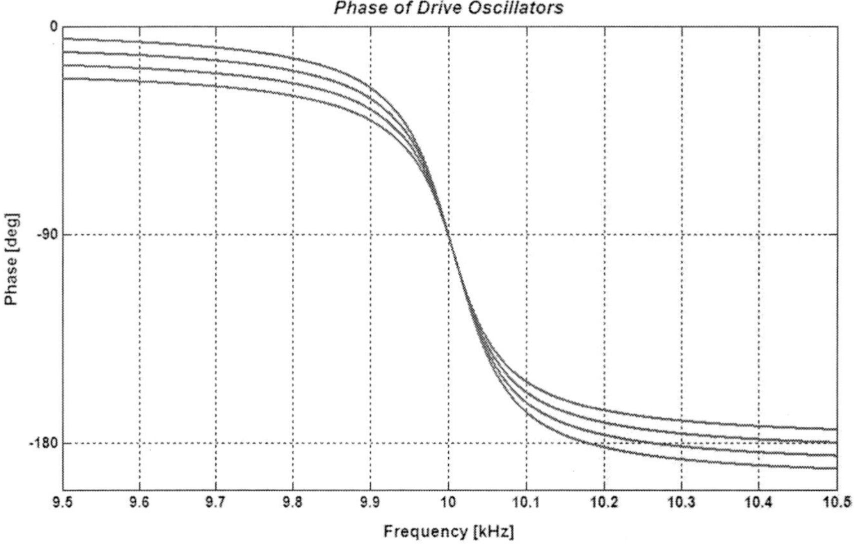

Fig. 8.4 For a system with nominal operation frequency of $\omega_d = 10$ kHz, a frequency spacing $\Delta\omega = 5$ Hz, and a quality factor $Q_d = 100$, the driving phase increment $\Delta\phi_d = 5.7092°$ between the resonators provides equal drive phase in the operation region.

The most important task of the drive-mode control system is to set the individual drive phases of the drive resonators. Since the relative location of the resonant frequency of each resonator relative to the drive frequency is different, using the same drive signal on every resonator results in a different drive phase in each resonator. In order to assure that the Coriolis force generated by the resonators can be superposed, the drive phase of the resonators at the drive frequency have to be equal. This can be achieved by setting incremental phase differences between the drive signals applied on each resonator, so that each resonator oscillates with the same phase in the operation region.

The driving phase increments between the resonators can be calculated from the frequency spacing between the resonators. If we denote the frequency spacing as $\Delta\omega$, nominal operation frequency as ω_d, and drive quality factor as Q_d; the driving phase increment becomes

$$\Delta\phi_d = -90° - \tan^{-1}\frac{\frac{1}{Q_d}\frac{\omega_d}{\omega_d+\Delta\omega}}{1 - \left(\frac{\omega_d}{\omega_d+\Delta\omega}\right)^2} \tag{8.5}$$

Thus, the drive force applied on the ith resonator is expressed as

$$F_{di} = F_0\sin(\omega_d + i\Delta\phi_d) \tag{8.6}$$

An example driving phase scheme is shown in Figure 8.4 for a system with nominal operation frequency of $\omega_d = 10$ kHz, a frequency spacing $\Delta\omega = 5$ Hz, and a quality factor $Q_d = 100$. With the expression above, the driving phase increment is found as $\Delta\phi_d = 5.70921°$. It is seen in Figure 8.4 that the same drive phase is achieved within the operation region when this driving phase increment is set between the resonators.

8.2.2.2 Frequency Spacing Design

It should be noticed that the resonance frequency separation of the oscillators are dictated by the bandwidth of the response, and thus by damping. In order to obtain a levelled operation region in the drive-mode, the frequency separation should be less than the bandwidth of a single oscillator.

If the separation of frequencies is large for low damping resonators, the resonance peaks become noticeable in the total response (Figure 8.5), and the levelled operation region will not be achieved in the response. On the contrary, the total response will converge to a 1-DOF resonance peak as the frequency separation approaches zero, where the highest possible gain is attained with the narrowest bandwidth.

8.3 Theoretical Analysis of the Trade-offs

The proposed design approach allows to widen the operation frequency range of the gyroscope drive-mode to achieve improved robustness, by trading off response amplitude. The optimal compromise between amplitude of the response and bandwidth (affecting sensitivity and robustness, respectively) can be obtained by selecting the frequency increments of the drive-mode oscillators.

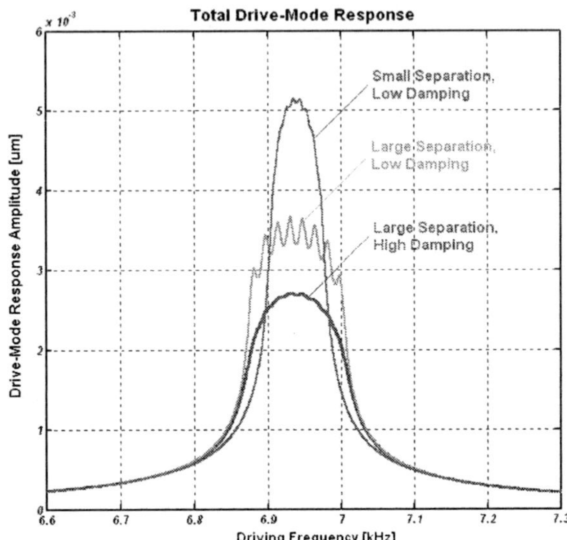

Fig. 8.5 The effect of damping and resonance frequency separation on the drive-mode response.

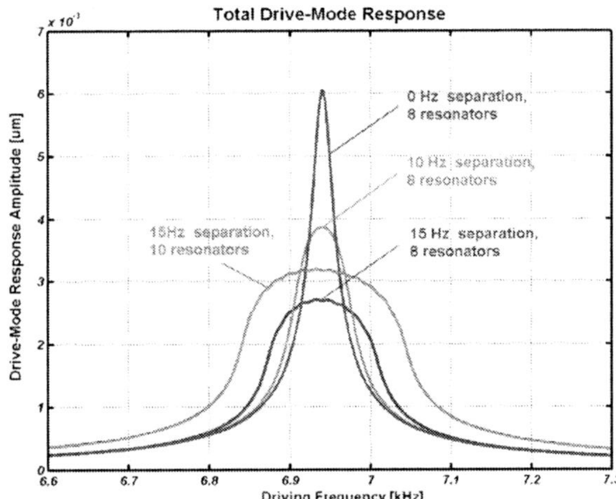

Fig. 8.6 The effect of frequency separation on the response gain and bandwidth (effecting sensitivity and robustness, respectively). The gain is maximized for zero frequency separation, and the overall bandwidth increases proportionally to spacing.

As a numerical example, the response of a device consisting of 8 drive-mode oscillators with resonance frequencies from 6.895 to 7 kHz and a frequency spacing of 15Hz will be considered. For 1 °/s input angular rate and a Q factor of 100 in the drive and sense modes, the supporting frame of the distributed-mass gyroscope will have an angular amplitude of response equal to $1.39 \times 10^{-6} rad$, which is equivalent to $2.8 \times 10^{-3} \mu m$ displacement at the sensing electrodes. If the frequency spacing of the drive-mode oscillators is decreased from 15Hz to 10Hz, the amplitude of the response in the sense direction will increase from $2.8 \times 10^{-3} \mu m$ to $3.9 \times 10^{-3} \mu m$; while the response bandwidth will decrease from 200 Hz to 140 Hz, which is still over an order of magnitude larger than the bandwidth of a single-mass conventional

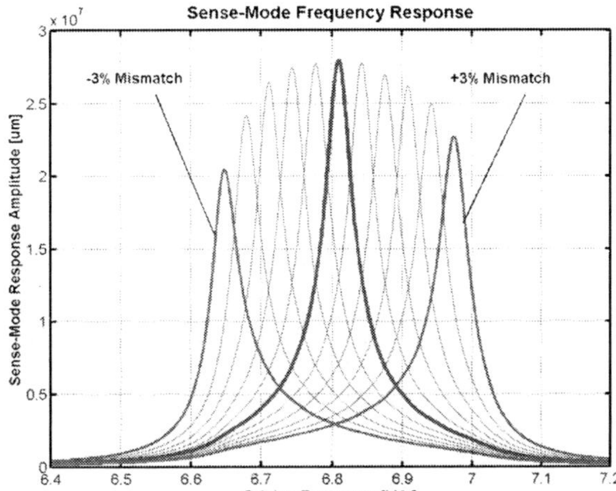

Fig. 8.7 The results of frequency mismatch on the overall system response. A mismatch of 3% results in only 16% error, while mismatches less than 1% have no significant effect.

gyroscope. The bandwidth can be further widened by increasing the number of oscillators. In Figure 8.6, the response of a gyroscope with 10 oscillators is modeled along with 8-oscillator systems with 0, 10, and 15Hz spacing; illustrating the effect of frequency separation and the number of oscillators. If the frequency separation is set to zero, the response gain will be at its maximum of $6.1 \times 10^{-3} \mu m$, with a bandwidth of 100Hz. The trade-offs between gain of the response (higher sensitivity) and the system bandwidth (increased robustness) are typically guided by application requirements.

The wide-bandwidth response in the drive-mode relaxes the mode-matching tolerances, and provides improved robustness against variations in the dynamical system parameters. Figure 8.7 illustrates the case of a potential relative shift in the sense-mode resonance frequency, e.g. due to temperature fluctuations, residual stresses, or fabrication variations. It is observed that the response amplitude is sustained at a constant value to a great extent without requiring feedback control or active tuning of resonance frequencies. For example, a 3% shift in the sense-mode resonance frequency results in only 16% error in the response amplitude, while mismatches less than 1% have no significant effect on the response. Without active compensation, a conventional 2-DOF gyroscope can exhibit over 60% error for the same 1% frequency shift under the same operation conditions.

8.4 Illustrative Example

8.4.1 Prototype Design

The wide-bandwidth design concept was analyzed experimentally on the bulk-micromachined prototype structures, fabricated in the UCI Integrated Nano-Systems

Research Facility. Two different prototype gyroscope structures utilizing the wide-bandwidth design concept were designed: one structure employing comb-drive actuation to achieve large drive amplitudes (Figure 8.8), and one structure employing parallel-plate actuation for a wide electrostatic tuning range (Figure 8.9).

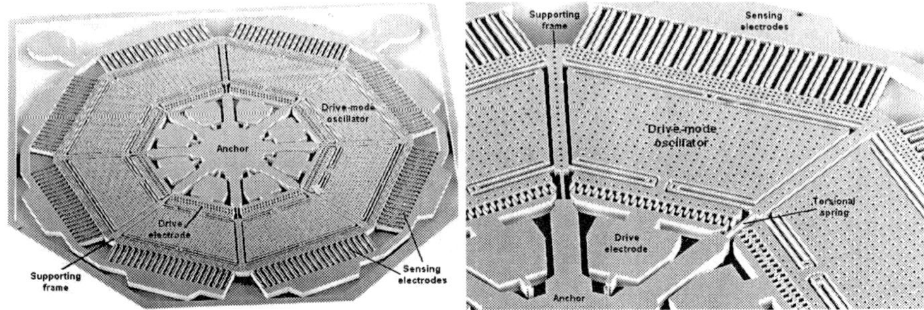

Fig. 8.8 SEM micrographs of the characterized structure employing comb-drive actuation for large drive amplitudes.

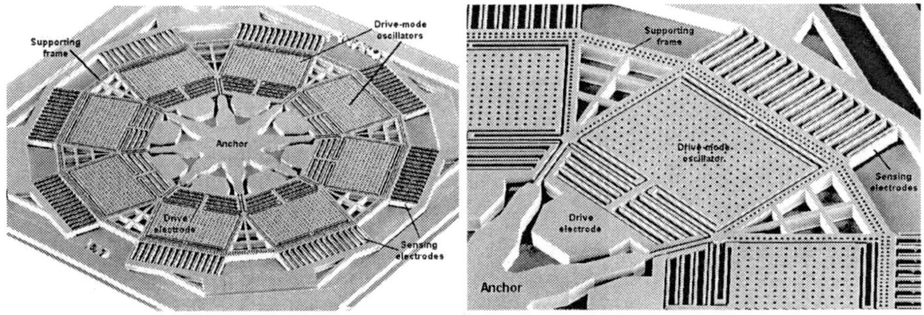

Fig. 8.9 SEM micrographs of the structure employing parallel-plate actuation for a wide electrostatic tuning range.

Each drive-mode oscillator was designed identically, although it will be shown in the next section that the natural frequency of each oscillator will be shifted due to material anisotropy and fabrication imperfections. This phenomenon is exploited to naturally provide the required frequency spacing for this demonstration.

In order to optimize the system parameters and verify the validity of the theoretical analysis assumptions, the operational modes of the system were simulated using the Finite Element Analysis package MSC Nastran/Patran. Each drive-mode mass of the analyzed prototype system is $1240\mu m \times 770\mu m$, suspended by four $350\mu m \times 7\mu m$ folded springs; yielding a resonance frequency estimation of 7.15kHz with an elastic modulus of 130GPa for single-crystal Silicon in (100) direction.

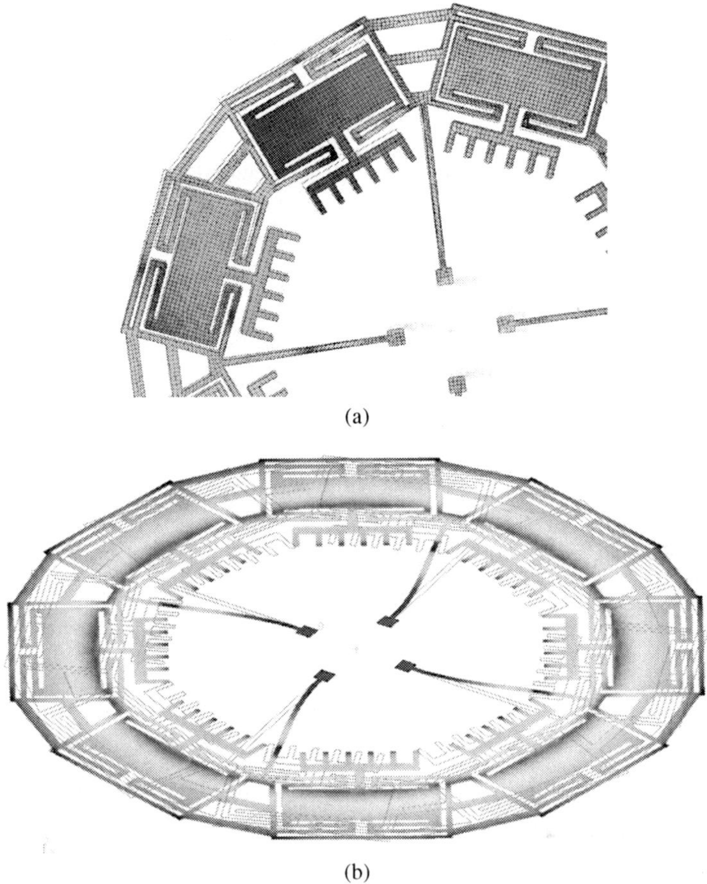

(a)

(b)

Fig. 8.10 The FEA simulation results. (a) Linear in-plane resonance mode of the drive oscillators, obtained at 6.98kHz; (b) The torsional sensing mode of the complete structure, optimized as $\omega_z =6.79$kHz.

Through FEA simulations, the resonance frequency of the drive-mode oscillators were obtained at 6.98kHz (Figure 8.10a). The torsional sense mode resonance frequency of the structure about the sense axis was then located at $\omega_z =6.79$kHz (Figure 8.10b) with four $296\mu m \times 10\mu m$ torsional suspension beams, by iteratively optimizing the beam length.

8.4.2 Experimental Characterization Results

The dynamic response of the linear drive-mode oscillators and the torsional sense-mode of the prototype gyroscope were characterized in a cryogenic MMR Probe

Station. The frequency response of the prototype devices were acquired under varying pressure values and at room temperature, using off-chip transimpedance amplifiers with a feedback resistor of R_A=1MΩ connected to an HP Signal Analyzer in sine-sweep mode. The drive-mode frequency responses were acquired utilizing one-port actuation and detection (Figure 8.11), where a single electrode was used for both driving and sensing at the same time. The driving AC signal plus the DC bias voltage was imposed on the gyroscope structure through the anchor, and the actuation and detection port was directly connected to the transimpedance amplifier.

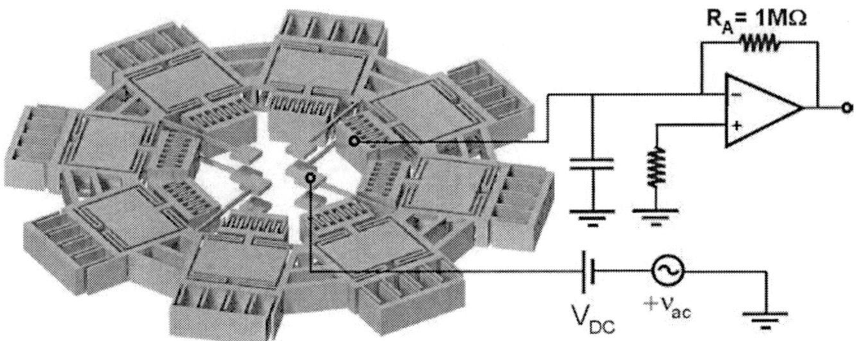

Fig. 8.11 The one-port actuation and detection scheme, where a single electrode is used for both driving and sensing.

The resonance frequencies of the drive-mode resonators were observed to be scattered between 4.546 kHz and 5.355 kHz within a 809Hz frequency band. The 16.36% maximum frequency deviation of the identically-designed drive-mode resonance frequencies results purely from the fabrication imperfections. The deviation of approximately 26% from the FEA results could be attributed to excessive lateral over-etching during DRIE, the resolution of the mask used in fabrication, and the exposure and development steps of the photolithography process. In the presence of this wide-band scatter, measuring the bandwidth of the drive-mode oscillators is crucial to assess the feasibility of the design concept.

8.4.2.1 System Identification

The dynamical parameters of the drive-mode oscillators can be identified by electrostatically acquiring the frequency responses. However, the output signal is corrupted by the feed-through of the excitation signal to the detected signal over a lumped parasitic capacitance C_p (e.g., between the bonding pads and the substrate, and between the drive and sense probes) and a finite substrate resistance R_p in parallel to the ideal system dynamics. The parasitic-free mechanical system response was extracted from the response utilizing the system identification algorithm presented in Chapter 5.

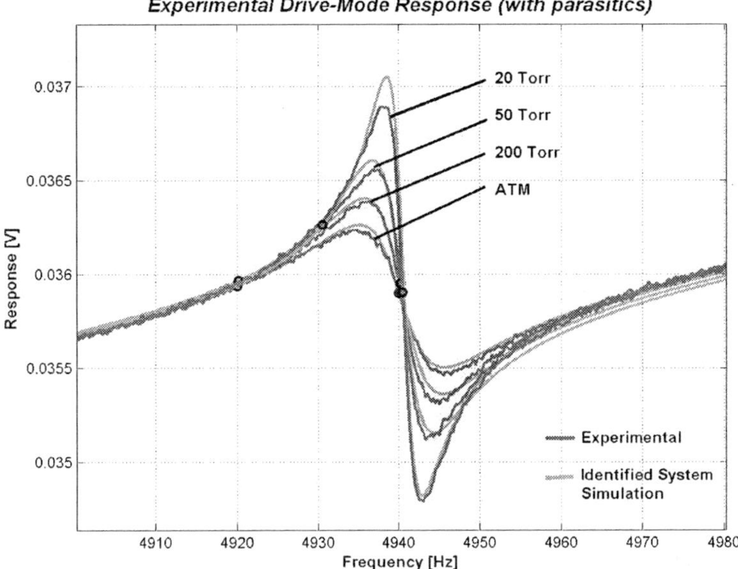

Fig. 8.12 The measured frequency response at different pressure values, corrupted by the drive feed-through signal. The system parameters and the parasitics are effectively identified.

Figure 8.12 shows the experimentally acquired response, and the response of the identified model, verifying the estimation accuracy of the system parameters and parasitics. For the oscillator mass of 1.03×10^{-7}kg, the identified parameters using the proposed algorithm are C_p=1.157pF, R_p=1.701MΩ, and $\alpha = 6.65 \times 10^{15}(VF/m)^2$.

Having identified the parasitic terms in the real and imaginary parts of the response, these terms were numerically filtered from the measured signal, by subtracting the evaluated parasitic term at each frequency from the acquired trace. The filtered frequency response reflects the actual mechanical dynamics of the drive-mode oscillator. Figure 8.13 presents the experimentally acquired frequency responses from atmospheric pressure to 4 Torr with numerical parasitic filtering, and the estimated Q factor and the bandwidth values.

8.4.2.2 Uniform Frequency Spacing with Tuning

The bandwidth of the drive-mode response even at atmospheric pressure was observed to be too narrow to achieve wide-band operation without electrostatic tuning of the drive-mode frequencies. Thus, the prototype with the parallel-plate actuated drive-mode oscillators (Figure 8.9) which provides a wider range of electrostatic tuning was tested, and the resonance frequency of each oscillator was electrostatically tuned to achieve uniform and smaller frequency separation.

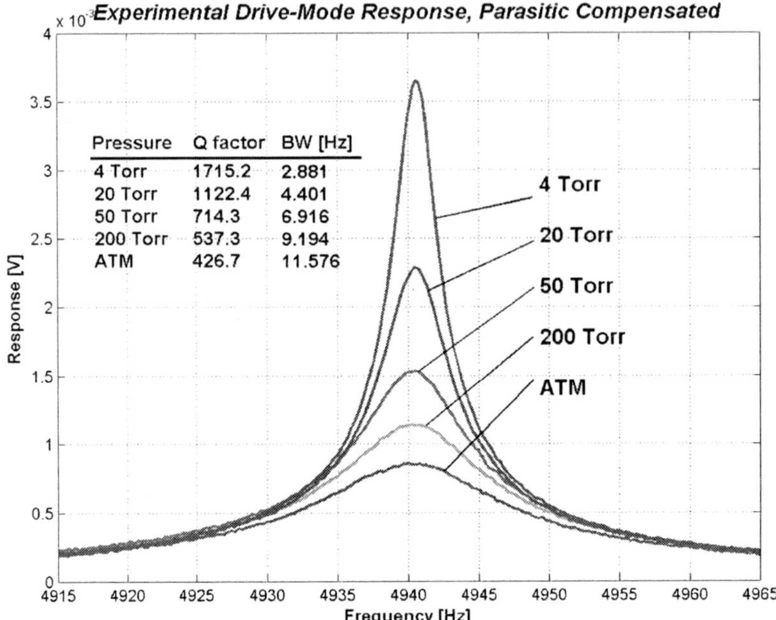

Pressure	Q factor	BW [Hz]
4 Torr	1715.2	2.881
20 Torr	1122.4	4.401
50 Torr	714.3	6.916
200 Torr	537.3	9.194
ATM	426.7	11.576

Fig. 8.13 Experimental measurements of the drive-mode frequency response of one of the oscilla-tors, with numerical filtering of the parasitics. The clean output signal reflects the actual mechanical dynamics.

Electrostatic frequency tuning using parallel-plate electrodes due to the non-linear electrostatic force profile was explained in Chapter 5. Taking the derivative of the electrostatic force F_{pp} with respect to displacement, the negative electrostatic spring constant was shown as

$$k_{el} = \frac{\partial F_{pp}}{\partial x} = -\frac{\varepsilon_0 A}{d_0^3} V_{DC}^2 \qquad (8.7)$$

where $\varepsilon_0 = 8.854 \times 10^{12}$ F/m is the dielectric constant, $A = 1080 \times 100 \mu$m is the to-tal actuation area, $d_0 = 15 \mu$m is the electrode gap, and V_{DC} is the DC bias voltage. The resonance frequencies of the parallel-plate oscillators were observed to shift with increasing DC bias following the theoretically estimated trend. Figure 8.14b presents the theoretical drive-mode tuning curve, obtained using the negative elec-trostatic spring model, and the experimental tuning data obtained from the same parallel-plate oscillator.

After electrostatic tuning of the parallel-plate oscillators for 10Hz spacing (Fig-ure 8.15a), the close spacing of the drive-mode resonance frequencies allowed all of the resonators to be excited together, to jointly generate a resultant Coriolis torque. The total Coriolis torque, which is estimated by summing the experimentally mea-

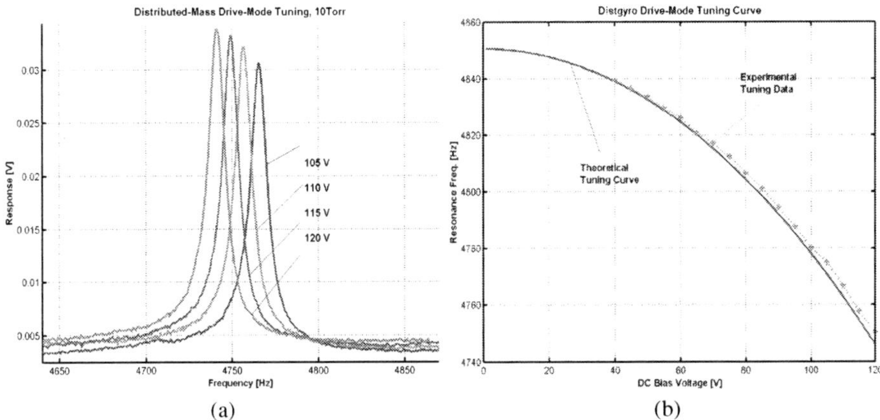

Fig. 8.14 (a) The resonance frequency shift of the parallel-plate oscillators with increasing DC bias; (b) The theoretical drive-mode tuning curve obtained using the negative electrostatic spring model, and the experimental tuning data.

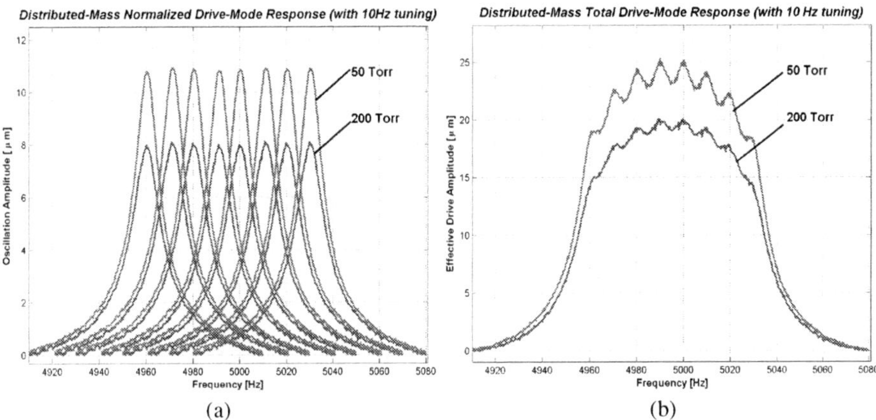

Fig. 8.15 (a) The normalized frequency response of the parallel-plate drive-mode oscillators with numerical parasitic filtering, after tuning for 10Hz spacing. (b) Experimental frequency response measurements of the total drive-mode response, obtained by summing the measured drive-mode response of the drive ports.

sured response of the eight drive-ports, was observed to provide a levelled range of over 90Hz (Figure 8.15b). When the experiments were repeated at reduced pressures, the resonance peaks in the levelled region of the overall response became more emphasized, as was theoretically illustrated in the previous section. Based on the experimental results, it was concluded that 200 to 300 Torr is the optimal pressure for the parallel-plate devices to achieve a levelled wide-bandwidth drive-mode response with 10Hz spacing. When the resonant frequencies were tuned for 5Hz spacing, the total drive-mode response gain was measured to be 65% larger gain, but the bandwidth was observed to drop to 50Hz (Figure 8.16).

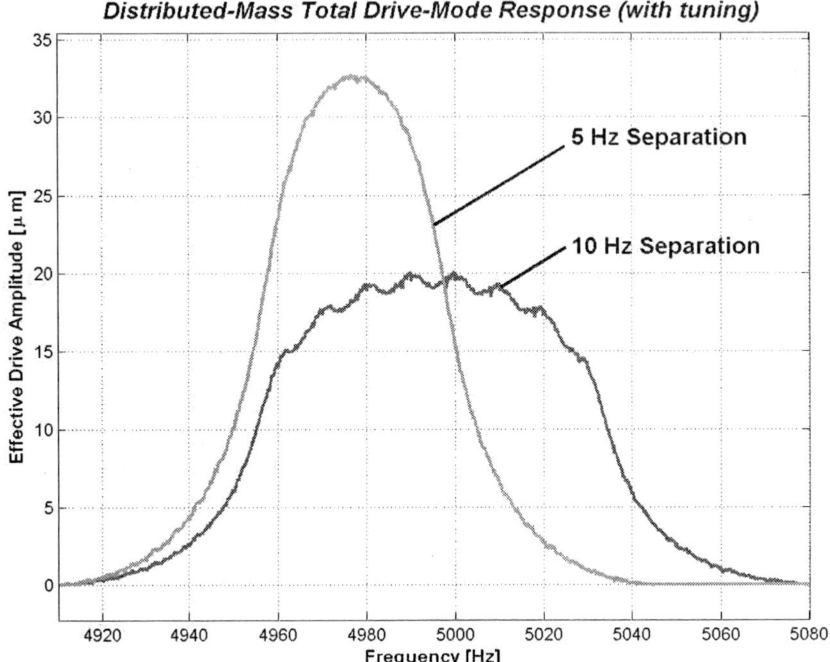

Fig. 8.16 The total drive-mode response measurements with 5Hz spacing of the resonant frequencies, providing 65% larger gain, with the expense of less than 50Hz bandwidth.

8.4.2.3 Narrow-Band Frequency Spacing Without Tuning

In order to minimize the effects of suspension width variation due to fabrication imperfections on random scattering of the drive-mode resonance frequencies, a new generation of devices with $10\mu m$ wide drive-mode beams were designed, and fabricated using a higher resolution mask. The resonance frequencies of the drive-mode resonators with the wider suspension beams were observed to be scattered between 6.490 kHz and 6.920 kHz in a 430Hz frequency band, with a 6.21% maximum frequency deviation (Figure 8.17a). The frequency separation of the resonators with $10\mu m$ wide beams was observed to provide over 600Hz operating frequency region with levelled output in atmospheric pressure (Figure 8.17b). However, the levelled region showed a maximum variation of 17.2% in the total response due to the non-uniform frequency separation.

8.4.2.4 Sense-Mode Characterization

The parasitics in the sense mode were modeled and identified similarly, yielding C_p=7.18pF and $\alpha = 5.34 \times 10^{13} (VF/m)^2$. In Figure 8.18a, the experimentally ac-

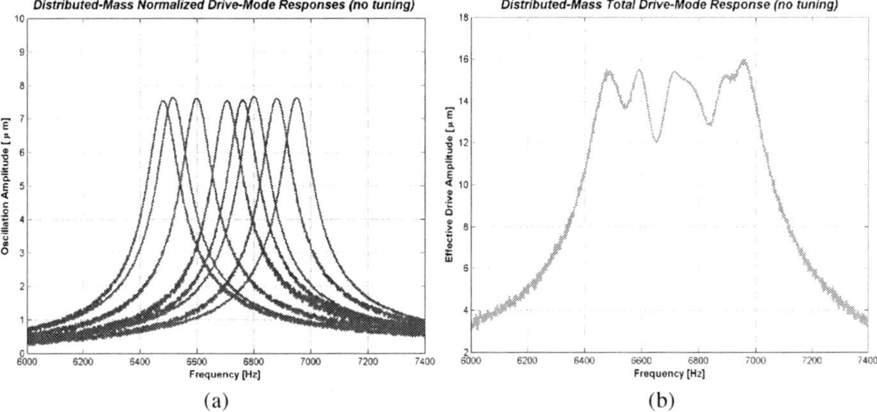

Fig. 8.17 (a) The natural frequency scatter of the drive-mode oscillators with $10\mu m$ wide drive-mode beams; (b) Experimental frequency response measurements of the total drive-mode response at atmospheric pressure, with a maximum gain variation of 17.2% in the 600Hz operating frequency region.

quired frequency responses of the torsional sense-mode and the identified system simulation are shown, verifying the estimation accuracy. Figure 8.18b presents the experimentally acquired responses with numerical filtering of parasitics, from atmospheric pressure to 10 Torr, and the estimated Q factor and the bandwidth values. The sense-mode resonance frequency of the frame was measured at 3.758kHz with 20V DC bias voltage.

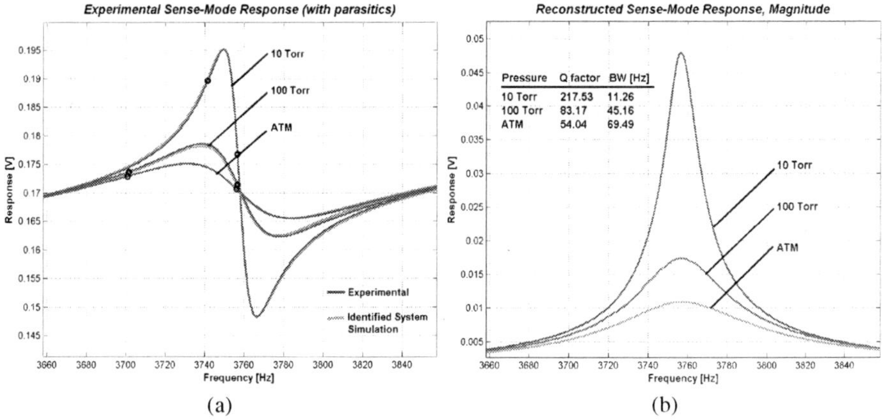

Fig. 8.18 (a) Experimental measurements of the torsional sense-mode frequency response, under different pressure conditions, with the parasitic feed-through. (b) The experimentally measured response amplitude with numerical filtering of the identified parasitics C_p and R_p.

Consequently, at pressures around 200 Torr, the parallel-plate devices with $7\mu m$ wide drive-mode suspension beams were observed to provide a levelled wide-bandwidth drive-mode response with individual tuning for 10Hz spacing, and sufficient off-resonance sense-mode gain. The devices with $10\mu m$ wide drive-mode beams provided over 600Hz operating frequency region with levelled output in atmospheric pressure; with a maximum variation of 17.2% in the total response due to the non-uniform frequency spacing.

8.5 Summary

A novel design approach based on utilizing multiple drive-mode oscillators with incrementally spaced resonance frequencies was presented, which provides wider drive-mode bandwidth in vibratory MEMS gyroscopes. The approach was theoretically illustrated, and experimentally verified. The linear drive-mode oscillators and the torsional sense-mode of the prototype gyroscope structures were characterized under varying pressure values. The resonance frequencies of the identically-designed drive-mode resonators were observed to be scattered within a 809Hz frequency band, due to the fabrication imperfections. The bandwidth of the drive-mode response even at atmospheric pressure was observed to be too narrow to achieve wide-band operation. After electrostatic tuning of the parallel-plate oscillators for 10Hz spacing, the close spacing of the drive-mode resonance frequencies allowed all of the resonators to be excited together, to jointly generate a resultant Coriolis torque with a levelled region of over 90Hz. At pressures around 200 Torr, the levelled wide-bandwidth drive-mode response was achieved together with sufficient off-resonance sense-mode gain, experimentally demonstrating the feasibility of the wide-bandwidth drive mode principle. The devices with $10\mu m$ wide suspension beams provided a levelled frequency region of 600Hz with a maximum variation of 17.2% due to non-uniform spacing, demonstrating that the natural frequency scatter due to material anisotropy could be utilized to provide the required frequency spacing for wide-bandwidth operation.

Chapter 9
Conclusions and Future Trends

9.1 Introduction

Micromachined gyroscopes could potentially provide reliable, robust and high performance angular-rate measurements leading to a wide range of applications including navigation and guidance systems, automotive safety systems, and consumer electronics. However, truly low-cost and high-performance devices are not on the market yet, and the current state of the art micromachined gyroscopes require orders of magnitude improvement in performance, stability, and robustness.

In the first part of this book, we reviewed the fundamental theory, design and implementation of micromachined vibratory rate gyroscopes. We emphasized that the bias and scale factor of gyroscopes with conventional 1-DOF drive and 1-DOF sense-mode oscillators are very sensitive to variations in dynamic system parameters. Shifts in resonant frequencies and damping due to environmental variations result in drastic changes in system characteristics. Frequency variations from device to device due to fabrication imperfections cause low yields, increasing the unit cost. Consequently, the drive and sense mode resonant frequencies are intentionally separated by design in most practical applications, in order to achieve a stable response and reduce device-to-device variation.

In the second part, we presented structural design concepts that explore the possibility of shifting the complexity from the control electronics to the mechanical sensing element design. The common aspect of the presented concepts was to achieve a gyroscope dynamical system that provide inherent robustness against structural and environmental parameter variations. The primary focus was on obtaining a gain and phase stable region in the drive or sense-mode frequency responses which provides improved bias stability, temperature stability, and immunity to environmental and fabrication variations.

In this chapter, we summarize the advantages and disadvantages of each presented design concept, and discuss the overall comparison of the design approaches. We experimentally compare the robustness of the wide-bandwidth system to a conventional gyroscope system to confirm the improvement in robustness. We then analyze the performance trade-offs with respect to a conventional system, and finally present future trends that address more demanding application requirements.

9.2 Comparative Analysis of the Presented Concepts

The observations and conclusions on the advantages and limitations of each design approach presented in the previous chapters are summarized below.

9.2.1 2-DOF Oscillator in the Sense-Mode

This approach was illustrated to potentially yield high-performance and high-robustness gyroscopes with conventional drive-mode control architectures. The main reason is the fact that the mass ratio in the 2-DOF sense-mode oscillator can be designed to be substantially large. This allows larger dynamical amplification and closer spaced resonance peaks in the 2-DOF oscillator response, leading to improved passive-mass response amplitude in the flat operation region. Furthermore, increasing the mass ratio means increasing the mass of the sense-mode active mass, which is the proof-mass utilized for generating the Coriolis force that excites the sense-mode oscillator. This increases the sense-mode response amplitude, and the gyroscope performance, even further.

Summary of Experimental Results

The prototype devices with 2-DOF sense-mode oscillator were successfully demonstrated to operate in the flat region of the sense-mode response. With this operation scheme, the gyroscope exhibited a measured noise-floor of $0.64°/s/\sqrt{Hz}$ over 50Hz bandwidth in atmospheric pressure. The sense-mode response in the flat operating region was also experimentally demonstrated to be inherently insensitive to pressure, temperature and DC bias variations. Rate table characterization of the device at 25°C and 75°C was performed, and only 1.62% change in the sensitivity was observed, verifying the improved robustness.

9.2.2 2-DOF Oscillator in the Drive-Mode

This approach allows to achieve large drive-mode oscillation amplitudes in the drive-mode passive-mass, while the oscillation amplitudes of the driven mass is drastically suppressed. Thus, parallel-plate actuators with small gaps could be utilized to produce large drive-mode amplitudes with low actuation voltages. The major limitation of the approach is that, a large active proof-mass in the drive-mode is required, which is not utilized in generating Coriolis force. Thus, increasing the drive-mode mass ratio to increase dynamical amplification requires very large foot-step areas for the active-mass, resulting in an increased overall device size.

Summary of Experimental Results

Using the prototype devices with 2-DOF drive-mode, dynamical amplification to achieve large drive-mode oscillation amplitudes was experimentally demonstrated. This allowed utilizing parallel-plate actuators with small gaps to produce large drive-mode amplitudes with low actuation voltages. The resulting gyroscope with dynamically amplified drive-mode exhibited a measured noise-floor of $3.05°/s/\sqrt{Hz}$ over 50Hz bandwidth in atmospheric pressure. The concept was also implemented in surface micromachining with a torsional gimbal-type structure, to achieve large actuation and detection capacitances with limited structural thickness.

9.2.3 Multiple Drive-Mode Oscillators

The major advantage of this approach is the versatility in allowing to optimize the response for more sensitivity or more robustness. By selecting the frequency spacing of the drive-mode oscillators, the optimal compromise between amplitude of the overall drive-mode response and the bandwidth (affecting sensitivity and robustness, respectively) can be obtained. The major challenge in production and commercialization of this device is the development and implementation of the drive-mode control strategies.

Summary of Experimental Results

With the distributed-mass design concept, the wide-bandwidth response in the drive-mode was experimentally demonstrated by utilizing multiple drive-mode oscillators with incrementally spaced resonance frequencies. The devices with $10\mu m$ wide suspension beams exhibited a levelled frequency region of 600Hz, demonstrating that the natural frequency scatter due to imperfections could be utilized to provide the required frequency spacing for wide-bandwidth operation. By minimizing the random scatter with process modifications, the oscillators could be ultimately designed with incrementally spaced resonant frequencies to provide the wide-bandwidth response.

9.3 Demonstration of Improved Robustness

Having presented the advantages and trade-offs offered by each system architecture, we will now focus on the 2-DOF sense-mode concept for demonstration of inherent robustness. In order to verify that robustness to parameter variations is achieved in the overall Coriolis response, a prototype gyroscope with 2-DOF sense-mode was characterized on a rate table in a thermally controlled environment.

As outlined in Chapter 6, the gyroscope system with 2-DOF sense-mode oscillator is operated at resonance in the drive-mode, and the wide-bandwidth frequency region is obtained in the sense-mode frequency response. The scanning electron micrograph of the characterized device is presented in Figure 9.1.

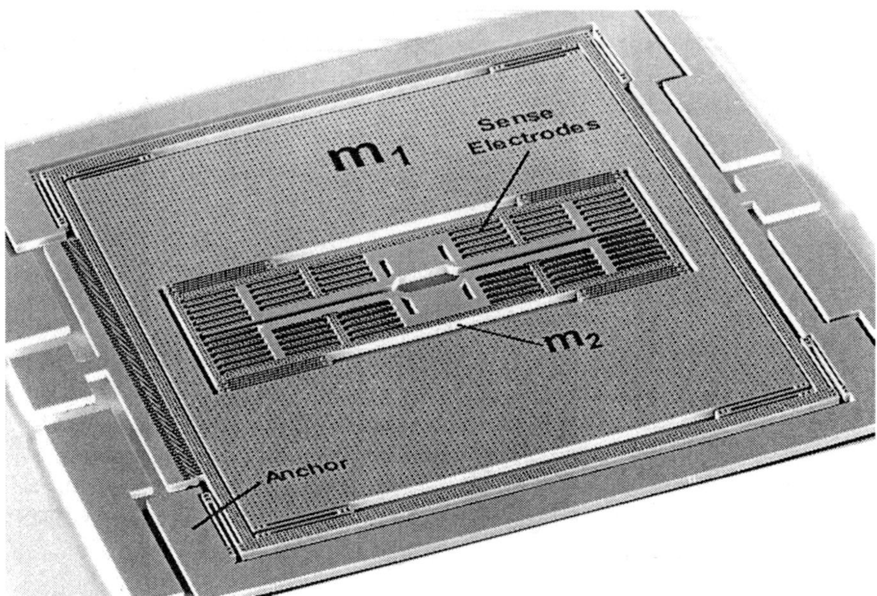

Fig. 9.1 Scanning electron micrograph of the prototype bulk-micromachined 3-DOF gyroscope with 2-DOF sense-mode.

9.3.1 Temperature Dependence of Drive and Sense-Modes

Variation of the 2-DOF sense-mode response with temperature was presented in Chapter 6. It was experimentally demonstrated that the effect of temperature variations on the gain and phase of the sense-mode response is significantly suppressed in the 2-DOF system. Within the flat region, the sense-mode response amplitude at the operation frequency of 752Hz was observed to change by 0.6% when the temperature was increased from 25°C to 75°C.

The temperature dependence of the 1-DOF drive-mode oscillator response was optically characterized during device operation on the rate table. A microscope and a digital video camera was mounted on the rate table, and slip rings were used for the electrical connections of the camera. The gyroscope was attached on a heated stage mounted on the rate table, to monitor the drive amplitude while increasing the temperature of the device. The drive motion envelope was optically measured with

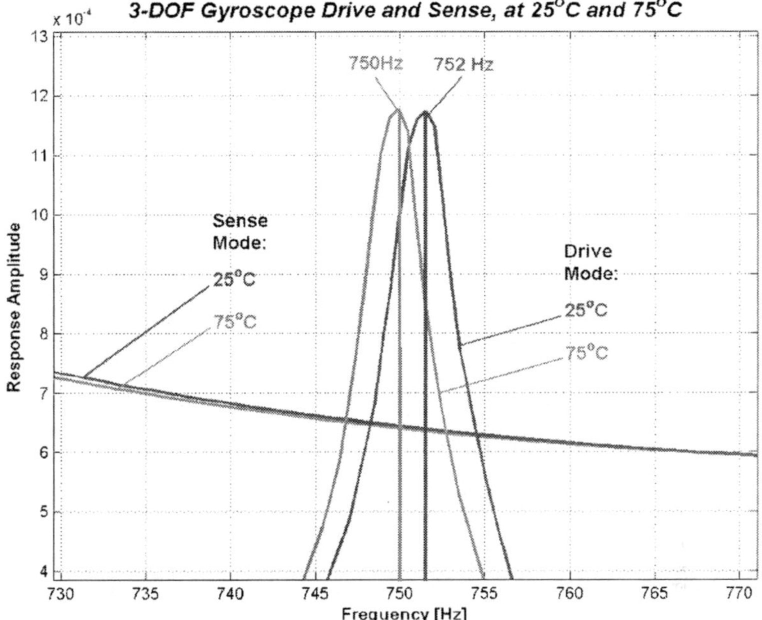

Fig. 9.2 The experimentally measured frequency response of the drive and sense-mode oscillators, showing the drive and sense-mode response gain at 25°C with 752 Hz drive, and at 75°C with 752 Hz drive.

a calibrated scale, while continuously monitoring the temperature of the gyroscope structure.

As the device temperature was increased from 25°C to 75°C, it was observed that the drive-mode resonant frequency shifted from 752 Hz to 750 Hz (Figure 9.2). When the excitation frequency was kept constant at 752Hz, the drive-mode amplitude was observed to drop from 5.8μm to 4.3μm at 75°C. By changing the drive frequency to 750Hz, the drive-mode amplitude was restored to 5.8μm. The drive-mode characterization results provided the frequency information needed to manually control the drive frequency during the rate-table experiment, in order to effectively isolate the temperature sensitivity of the sense-mode. An automatic drive-mode control with a self-oscillation loop and AGC loop could also be utilized, which achieves the same result in an actual device implementation.

9.3.2 Rate-Table Characterization Results

Synchronous demodulation technique described in Chapter 5 was utilized to extract the angular rate response of the 3-DOF system with 2-DOF sense-mode at different temperatures. The drive-mode amplitude was continuously monitored during the operation of the device using a microscope attached to the rate-table platform, as explained in the previous section. The temperature of the device also continuously monitored using a solid-state temperature sensor attached to the sample.

Fig. 9.3 The experimental setup utilized for rate-table characterization of prototype gyroscopes. Microscope for drive amplitude monitoring and heating stage for temperature control are mounted on the rate table platform.

A lock-in amplifier was utilized to provide the driving AC signal applied on the comb-drives, and to synchronously demodulate the Coriolis signal at the drive frequency in the internal reference mode. The output of the lock-in amplifier is low-pass filtered to provide a stable output reading proportional to the applied input angular rate. The experimental setup for implementing this characterization scheme is presented in Figure 9.3.

When the temperature of the gyroscope was increased from 25°C to 75°C while keeping the excitation frequency constant at 752Hz, the sensitivity of the gyroscope was observed to drop from 0.0308 mV/°/s to 0.0234 mV/°/s. This translates into 24.1% drop in the response gain. When the change in the drive-mode amplitude from 25°C to 75°C is investigated, it is seen that it changes from 5.8μm to 4.3μm; yielding a 25.9% change. Thus, it is demonstrated that the change in the gyroscope sensitivity is almost exactly equal to the drive-mode amplitude change, verifying the insensitivity of the sense-mode response to temperature variations.

In order to confirm this result, the rate table characterization at 75°C was repeated, this time changing the drive frequency to 750Hz. At this frequency, the drive-mode amplitude was restored to 5.8μm, and a sensitivity of 0.0303 mV/°/s was measured (Figure 9.4). Consequently, it was experimentally demonstrated that, a temperature variation from 25°C to 75°C results in only 1.62% change in the output of the wide-bandwidth gyroscope approach, verifying the improved robustness. At elevated temperatures, the linearity of the response was also observed to be preserved.

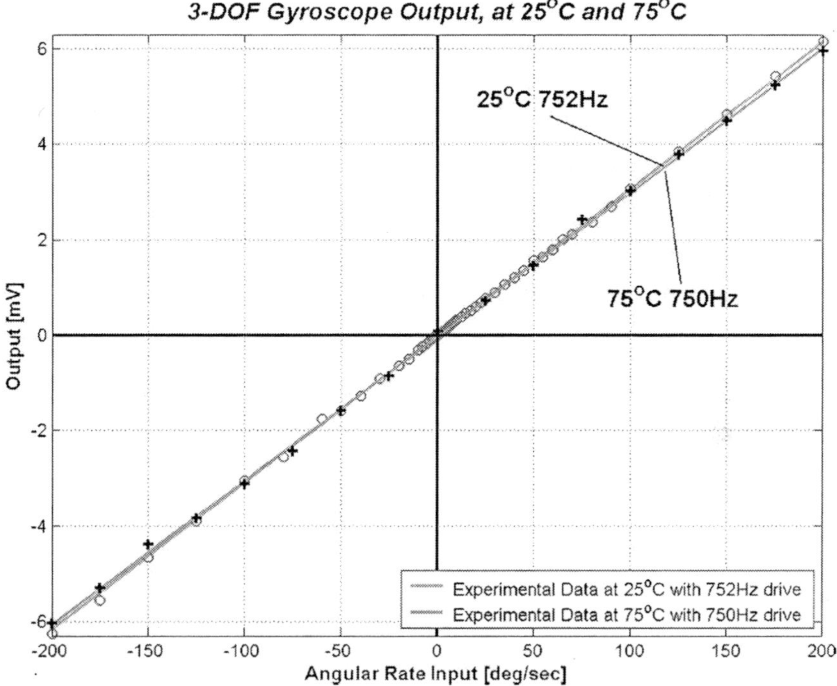

Fig. 9.4 The measured angular-rate response of the 3-DOF gyroscope at 25°C and 75°C, when the drive frequency is changed from 752Hz to 750Hz to maintain resonance in drive-mode.

9.3.3 Comparison of Response with a Conventional Gyroscope

In order to compare the improved robustness of the proposed wide-bandwidth approach, a micromachined gyroscope with a conventional 1-DOF sense oscillator was characterized under the same temperature variations and using the same signal conditioning electronics. The characterized conventional device (Figure 9.5) employs a structural mode-decoupling mechanism, and utilizes the post-release capacitance enhancement technique on the comb-drives and the sensing electrodes.

With the conventional gyroscope operated at the drive resonant frequency of 1332Hz with 12μm drive amplitude, a scale factor of 0.91 mV/°/s and a noise floor of $0.25°/\text{s}/\sqrt{\text{Hz}}$ was measured at 50Hz bandwidth in atmospheric pressure at 25°C.

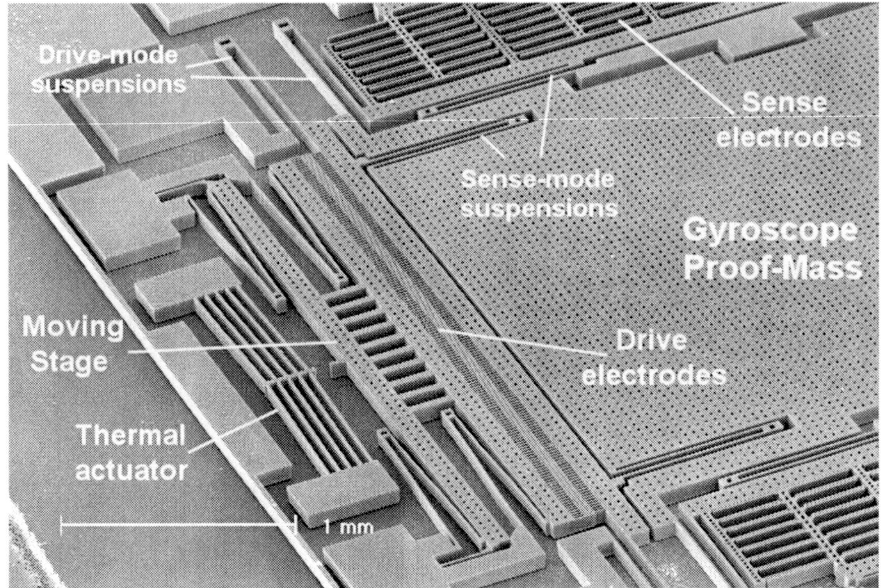

Fig. 9.5 The scanning-electron-micrograph of the characterized conventional gyroscope with enhanced structural decoupling, and post-release capacitance enhancement technique applied on the comb-drives and the sensing electrodes.

When the temperature of the tested conventional gyroscope was increased from $25°C$ to $75°C$ while restoring the drive-mode amplitude to 12μm, the sensitivity was observed to drop from 0.91 mV/°/s to 0.73 mV/°/s. Thus, a $50°C$ temperature increase was observed to result in 19.8% sensitivity change in the conventional gyroscope, which is over 12.2 times larger than the wide-bandwidth gyroscope approach.

Figure 9.6 shows the measured angular-rate response of the 2-DOF sense gyroscope and the conventional gyroscope at $25°C$ and $75°C$ on the same plot, with manual frequency tracking. For clarity, the response of the conventional gyroscope has been scaled so that the scale factor at $25°C$ is equal to that of the 2-DOF sense gyroscope. It is clearly observed that the scale factor variation of the 2-DOF sense gyroscope is over an order of magnitude lower than the conventional device.

9.4 Scale Factor Trade-off Analysis

Having experimentally demonstrated the inherent robustness achieved with the 2-DOF sense-mode approach, the trade-off in sensitivity due to operation in the flat region has to be analyzed. It is clear that the 2-DOF sense-mode approach sacrifices sense-mode gain by operating away from resonance. However, it should be remem-

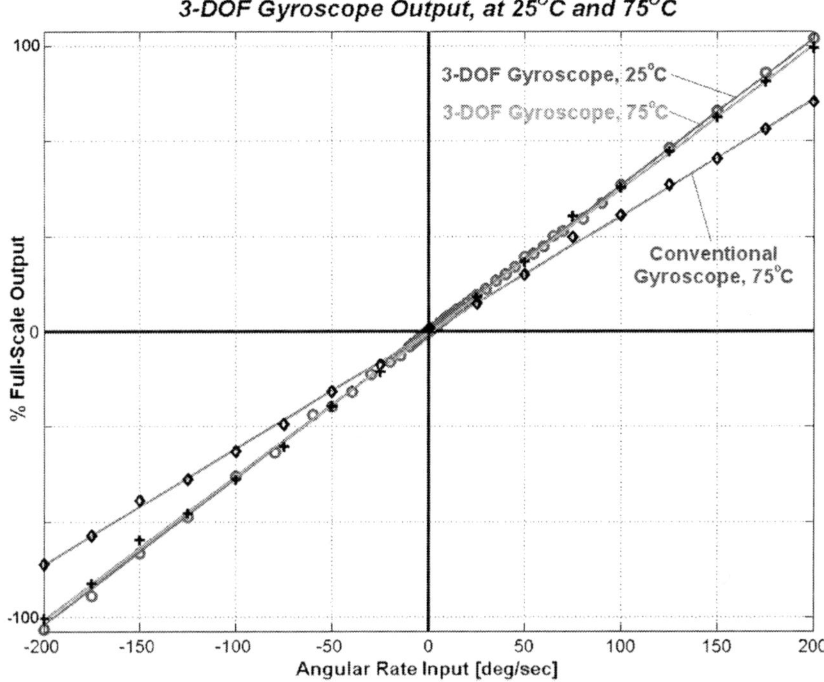

Fig. 9.6 The measured angular-rate response of the 2-DOF sense gyroscope and a conventional 1-DOF sense gyroscope at 25°C and 75°C.

bered that vast majority of conventional devices in practice are mode mismatched, and operated off-resonance in the sense-mode to minimize scale factor and bias variation over temperature and from device to device. Thus, a realistic comparison can only be made with a mode-mismatched conventional gyroscope.

As an example, let us take a conventional gyroscope system operated at the drive resonant frequency of 2 kHz, with a drive amplitude of 10μm. For 1 °/s angular rate input, with a sense-mode Q factor of 100, the sense-mode response amplitude of a mode-matched system, i.e. $\Delta f = 0$ Hz, is 2.778 nm. When there is a 10 Hz relative variation between the drive and sense frequencies, the response amplitude drops to 1.959 nm, which results in 29% drop in scale factor. As the initial Δf increases, the scale factor error due to relative frequency variation decreases. For example, when $\Delta f = 100$ Hz, a 10 Hz relative variation results in 10% scale factor error.

At higher Q values, the scale factor error due to relative frequency variations are more drastic. For a Q factor of 1000, the scale factor error due to 10 Hz variation in a mode-matched system is 90%. It should also be noticed that the sense-mode phase error due to frequency variation also reduces for higher Δf values. The scale factor and phase error for 10 Hz relative frequency variation at increasing Δf values are plotted for Q factors of 100 and 1000 in Figure 9.7.

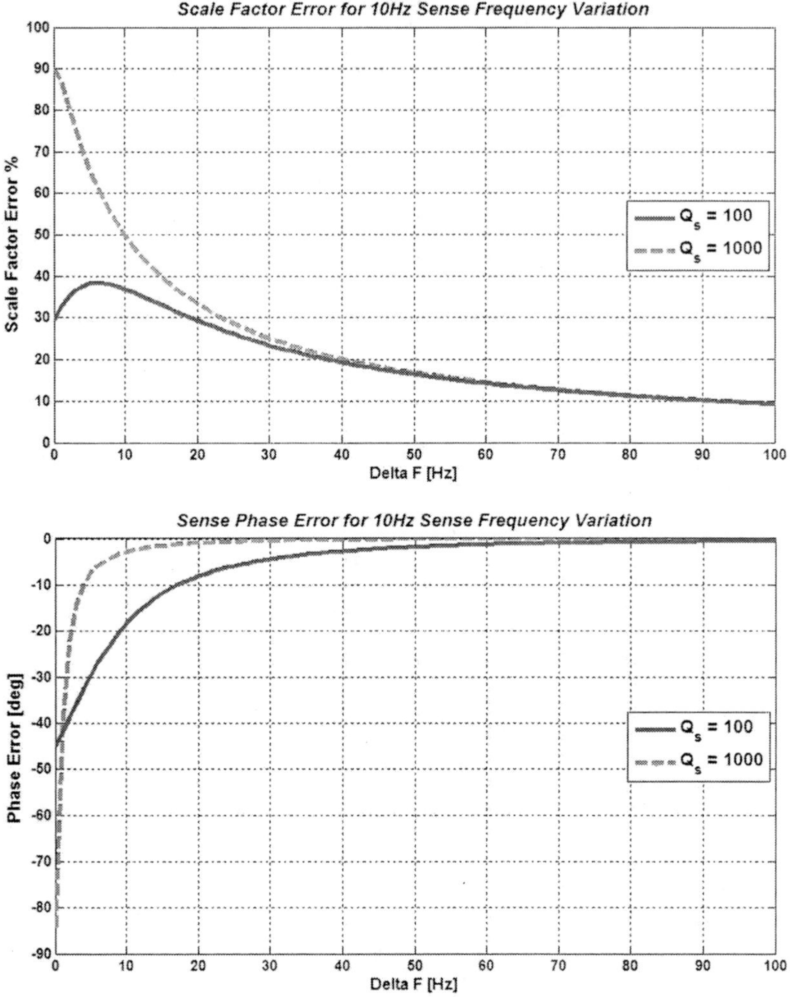

Fig. 9.7 The scale factor and phase error for 10 Hz relative frequency variation in a conventional gyroscope system.

Even though increasing the Δf in a conventional system improves robustness, it also decreases the scale factor. For example, when the Δf is increased to 100 Hz by setting the sense-mode frequency to 2100 Hz for the system discussed above, the sense-mode amplitude drops from 2.778 nm to 0.266 nm. This translates into a scale factor reduction of over 90%. Figure 9.8 shows the reduction in scale factor for increasing frequency mismatch. Consequently, the Δf value should be selected as to provide sufficient angular rate sensitivity and at the same time sufficiently low sensitivity to frequency variations, guided by application specifications and process variation data.

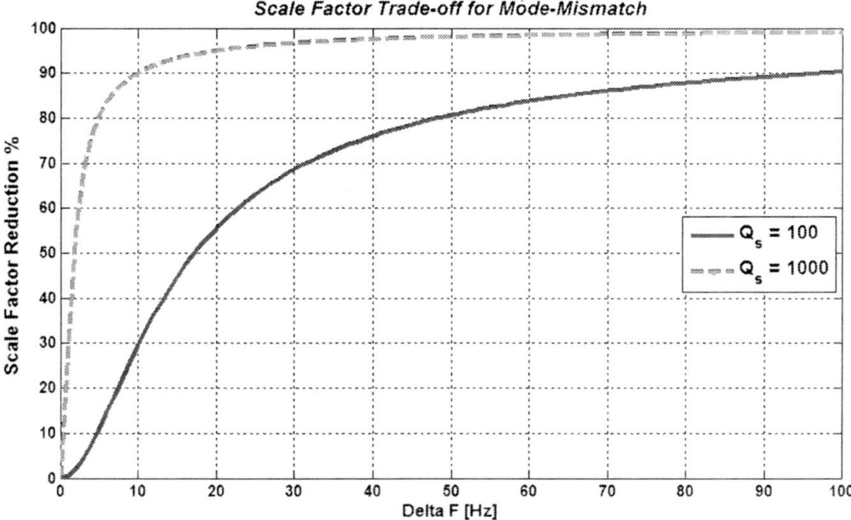

Fig. 9.8 Reduction in scale factor for increasing frequency mismatch in a conventional gyroscope system.

Given the requirement for a finite frequency mismatch Δf in a conventional gyroscope system, the 2-DOF sense-mode architecture can actually improve both robustness and performance. In Figure 9.9, the frequency responses of a 2-DOF sense gyroscope and a conventional 1-DOF sense gyroscope with $\Delta f = 100$ Hz are presented. Both systems have identical drive oscillators, operated at the drive resonant frequency of 2 kHz, with a drive amplitude of 10μm. In the sense-mode, the primary mass m_1 of 2-DOF oscillator is equal to the mass of the conventional device, and the mass ratio is $m_1/m_2 = 20$. The isolated primary and secondary resonant frequencies are 1975 Hz, which locates the flat region at the 2 kHz operation frequency. Damping conditions in both devices are also identical, resulting in a Q factor of 100 in the conventional device.

For 1 °/s angular rate input, the sense-mode response amplitude of the conventional system is 0.266 nm, as presented above. The response amplitude of the 2-DOF system is 0.5480 nm, which is over two times larger than the conventional system (Figure 9.9). When there is a 10 Hz variation in operation frequency, the response amplitude becomes 0.5492, changing by less than 0.2%. Thus, the 2-DOF system provides over two times larger scale factor and fifty times higher robustness to 10 Hz frequency variation.

The scale factor of the conventional system becomes higher than the 2-DOF sense system when the Δf is less than 50 Hz. For this range of Δf, the scale factor error for 10 Hz variation is more than 16%. Thus, in applications that require higher robustness or fabrication processes with larger relative frequency variation, the 2-DOF sense-mode gyroscope system achieves better performance and robustness.

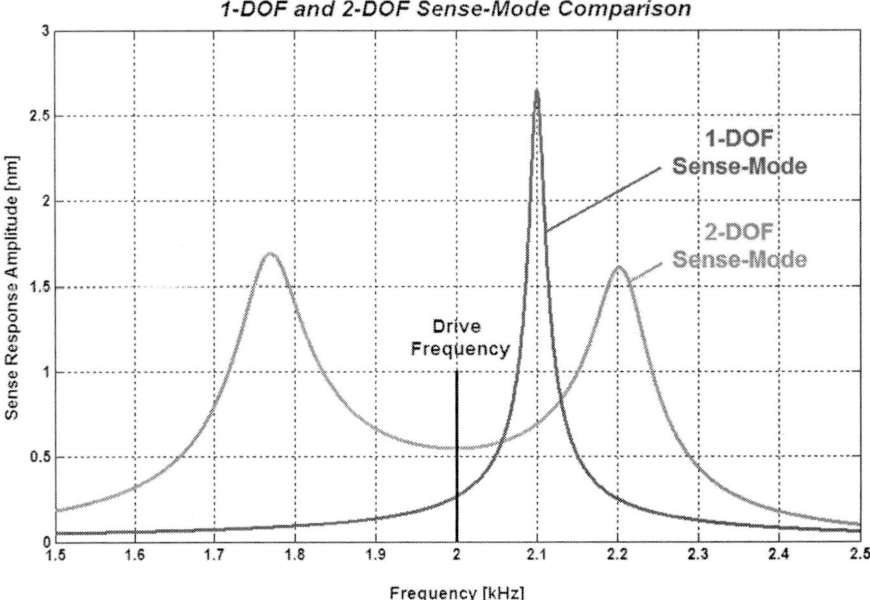

Fig. 9.9 Comparison of response between the 1-DOF and 2-DOF sense-mode oscillators. The 2-DOF sense-mode oscillator can even improve the scale factor and rate sensitivity compared to a conventional 1-DOF sense oscillator with mismatched modes.

9.5 Future Trends

Many emerging applications are known to impose quite harsh environmental conditions such as vibration, shock, temperature and thermal cycling on MEMS devices. Meeting performance specifications in demanding applications with low-cost and high-yield devices imposes even tighter robustness requirements on the MEMS sensing element. In the following sections, we outline design concepts developed as extension of the 2-DOF sense-mode architecture, addressing imperative design challenges such as vibration rejection and resonant frequency location.

9.5.1 Anti-Phase 2-DOF Sense Mode Gyroscope

A common design technique used to alleviate ambient vibrations and shock issues [145, 148] is the use of tuning fork architectures which allows the sensor to reject common mode inputs while preserving the angular rate signal [142]. Since the 2-DOF sense-mode concept presented in the previous chapters also used mechanical design to improve temperature robustness, a hybrid device promised a sensor with the benefits of both concepts. Thus, the multi-DOF sense mode tuning fork gyroscope was introduced intending to combine the robustness of multi-DOF sense systems with the common mode rejection capability of tuning fork designs [149].

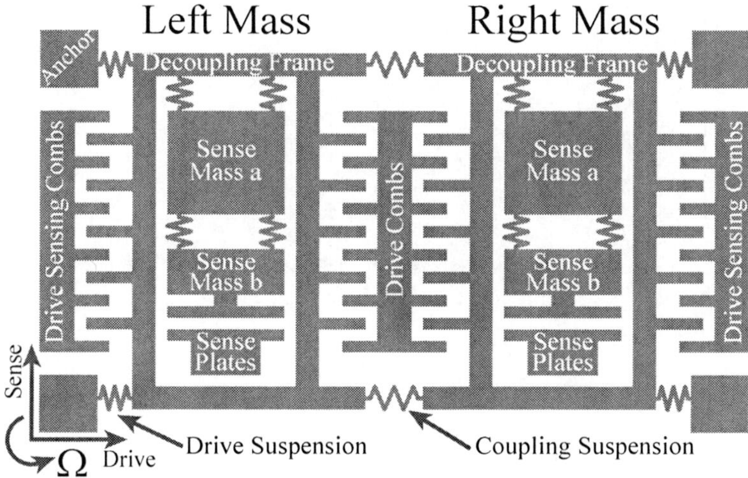

Fig. 9.10 Schematic of the anti-phase 2-DOF sense-mode gyroscope.

9.5.1.1 Design Concept

A schematic of the multi-DOF sense mode tuning fork gyroscope concept is shown in Figure 9.10. It consists of two identical multi-DOF systems, labeled left mass and right mass, connected via a coupling suspension forming a 2-DOF dynamic system in the drive mode. These systems are equivalent to the tines of a tuning fork; the major difference between conventional tuning fork gyroscopes, however, is that both tines are multi-DOF devices which contain decoupled 2-DOF sense modes. The frequency response of an ideal multi-DOF tine is identical to the regular multi-DOF gyroscope, except that the operational frequency of the tuning fork device is the anti-phase natural frequency of the 2-DOF drive mode.

The operation of the multi-DOF sense mode tuning fork, as with conventional ones, is based upon anti-phase motion of the drive masses. By driving in anti-phase,

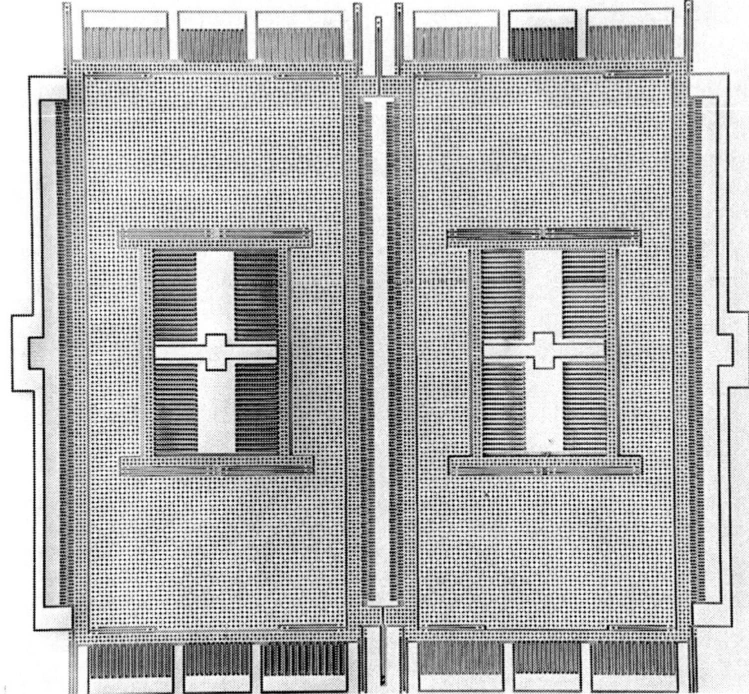

Fig. 9.11 SEM image of fabricated anti-phase 2-DOF sense-mode gyroscope. Courtesy of Adam R. Schofield.

the Coriolis responses of the corresponding sense systems will also be in anti-phase. Thus, a differential of the sense signals will preserve the angular rate signal while rejecting common mode inputs such as vibration and shock. In order to excite the anti-phase mode, a central driving electrode, Figure 9.10, is used to electrostatically force both the left and right masses simultaneously so that the ideal, linear time invariant (LTI) equations of motion for the drive mode become,

$$M_\delta \ddot{x}_L + c_\delta \dot{x}_L + k_\delta x_L - k_c x_R = -F(\omega_d),$$
$$M_\delta \ddot{x}_R + c_\delta \dot{x}_R + k_\delta x_R - k_c x_L = +F(\omega_d), \qquad (9.1)$$

where x_L and x_R denote the displacements of the left and right masses respectively in the drive direction.

Each drive mass, $M_\delta = m_f + m_a + m_b$, consists of the decoupling frame, m_f, and both sense masses, m_a and m_b, moving in unison along opposite drive directions due to the electrostatic forcing, $F(\omega_d)$, at the operational frequency. Each mass is suspended relative to the anchor via the drive suspension, k_δ, while the coupling suspension, k_c, connects the two tines forming the 2-DOF drive mode dynamic system. These two stiffnesses determine the operational frequency of the multi-DOF tuning fork according to the drive mode eigenvalue equation,

$$\omega^4 - 2\omega^2\omega_\delta^2 + \omega_\delta^4 - \omega_c^4 = 0, \tag{9.2}$$

where $\omega_\delta^2 = (k_\delta + k_c)/M_\delta$ and $\omega_c^2 = k_c/M_\delta$. The solution to (9.2) defines the in-phase and anti-phase modes of the multi-DOF tuning fork drive mode; as mentioned above, the operational frequency, ω_d, is the anti-phase natural frequency which can be expressed as

$$\begin{aligned} \omega_d^2 &= \omega_\delta^2 + \omega_c^2, \\ &= \frac{k_\delta + 2k_c}{M_\delta}. \end{aligned} \tag{9.3}$$

Despite having significantly different drive modes, the sense modes of the multi-DOF sense mode tuning fork are exactly the same as the 2-DOF sense-mode device presented in the previous chapters. While the anti-phase forcing provides mechanical rejection of common mode signals, the multi-DOF sense modes inherently provide the device with improved bandwidth and temperature robustness. The sense mode resonant frequencies are designed to be symmetric about the operational frequency, which in this case, is the anti-phase natural frequency defined above in (9.3). The region between the sense mode resonances not only provides a region of constant amplitude, but also a region of constant phase. This allows for easier phase matching of the sense modes which enhances the ability of the sensor to reject common modes.

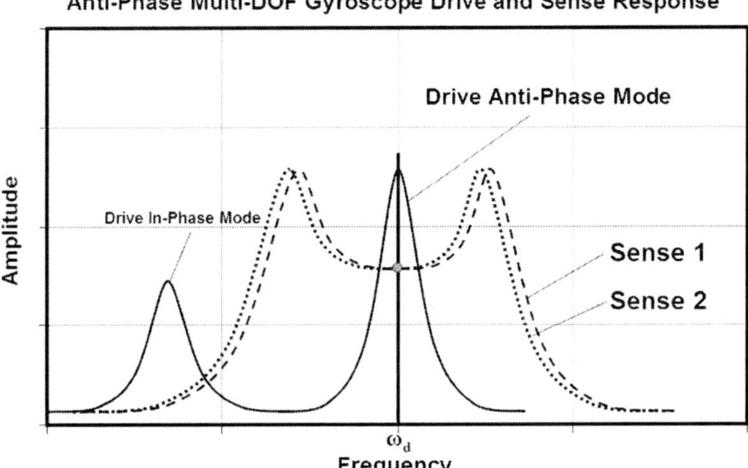

Fig. 9.12 The drive and sense-mode frequency responses of the anti-phase multi-DOF gyroscope system. The 2-DOF drive mode results in one in-phase and one anti-phase resonant mode. The anti-phase resonant mode is located within the two overlapping sense-mode flat regions.

In order to demonstrate the combined advantages of the multi-DOF sense mode tuning fork gyroscope, prototypes were fabricated using a wafer scale SOI process with a 50 μm conductive device layer, 5 μm buried oxide layer, and a 500 μm thick substrate. Figure 9.11 shows an SEM image of a fabricated multi-DOF tuning fork device. A central lateral comb drive provides the anti-phase drive actuation at the operational frequency while mechanical mode decoupling is used to isolate the drive and sense modes to minimize quadrature. Each sense system consists of two masses, a larger mass m_a and a smaller mass m_b which has differential parallel plate capacitors for detection. The following is a summary of results obtained by Adam R. Schofield at the UCI Microsystems Laboratory.

9.5.1.2 Common Mode Rejection

The common mode rejection capability of the multi-DOF sense mode tuning fork gyroscope was demonstrated using a shock input applied along the sense mode at atmospheric pressure while monitoring the response of both sense systems. No AC driving voltages were used, however a 10 V DC probing bias was applied to the gyroscope mass for detection. Each of the sense outputs were amplified via trans-impedance amplifiers and collected using a LeCroy WaveSurfer Model 452 Digital Oscilloscope.

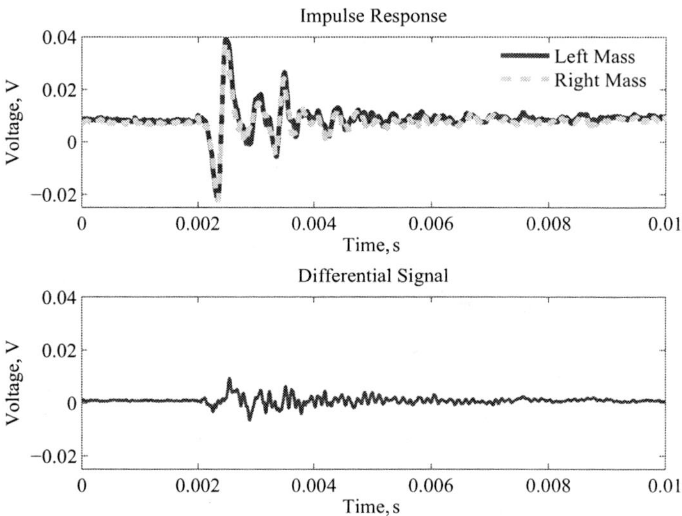

Fig. 9.13 Experimentally obtained time history of right and left sense mode output due to shock input.

Figure 9.13 (top) shows the experimentally obtained time histories of each sense system trigged due to a mechanical impulse applied at 0.002 seconds. The output of both sense masses track each other showing that the masses indeed responded in a common mode, as expected. The differential signal is shown in Figure 9.13

(bottom) which results in an uncalibrated 12 dB reduction in amplitude versus the individual signals. The amount of rejection can be increased by using more precise amplification or capacitance matching between the two outputs.

9.5.1.3 Coriolis Response

Having demonstrated the common mode response due to shock inputs, the multi-DOF tuning fork gyroscope was then characterized using constant angular rates to verify the anti-phase nature of the Coriolis response. This was accomplished by monitoring the sense signal of each mass individually, so that the scale factors of each could be observed; the final device output, however, would be the difference of these two signals.

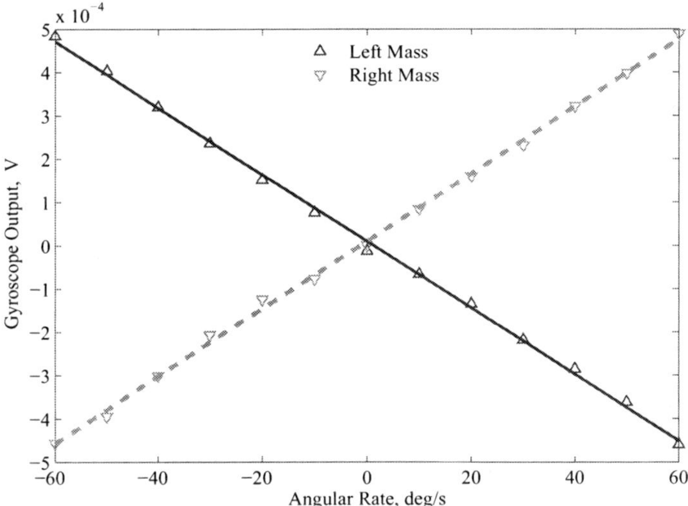

Fig. 9.14 Constant rate response of both right and left masses of the MDTF gyroscope.

The motional sense currents of both the right and left masses were amplified via trans-impedance amplifiers and demodulated to produce the DC output voltages shown in Figure 9.14. As expected, the rate responses of the two masses are in anti-phase as demonstrated by the sign difference of the scale factors. A linear fit was performed resulting in a scale factor of -7.68 μV/deg/s, zero offset of 10 μV, and linearity of 2.2% FSO for the left mass and a scale factor of 7.78 μV/deg/s, zero offset of 10 μV, and linearity of 2.2% FSO for the right mass. Since the actual output of the device would be the differential of these two signals, the overall scale factor is expected to increase by a factor of two.

9.5.2 2-DOF Sense Mode Gyroscope with Scalable Peak Spacing

In this section we review an alternative MEMS vibratory rate gyroscope design with
2-DOF sense mode recently introduced by Alexander A. Trusov in [151]. The proposed architecture, Figure 9.15, utilizes a single degree-of-freedom (DOF) drive-
mode and a fully coupled 2-DOF sense-mode, and yields robust, wide-bandwidth
devices without sacrifice in response gain. Due to the flexibility of the extended
design space of the architecture, the sense-mode resonant peak spacing and bandwidth can be adjusted independently of the operational frequency. For arbitrary
given proof mass m_p and detection mass m_d the three design stiffnesses are obtained
as

$$\begin{cases} k_1 = m_p \Phi^2 - k_2, \\ k_2 = \Delta\Phi \sqrt{m_p m_d} \sqrt{\Phi^2 - 0.25\Delta\Phi^2}, \\ k_3 = m_d \Phi^2 - k_2, \end{cases} \tag{9.4}$$

where Φ and $\Delta\Phi$ are respectively the desired operational frequency and the sense-
mode resonant peak spacing. Inherent to the design, the drive-mode resonance is
automatically positioned between the sense-mode peaks, eliminating the need to
trim and tune drive- and sense-modes.

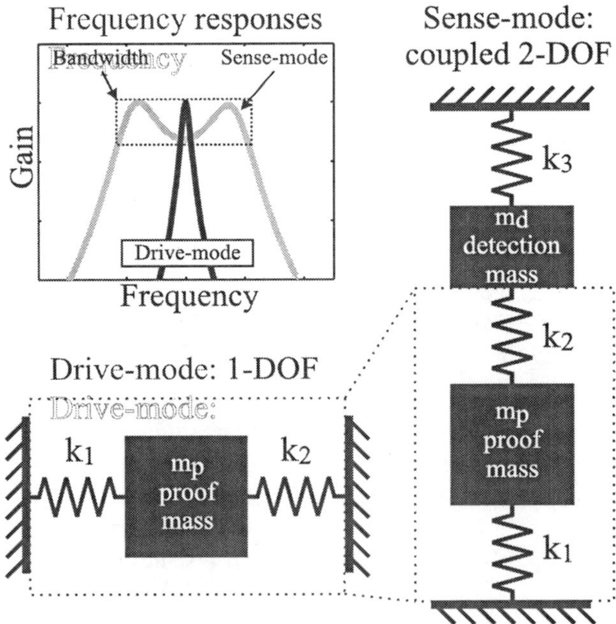

Fig. 9.15 Lumped models of the drive and sense-modes.

A micromachined implementation of the proposed gyroscope is shown in Figure 9.16. The device consists of an anchored outer frame, two drive-mode and two sense-mode shuttles, a proof mass m_p, a detection mass m_d, and a central anchor. Each of the two drive-mode and two sense-mode shuttles is suspended relative to the fixed frame by two springs constraining shuttles' motion to the respective axes. Similar suspension elements couple the four shuttles to the proof mass m_p, forming a symmetrically decoupled suspension [152]. Using the capacitive electrodes on the drive-mode shuttles, the proof mass m_p is driven into a drive-mode oscillation to form a z-axis sensitive Coriolis element. Unlike the conventional case [153, 154], the Coriolis-induced motion is not directly picked-up from the proof mass, m_p; instead, the proof mass is coupled to the detection mass m_d using a bi-directional flexure with equal x and y stiffnesses. The detection mass is also coupled to the substrate with an inner suspension. During rotation, the Coriolis acceleration of the proof mass is transferred to the detection mass, which responds in a wide frequency bandwidth due to the coupled dynamics of the proposed 2-DOF sense-mode.

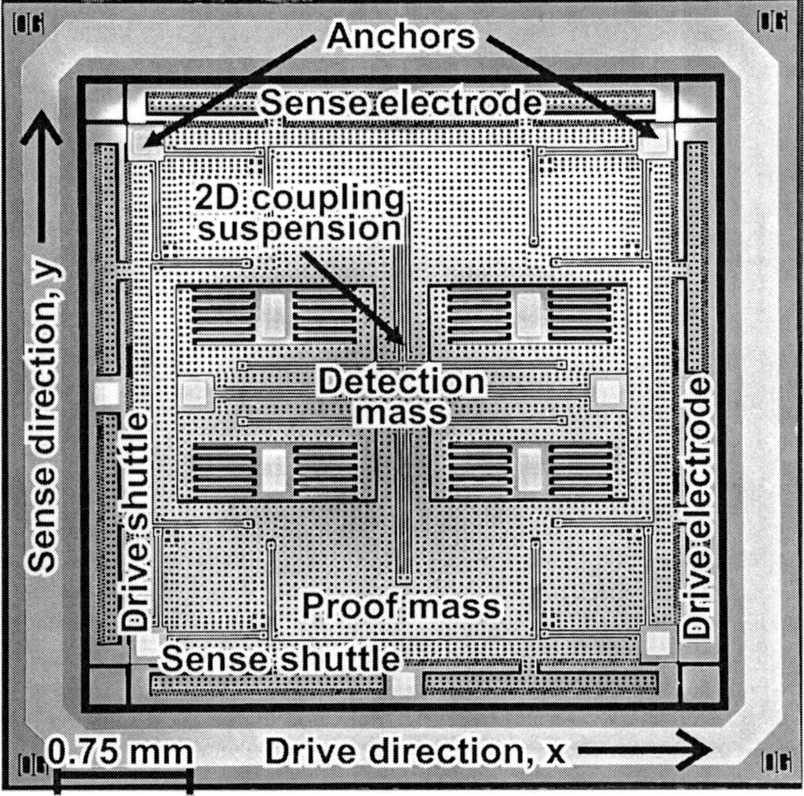

Fig. 9.16 SEM image of a fabricated sensor element with 1-DOF drive-mode and a fully coupled 2-DOF sense-mode. Courtesy of Alexander A. Trusov.

Fabricated prototypes were packaged and characterized under ambient pressure conditions. The measured drive-mode resonance was at 2.58 kHz, located in-between the 2-DOF sense-mode resonances at 2.47 and 2.73 kHz. The 2-DOF sense-mode exhibited a 250 Hz 3 dB bandwidth providing an 8 times improvement in temperature drift compared to the identical conventional mode-matched device. Figure 9.17 shows rate response experimentally obtained at room temperature and estimated responses for -55 °C and 125 °C temperatures. Analysis of the gyroscope's noise performance is shown in Figure 9.18. With the off-chip detection electronics, the measured resolution was 0.09 °/s/$\sqrt{\text{Hz}}$ and the bias drift was 0.08 °/s.

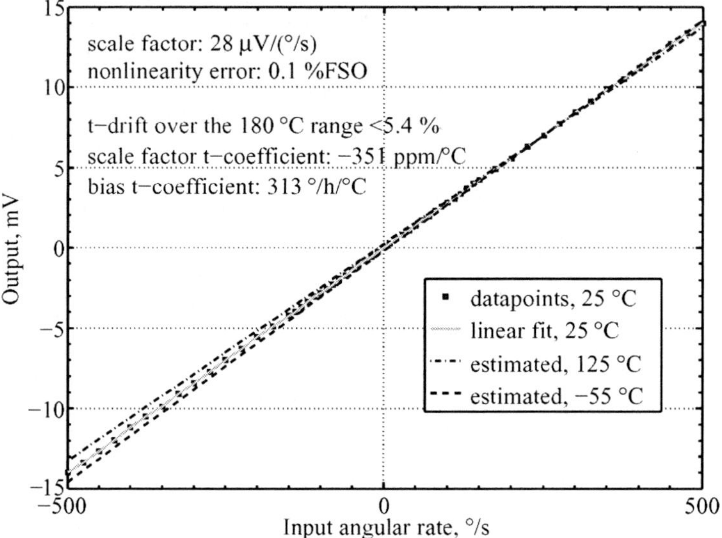

Fig. 9.17 Rate responses at different temperatures.

For the described gyroscope design architecture, the resonant peak spacing of the fully coupled 2-DOF sense-mode defines the bandwidth of the gyroscope and can be adjusted independently of the operational frequency. The rate sensitivity and quadrature of the gyroscope are comparable to the best reported performance characteristics for MEMS gyroscopes operated in air, e.g., [155]. At the same time, the gyroscope provides much larger bandwidth than the state of the art rate gyroscopes, as well as excellent robustness to fabrication imperfections and in-operation temperature variations.

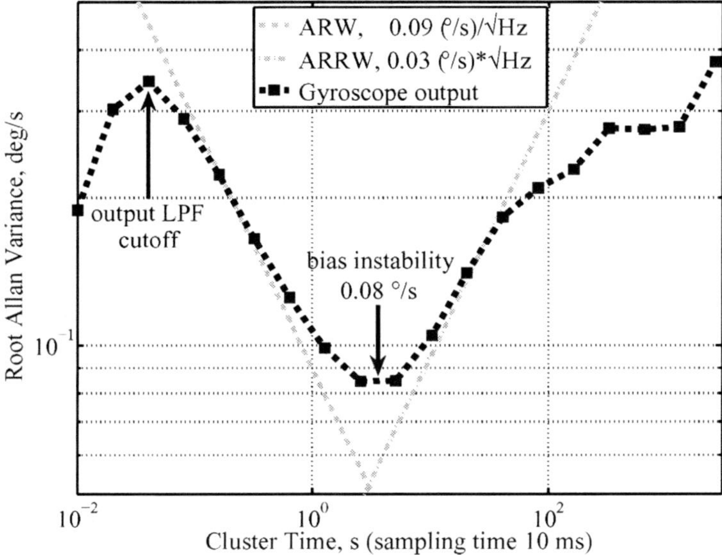

Fig. 9.18 Allan Variance characterization of noise modes at zero rate.

9.6 Conclusion

Commercialization of reliable vibratory micromachined gyroscopes for high-volume applications have proven to be extremely challenging, primarily due to the high sensitivity of the dynamical system response to fabrication and environmental variations. In this book, we provided a solid foundation on the fundamental operational principles and the primary implementation challenges of micromachined vibratory gyroscopes. We then presented and compared several promising structural design concepts for micromachined vibratory gyroscopes, which provide inherent robustness against structural parameter variations. We demonstrated that an inherently robust mechanical sensing element of a micromachined gyroscopes system could improve bias stability, temperature stability, linearity and immunity to environmental variations. Thus, the presented structural concepts are expected to provide reliable, robust, high-yield, low-cost and high performance micromachined vibratory rate gyroscopes suitable for many high-volume applications.

References

1. G.G. Coriolis. Mémoire sur les équations du mouvement relatif des systémes de corps. *Journal de l'école Polytechnique*, Vol 15, pp. 142154, 1835.
2. IEEE Std 1431-2004. IEEE Standard Specification Format Guide and Test Procedure for Coriolis Vibratory Gyros.
3. IEEE Std 528-2001. IEEE Standard for Inertial Sensor Terminology.
4. N. Yazdi, F. Ayazi, and K. Najafi. Micromachined Inertial Sensors. *Proc. of IEEE, Vol. 86, No. 8*, August 1998.
5. H. Xie and G. K. Fedder. Integrated Microelectromechanical Gyroscopes. *Journal of Aerospace Engineering, Vol. 16, No. 2*, April 2003, pp. 65-75.
6. P. Greiff, B. Boxenhorn, T. King, and L. Niles. Silicon monolithic micromechanical gyroscope. *Tech. Dig. 6th Int. Conf. Solid-State Sensors and Actuators (Transducers'91), San Francisco, CA*, June 1991, pp. 966-968.
7. S. D. Orlosky and H. D. Morris. Quartz rotation (rate) sensor. *Proc. Sensor Expo, Cleveland, OH*. 1994, pp. 171-177.
8. R. R. Ragan and D. D. Lynch. Inertial technology for the future, Part X: Hemispherical resonator gyro. *IEEE Trans. Aerosp. Electron. Syst., vol. AES-20*, July 1984. p. 432
9. J. Bernstein, S. Cho, A. T. King, A. Kourepenis, P. Maciel, and M.Weinberg. A micromachined comb-drive tuning fork rate gyroscope. *Proc. IEEE Microelectromechanical Systems, Fort Lauderdale, FL*, Feb. 1993, pp. 143-148.
10. M. W. Putty and K. Najafi. A micromachined vibrating ring gyroscope. *Tech. Dig. Solid-State Sens. Actuator Workshop, Hilton Head Island, SC*, June 1994, pp. 213-220.
11. Hopkin, I. D. Vibrating gyroscopes (automotive sensors). *IEEE Colloquium on Automative Sensors. Digest No. 1994/170. Solihull, UK*, Sept. 1994, pp. 1-4.
12. Tanaka, K., Mochida, Y., Sugimoto, S., Moriya, K., Hasegawa, T., Atsuchi, K., and Ohwada, K. A micromachined vibrating gyroscope. *Proc., Eighth IEEE Int. Conf. on Micro Electro Mechanical Systems (MEMS '95), Amsterdam, Netherlands*, Jan. 1995, pp. 278-281.
13. Clark, W. A., Howe, R. T., and Horowitz, R. Surface micromachined Z-axis vibratory rate gyroscope. *Technical Digest. Solid-State Sensor and Actuator Workshop, Hilton Head Island, S.C.*, June 1996, pp. 283-287.
14. Juneau, T., Pisano, A. P., and Smith, J. H. Dual axis operation of a micromachined rate gyroscope. *Proc., IEEE 1997 Int. Conf. on Solid State Sensors and Actuators (Tranducers '97), Chicago*, June 1997, pp. 883-886.
15. M. Lutz, W. Golderer, J. Gerstenmeier, J. Marek, B. Maihofer, S. Mahler, H. Munzel, and U. Bischof. Aprecision yaw rate sensor in silicon micromachining, *Transducers 1997, Chicago, IL*, June 1997, pp. 847-850.
16. T. K. Tang, R. C. Gutierrez, C. B. Stell, V. Vorperian, G. A. Arakaki, J. T. Rice, W. J. Li, I. Chakraborty, K. Shcheglov, J. Z. Wilcox, and W. J. Kaiser. A packaged silicon MEMS vibratory gyroscope for microspacecraft. *Proc. IEEE Micro Electro Mechanical Systems Workshop (MEMS'97), Japan*, 1997, pp. 500-505.

17. D. R. Sparks, S. R. Zarabadi, J. D. Johnson, Q. Jiang, M. Chia, O. Larsen, W. Higdon, and P. Castillo-Borelley. A CMOS integrated surface micromachined angular rate sensor: It's automotive applications. *Tech. Dig. 9th Int. Conf. Solid-State Sensors and Actuators (Transducers'97), Chicago, IL,* June 1997, pp. 851-854.

18. Y. Oh, B. Lee, S. Baek, H. Kim, J. Kim, S. Kang, and C. Song. A surface-micromachined tunable vibratory gyroscope. *Proc. IEEE Micro Electro Mechanical Systems Workshop (MEMS'97), Japan,* 1997, pp. 272-277.

19. K. Y. Park, C. W. Lee, Y. S. Oh, and Y. H. Cho. Laterally oscillated and force-balanced micro vibratory rate gyroscope supported by fish hook shape springs. *Proc. IEEE MicroElectro Mechanical Systems Workshop (MEMS'97), Japan,* 1997, pp. 494-499.

20. R. Voss, K. Bauer, W. Ficker, T. Gleissner, W. Kupke, M. Rose, S. Sassen, J. Schalk, H. Seidel, and E. Stenzel. Silicon angular rate sensor for automotive applications with piezoelectric drive and piezoresistive read-out. *Tech. Dig. 9th Int. Conf. Solid-State Sensors and Actuators (Transducers'97), Chicago, IL,* June 1997, pp. 879-882.

21. R. Hulsing. MEMS inertial rate and acceleration sensor. *IEEE AES Systems Magazine.* November 1998, pp. 17-23.

22. A. Kourepenis, J. Bernstein, J. Connely, R. Elliot, P. Ward, and M.Weinberg. Performance of MEMS inertial sensors. *Proc. IEEE Position Location and Navigation Symposium,* 1998, pp. 1-8.

23. Mochida, Y., Tamura, M., and Ohwada, K. A micromachined vibrating rate gyroscope with independent beams for the drive and detection modes. *Proc., Twelfth IEEE Int. Conf. on Micro Electro Mechanical Systems (MEMS '99), Orlando, FL,* Jan. 1999, pp. 618-623.

24. Funk, K., Emmerich, H., Schilp, A., Offenberg, M., Neul, R., and Larmer, F. A surface micromachined silicon gyroscope using a thick polysilicon layer. *Proc., Twelfth IEEE Int. Conf. on Micro Electro Mechanical Systems (MEMS '99), Orlando, FL,* Jan. 1999, pp. 57-60.

25. Park, K. Y., Jeong, H. S., An, S., Shin, S. H., and Lee, C. W. Lateral gyroscope suspended by two gimbals through high aspect ratio ICP etching. *Proc., IEEE 1999 Int. Conf. on Solid State Sensors and Actuators (Tranducers '99), Sendai, Japan,* June 1999, pp. 972-975.

26. S. Lee, S. Park, J. Kim, and D. Cho. Surface/bulk micromachined single-crystalline-silicon micro-gyroscope. *IEEE/ASME Journal of Microelectromechanical Systems, Vol.9, No.4* Dec. 2000, pp. 557-567.

27. Jiang, X., Seeger, J. I., Kraft, M., and Boser, B. E. A monolithic surface micromachined Z-axis gyroscope with digital output. *Digest of Technical Papers, 2000 Symposium on VLSI Circuits, Honolulu,* June 2000, pp. 16-19.

28. H. Xie and G. K. Fedder. A CMOS-MEMS lateral-axis gyroscope. *Proc. 14th IEEE Int. Conf. Microelectromechanical Systems, Interlaken, Switzerland,* Jan. 2001, pp. 162-165.

29. H. Luo, X. Zhu, H. Lakdawala, L. R. Carley, and G. K. Fedder. A copper CMOS- MEMS z-axis gyroscope. *Proc. 15th IEEE Int. Conf. Microelectromechanical Systems, Las Vegas, NV,* Jan. 2001, pp. 631-634.

30. W. Geiger, W.U. Butt, A. Gaisser, J. Fretch, M. Braxmaier, T. Link, A. Kohne, P. Nommensen, H. Sandmaier, and W. Lang. Decoupled Microgyros and the Design Principle DAVED. *Sensors and Actuators A (Physical), Vol. A95, No.2-3,* Jan. 2002, pp. 239-249.

31. J. A. Geen, S. J. Sherman, J. F. Chang, and S. R. Lewis. Single-chip surface-micromachining integrated gyroscope with 50 deg/hour root Allan variance. *Dig. IEEE Int. Solid-State Circuits Conf., San Francisco, CA,* Feb. 2002, pp. 426-427.

32. Seshia, A. A., Howe, R. T., and Montague, S. An integrated microelectromechanical resonant output gyroscope. *Proc., The Fifteenth IEEE Int. Conf. on Micro Electro Mechanical Systems (MEMS 2002), Las Vegas,* Jan. 2002, pp. 722-726.

33. G. He and K. Najafi. A single crystal silicon vibrating ring gyroscope. *Proc., The Fifteenth IEEE Int. Conf. on Micro Electro Mechanical Systems (MEMS 2002), Las Vegas,* Jan. 2002, pp. 718-721.

34. H. Xie and G. K. Fedder. Fabrication, Characterization, and Analysis of a DRIE CMOS-MEMS Gyroscope. *IEEE Sensors Journal, Vol. 3, No. 5,* 2003, pp. 622-631.

35. Karnick, D.; Ballas, G.; Koland, L.; Secord, M.; Braman, T.; Kourepenis, T. Honeywell gun-hard inertial measurement unit (IMU) development. *Position Location and Navigation Symposium, PLANS 2004.* 26-29 April 2004 Page(s):49 - 55.

36. Byoung-doo Choi; Sangjun Park; Hyoungho Ko; Seung-Joon Paik; Yonghwa Park; Geunwon Lim; Ahra Lee; Sang Chul Lee; Carr, W.; Setiadi, D.; Mozulay, R.; Dong-il Cho. The first sub-deg/hr bias stability, silicon-microfabricated gyroscope. *Solid-State Sensors, Actuators and Microsystems, 2005. TRANSDUCERS '05.* 5-9 June 2005, pp. 180 - 183.

37. A. Sharma, M.F. Zaman, and F.A. Ayazi. 0.2 deg/hr Micro-Gyroscope with Automatic CMOS Mode Matching. *IEEE International Solid-State Circuits Conference, 2007. ISSCC 2007.* pp. 386 - 610.

38. M.F. Zaman, A. Sharma, and F.A. Ayazi. High Performance Matched-Mode Tuning Fork Gyroscope. *Tech. Dig 19th IEEE International Conference on Micro-Electormechanical Systems Conference 2006 (MEMS 2006).* Istanbul, Turkey, Jan. 2006, pp. 66-69.

39. T. J. Brosnihan, J. M. Bustillo, A. P. Pisano, and R. T. Howe. Embedded interconnect and electrical isolation for high-aspect-ratio, SOI inertial instruments. *Transducers 1997, Chicago, IL,* June 1997, pp. 637-640.

40. Egyptian Museum, Egyptian Center for Documentation of Cultural and Natural Heritage. www.eternalegypt.org

41. A. Persson, The Coriolis Effect A Conflict Between Common Sense and Mathematics. The Swedish Meteorological and Hydrological Institute, Norrkping, Sweden.

42. N. Barbour, and G. Schmidt. Inertial Sensor Technology Trends. *IEEE Sensors Journal, Vol. 1, No. 4,* Dec 2001, pp. 332-339.

43. W.A. Clark, R.T. Howe, and R. Horowitz. Surface Micromachined Z-Axis Vibratory Rate Gyroscope. *Proceedings of Solid-State Sensor and Actuator Workshop,* June 1994.

44. A. Shkel, R.T. Howe, and R. Horowitz. Micromachined Gyroscopes: Challenges, Design Solutions, and Opportunities. *Int. Workshop on Micro-Robots, Micro-Machines and Systems, Moscow, Russia,*1999.

45. A. Shkel, R. Horowitz, A. Seshia, S. Park and R.T. Howe. Dynamics and Control of Micromachined Gyroscopes *American Control Conference, CA* ,1999.

46. S. Park and R. Horowitz. Adaptive Control for Z-Axis MEMS Gyroscopes. *American Control Conference, Arlington, VA,* June 2001.

47. R.P. Leland. Adaptive Tuning for Vibrational Gyroscopes. *Proceedings of IEEE Conference on Decision and Control, Orlando, FL,* Dec. 2001.

48. US 5,992,233. W.A. Clark. Micromachined z-Axis Vibratory Gyroscope. Nov. 1999.

49. US 6,122,961. J.A. Geen, and D.W. Carow. Micromachined Gyros. Sept. 2000.

50. US 6,089,089. Y.W. Hsu. Multi-Element Micro Gyro. April 1999.

51. C.W. Dyck, J. Allen, and R. Hueber. Parallel Plate Electrostatic Dual Mass Oscillator. *Proceedings of SPIE SOE, CA,* 1999.

52. X. Li, R. Lin, and K.W. Leow. Performance-Enhanced Micro-Machined Resonant Systems with Two-Degrees-of-Freedom Resonators. *Journal of Micromech, Microeng., Vol. 10,* 2000, pp. 534-539.

53. T. Usada. Operational Characteristics of Electrostatically Driven Torsional Resonator with Two-Degrees-of-Freedom. *Sensors and Actuators A,* Vol. 64, 1998, pp. 255-257.

54. W. Geiger, B. Folkmer, U. Sobe, H. Sandmaier, and W. Lang. New Designs of Micromachined Vibrating Rate Gyroscopes with Decoupled Oscillation Modes. *Sensors and Actuators A,* Vol. 66, 1998, pp. 118-124.

55. E. Netzer, and I. Porat. A Novel Vibratory Device for Angular Rate Measurement. *Journal of Dynamic Systems, Measurement and Control,* Dec. 1995.

56. A. Seshia. Design and Modeling of a Dual Mass SOI-MEMS Gyroscope. *M.S. Thesis, BSAC, U.C. Berkeley,* 1999.

57. D.A. Koester, R. Mahadevan, B. Hardy, and K.W. Markus. MUMPs Design Handbook, Revision 5.0. *Cronos Integrated Microsystems,* 2000.

58. L. Lin, R.T. Howe, and A.P. Pisano. Microelectromechanical Filters for Signal Processing. *Journal of Microelectromechanical Systems, Vol. 7,* Sept. 1998.

59. Lemkin. Micro Accelerometer Design with Digital Feedback Control. *Ph.D. Thesis, BSAC, U.C. Berkeley*, 1997.

60. H. Xie and G.K. Fedder. A DRIE CMOS-MEMS Gyroscope. *IEEE Sensors 2002 Conference, June 2002, Orlando, FL.*

61. A. Shkel, R.T. Howe, and R. Horowitz. Modeling and Simulation of Micromachined Gyroscopes in the Presence of Imperfections. *International Conference on Modeling and Simulation of Microsystems*, Puerto Rico, 1999, pp. 605-608.

62. W.A. Clark. Micromachined Vibratory Rate Gyroscope. *Ph.D. Thesis, BSAC, U.C. Berkeley*, 1994.

63. C. Acar, and A. Shkel. Inherently Robust Micromachined Gyroscopes with 2-DOF Sense-Mode Oscillator. *Journal of Microelectromechanical Systems.* Vol. 15, No. 2, pp. 380-387, 2006.

64. C. Acar, and A. Shkel. An Approach for Increasing Drive-Mode Bandwidth of MEMS Vibratory Gyroscopes. *Journal of Microelectromechanical Systems.* Vol. 14, No. 3, pp. 520-528, 2005.

65. C. Acar, and A. Shkel. Structurally Decoupled Micromachined Gyroscopes with Post-Release Capacitance Enhancement. *Journal of Micromechanics and Microengineering.* Vol.15, pp. 1092-1101, 2005.

66. C. Acar, and A. Shkel. Structural Design and Experimental Characterization of Torsional Micromachined Gyroscopes with Non-Resonant Drive-Mode. *Journal of Micromechanics and Microengineering.* Vol.14, pp.15-25, 2003.

67. C. Acar, and A. Shkel. Experimental Evaluation and Comparative Analysis of Commercial Variable-Capacitance MEMS Accelerometers. *Journal of Micromechanics and Microengineering.* Vol.13, pp.634-645, 2003.

68. C. Acar, and A. Shkel. Four Degrees-of-Freedom Micromachined Gyroscopes. *Journal of Modeling and Simulation of Microsystems, Vol. 2, No. 1,* pp. 71-82, 2001.

69. C. Acar, A. Shkel. Non-Resonant Micromachined Gyroscopes with Structural Mode-Decoupling. *IEEE Sensors Journal, Vol. 3, No. 4, pp. 497-506,* 2003.

70. US Patent 7,377,167. Cenk Acar and Andrei Shkel. Non-resonant micromachined gyroscopes with structural mode-decoupling.

71. US Patent 7,284,430. Cenk Acar and Andrei Shkel. Robust micromachined gyroscopes with two degrees of freedom sense-mode oscillator.

72. US Patent 7,279,761. Cenk Acar and Andrei Shkel. Post-release capacitance enhancement in micromachined devices and a method of performing the same.

73. US Patent 7,100,446. Cenk Acar and Andrei Shkel. Distributed-mass micromachined gyroscopes operated with drive-mode bandwidth enhancement.

74. US Patent 6,845,669. Cenk Acar and Andrei Shkel. Non-resonant four degrees-of-freedom micromachined gyroscope.

75. Cenk Acar, Adam Schofield, Andrei Shkel, Lynn Costlow, Asad Madni. Robust Six Degree-of-Freedom Micromachined Gyroscope with Anti-Phase Drive Scheme Patent Pending, UC Office of Technology Transfer.

76. C. Acar and A. Shkel. Torsional z-Axis Surface-Micromachined Gyroscopes with Non-Resonant Actuation. *Patent pending, UC Office of Technology Transfer.*

77. C. Acar, A. M. Shkel, L. Costlow, and A. M. Madni, "Inherently robust micromachined gyroscopes with 2-DOF sense-mode oscillator," in *Proc. of IEEE Sensors*, Irvine, CA, USA, Oct. 31– Nov. 3, 2005, pp. 664–667.

78. C. Acar, A. Shkel. Enhancement of Drive-Mode Bandwidth in MEMS Vibratory Gyroscopes Utilizing Multiple Oscillators. *Solid-State Sensor and Actuator Workshop, Hilton Head Island, SC,* June 2004.

79. C. Acar, A. Shkel. A Design Approach for Robustness Improvement of Rate Gyroscopes. *Modeling and Simulation of Microsystems Conference*, March 2001.

80. C. Acar, A. Shkel. Microgyroscopes with Dynamic Disturbance Rejection. *SPIE Conference on Smart Electronics and MEMS*, March 2001.

81. C. Acar, S. Eler, and A. Shkel. Concept, Implementation, and Control of Wide Bandwidth MEMS Gyroscopes. *American Control Conference*, June 2001.

82. C. Acar, A. Shkel. Design Concept and Preliminary Experimental Demonstration of MEMS Gyroscopes with 4-DOF "Master-Slave" Architecture. *SPIE Conference on Smart Electronics and MEMS*, March 2002.

83. C. Acar, and A. Shkel. A Class of MEMS Gyroscopes with Increased Parametric Space. *Proceedings of IEEE Sensors Conference*, Orlando, Florida, 2002, pp. 854-859.

84. C. Acar, A. Shkel. Distributed-Mass Micromachined Gyroscopes for Enhanced Mode-Decoupling. *IEEE Sensors Conference, September 2003, Toronto, Canada*.

85. J.A. Geen. A Path to Low Cost Gyroscopy. *Solid-State Sensor and Actuator Workshop*, Hilton-Head, SJ, 1998, pp. 51-54.

86. W. Geiger, W.U. Butt, A. Gaisser, J. Fretch, M. Braxmaier, T. Link, A. Kohne, P. Nommensen, H. Sandmaier, and W. Lang. Decoupled Microgyros and the Design Principle DAVED. *IEEE Sensors Journal*, 2001, pp. 170-173.

87. Y. Mochida, M. Tamura, and K. Ohwada. A Micromachined Vibrating Rate Gyroscope with Independent Beams for Drive and Detection Modes. *Sensors and Actuators A*, Vol. 80, 2000, pp. 170-178.

88. M. Niu, W. Xue, X. Wang, J. Xie, G. Yang, and W. Wang. Design and Characteristics of Two-Gimbals Micro-Gyroscopes Fabricated with Quasi-LIGA Process. *International Conference on Solid-State Sensor and Actuators*, 1997, pp. 891-894.

89. U. Breng, W. Guttman, P. Leinfelder, B. Ryrko, S. Zimmermann, D. Billep, T. Gessner, K. Hillner, and M. Weimer. A Bulk Micromachined Gyroscope Based on Coupled Resonators. *International Conference on Solid-State Sensor and Actuators*, Japan, 1999, pp. 1570-1573.

90. H.T. Lim, J.W. Song, J.G. Lee, and Y.K. Kim. A Few deg/hr Resolvable Low Noise Lateral Microgyroscope. *Proceedings of IEEE MEMS Conference, NV*, 2002, pp. 627-630.

91. S.E. Alper, and T. Akin. A Symmetric Surface Micromachined Gyroscope with Decoupled Oscillation Modes. *Sensors and Actuators A*, Vol. 97, 2002, pp. 347-358.

92. Y.S. Hong, J.H. Lee, and S.H. Kim. A Laterally Driven Symmetric Micro-Resonator for Gyroscopic Applications. *Journal of Micromechanics and Microengineering*, Vol. 10, 2000, pp. 452-458.

93. C.W. Dyck, J. Allen, R. Hueber. Parallel Plate Electrostatic Dual Mass Oscillator. *Proceedings of SPIE Conference on Micromachining and Microfabrication, Vol. 3876, CA*, 1999, pp. 198-209.

94. US 6,691,571. Willig et al. Rotational Speed Sensor. February 2004.

95. Funk, K. Emmerich, H. Schilp, A. Offenberg, M. Neul, R. Larmer, F.. A Surface Micromachined Silicon Gyroscope Using a Thick Polysilicon Layer. *Twelfth IEEE International Conference on Micro Electro Mechanical Systems*, 1999, pp. 57-60.

96. J. J. Wortman and R. A. Evans. Young's Modulus, Shear Modulus, and Poisson's Ratio in Silicon and Germanium. *Journal of Applied Physics*, 1965. 36, 153.

97. Kuisma, H.; Ryhanen, T.; Lahdenpera, J.; Punkka, E.; Ruotsalainen, S.; Sillanpaa, T.; Seppa, H.. A Bulk Micromachined Silicon Angular Rate Sensor *International Conference on Solid State Sensors and Actuators, 1997*. Volume 2, Chicago 16-19 Jun 1997 pp. 875 - 878

98. S. V. Iyer, Y. Zhou and T. Mukherjee. Analytical Modeling of Cross-axis Coupling in Micromechanical Springs *International Conference on Modeling and Simulation of Microsystems (MSM)*. pp. 632-635, April 19-21, 1999, San Juan, PR, USA.

99. Geiger, W.; Folkmer, B.; Merz, J.; Sandmaier, H.; Lang, W.. A New Silicon Rate Gyroscope *The Eleventh Annual International Workshop on Micro Electro Mechanical Systems, 1998.* 25-29 Jan 1998 pp. 615 - 620

100. S. Gnthner, K. Kapser, M. Rose, B. Hartmann, M. Kluge, U. Schmid, and H. Seidel. Analysis of Piezoresistive Read-out Signals for a Silicon Tuning Fork Gyroscope *IEEE Sensors Conference, 2004*. Vienna, Austria, Oct. 2004, pp. 14111414.

101. Weinberg, M.S.; Kourepenis, A. Error sources in in-plane silicon tuning-fork MEMS gyroscopes *Journal of Microelectromechanical Systems*. Volume 15, Issue 3, June 2006, pp. 479 - 491.

102. A. Duwel, M. Weinstein, J. Gorman, J. Borenstein, P. Ward. Quality Factors of MEMS Gyros and the Role of Thermoelastic Damping *International Conference on Micro Electro Mechanical Systems, 2002*. Las Vegas, NV, January 2002, pp. 214219.

103. C. Zener. Internal Friction in Solids II. General Theory of Thermoelastic Internal Friction. *Physical Review, 1938.* Vol. 53, pp. 90-99.

104. W. C. Tang. Electrostatic Comb-Drive for Resonant Sensor and Actuator Applications. *Ph.D. Thesis, University of California at Berkeley*, 1990.

105. B.E. Boser. Electronics For Micromachined Inertial Sensors. *Proceedings of Transducers,* 1997.

106. W.C. Young. Roark's Formulas for Stress and Strain. *McGraw-Hill, Inc.*, pp. 93-156, 1989.

107. M.J. Madou. Fundamentals of Microfabrication: The Science of Miniaturization. *CRC Press.*, Second Edition, 2002.

108. W. Kuehnel. Modeling of the Mechanical Behavior of a Differential Capacitor Acceleration Sensor. *Sensors and Actuators A*, Vol. 48, 1995, pp. 101-108.

109. T. Veijola, and M. Turowski. Compact Damping Models for Laterally Moving Microsctructures with Gas-Rarefaction Effects. *Journal of Microelectromechanical Systems*, Vol. 10, No.2, June 2001, pp. 263-273.

110. T. Veijola, H. Kuisma, J. Lahdenpera and T. Ryhenen. Equivalent Circuit Model of the Squeezed Gas Film in a Silicon Accelerator. *Sensors and Actuators A*, Vol. 48, 1995, pp. 239-248.

111. J.J. Blech. On Isothermal Squeeze Films. *Journal of Lubrication Technology*, Vol. 105, 1983, pp. 615-620.

112. Y.H. Cho, B.M. Kwak, A.P. Pisano, and R.T. Howe. Slide Film Damping in Laterally Driven Microsctructures. *Sensors and Actuators A*, Vol. 40, 1994, pp. 31-39.

113. C. Bourgeois, F. Parret, and A. Hoogerwerf. Analytical Modeling of Squeeze-Film Damping in Accelerometers. *International Conference on Solid-State Sensors and Actuators*, Chicago, IL, June 1997, pp. 1117-1120.

114. http://www.analogdevices.com.

115. T.N. Juneau, A.P. Pisano, J.H. Smith. Dual Axis Operation of a Micromachined Rate Gyroscope. *Ninth International Conference on Solid-State Sensors and Actuators*, Chicago, IL, June 1997.

116. S.E. Alper, and T. Akin. A Planar Gyroscope Using Standard Surface Micromachining Process. *Conference on Solid-State Transducers*, Copenhagen, Denmark, 2000, pp. 387-390.

117. http://www.sensofar.com.

118. http://www.saesgetters.com.

119. http://www.polytecpi.com.

120. http://www.mmr.com/ltmpd.htm

121. M. Handtmann, R. Aigner, A. Meckes, and G.K.M. Wachutka. A Sensitivity Enhancement of MEMS Inertial Sensors Using Negative Springs and Active Control. *Sensors and Actuators A*, Vol. 97, 2002, pp. 153-160.

122. A. Lawrence. Modern Inertial Technology. *Springer*, 1998.

123. T. Hirano, T. Furuhata, K.J. Gabriel, and H. Fujita. Design, Fabrication and Operation of Submicron Gap Comb-Drive Microactuators. *Journal of Microelectromechanical Sytems*, Vol. 1, 1992, pp. 52-59.

124. F. Ayazi, and K. Najafi. High Aspect-Ratio Combined Poly and Single-Crystal Silicon (HARPSS) MEMS Technology *Journal of Microelectromechanical Sytems*, Vol. 9, 2000, pp. 288-294.

125. S. Pourkamali, F. Ayazi. SOI-Based HF and VHF Single-Crystal Silicon Resonators With Sub-100 Nanometer Vertical Capacitive Gaps. *Proceedings of Solid-State Sensor and Actuator Workshop*, 2003, pp. 837-840.

126. http://www.microsensors.com

127. http://www.mems-exchange.org

128. http://www.suss.com

129. Smith, J.H.; Montague, S.; Sniegowski, J.J.; Murray, J.R.; McWhorter, P.J. Embedded micromechanical devices for the monolithic integration of MEMS with CMOS. *International Electron Devices Meeting, 1995.*, pp.609-612, 10-13 Dec 1995.

130. T. A. Core, W. K. Tsang, and S. J. Sherman. Fabrication technology for an integrated surface-micromachined sensor. *Solid State Technology.*, vol. 36, no. 10, pp. 3940, 42, 44, 4647, Oct. 1993.

131. K. Y. Park, C. W. Lee, Y. S. Oh, and Y. H. Cho. Laterally oscillated and force-balanced micro vibratory rate gyroscope supported by fish-hook shape springs. *Proc. IEEE Micro Electro Mechanical Systems Workshop (MEMS97)*, Japan, 1997, pp. 494499.

132. W. C. Tang, C.T.-C. Nguyen, M. W. Judy, and R. T. Howe. Electrostatic-comb drive of lateral polysilicon resonators. *Sensors and Actuators*, vol. A21-23, 1990, pp. 328-331.

133. C.T.-C. Nguyen. Micromachined Signal Processors. *Ph.D. Thesis, University of California, Berkeley*, 1994.

134. Laermer, F.; Urban, A.. Milestones in deep reactive ion etching. *International Conference on Solid State Sensors and Actuators Transducers '05* , Volume 2, pp. 1118-1121, 2005.

135. Lutz, M.; Golderer, W.; Gerstenmeier, J.; Marek, J.; Maihofer, B.; Mahler, S.; Munzel, H.; Bischof, U. A precision yaw rate sensor in silicon micromachining. *International Conference on Solid State Sensors and Actuators Transducers '97 Chicago*, Volume 2, pp. 847 - 850, 1997.

136. R.N. Candler, W.T. Park, H.M. Li, G. Yama, A. Partridge, M. Lutz, and T.W. Kenny. Single Wafer Encapsulation of MEMS Devices. *IEEE Trans. Advanced Packaging*, 26, 227, 2003.

137. Y. Dong, M. Kraft and W. Redman-White. Micromachined vibratory gyroscopes controlled by a high order band-pass sigma delta modulator. *IEEE Sensors Journal*, 7 (1). pp. 59-69, 2007.

138. Y. Dong, M. Kraft and W. Redman-White. High Order Bandpass Sigma Delta Interface for Vibratory Gyroscopes. *2005 IEEE Sensors Conference*, pp. 1080 - 1083, 2005

139. Rodjegard, H. Sandstrom, D. Pelin, P. Hedenstierna, N. Eckerbert, D. Andersson, G.I. A digitally controlled MEMS gyroscope with 3.2 deg/hr stability. *Solid-State Sensors, Actuators and Microsystems, Transducers 2005*, Vol 1, pp. 535- 538, 2005.

140. A. Madni and L. Costlow, "A third generation, highly monitored, micromachined quartz rate sensor for safety-critical vehicle stability control," in *Proc. Aerospace Conference IEEE*, vol. 5, 2001, pp. 2523–2534.

141. A. Madni, L. Costlow, and S. Knowles, "Common design techniques for BEI GyroChip quartz rate sensors for both automotive and aerospace/defense markets," *IEEE Sensors J.*, vol. 3, no. 5, pp. 569–578, 2003.

142. M. Weinberg and A. Kourepenis, "Error sources in in-plane silicon tuning-fork MEMS gyroscopes," *J. Microelectromech. Syst.*, vol. 15, no. 3, pp. 479–491, Jun. 2006.

143. U.-M. Gomez, B. Kuhlmann, J. Classen, W. Bauer, C. Lang, M. Veith, E. Esch, J. Frey, F. Grabmaier, K. Offterdinger, T. Raab, H.-J. Faisst, R. Willig, and R. Neul, "New surface micromachined angular rate sensor for vehicle stabilizing systems in automotive applications," in *Proc. of International Conference on Solid-State Sensors, Actuators and Microsystems (TRANSDUCERS '05)*, vol. 1, 2005, pp. 184–187.

144. R. Neul, U. Gomez, K. Kehr, W. Bauer, J. Classen, C. Doring, E. Esch, S. Gotz, J. Hauer, B. Kuhlmann, C. Lang, M. Veith, and R. Willig, "Micromachined gyros for automotive applications," in *Proc. IEEE Sensors*, 2005.

145. R. Neul, U.-M. Gomez, K. Kehr, W. Bauer, J. Classen, C. Doring, E. Esch, S. Gotz, J. Hauer, B. Kuhlmann, C. Lang, M. Veith, and R. Willig, "Micromachined angular rate sensors for automotive applications," *IEEE Sensors J.*, vol. 7, no. 2, pp. 302–309, Feb. 2007.

146. J. Geen, "Progress in integrated gyroscopes," in *Proc. Position Location and Navigation Symposium (PLANS 2004)*, 2004, pp. 1–6.

147. J. Geen, "Very low cost gyroscopes," in *Proc. IEEE Sensors*, 2005.

148. J. A. Geen, S. J. Sherman, J. F. Chang, and S. R. Lewis, "Single-chip surface micromachined integrated gyroscope with 50 deg/h allan deviation," *IEEE J. Solid-State Circuits*, vol. 37, no. 12, pp. 1860–1866, Dec. 2002.

149. A. R. Schofield, A. A. Trusov, C. Acar, and A. M. Shkel, "Anti-phase driven rate gyroscope with multi-degree of freedom sense mode," in *Proc. International Solid-State Sensors, Actuators and Microsystems Conference, TRANSDUCERS 2007*, Lyon, France, Jun. 10–14, 2007, pp. 1199–1202.

150. A. R. Schofield, A. A. Trusov, and A. M. Shkel, "Multi-degree of freedom tuning fork gyroscope demonstrating shock rejection," in *Proc. IEEE Sensors*, Atlanta, GA, USA, Oct. 28–31, 2007, pp. 120–123.

151. A. A. Trusov, A. R. Schofield, and A. M. Shkel, "New architectural design of a temperature robust MEMS gyroscope with improved gain-bandwidth characteristics," Hilton Head Workshop in Solid State Sensors, Actuators, and Microsystems, Jun. 1–5, 2008.

152. M. S. Kranz and G. K. Fedder, "Micromechanical vibratory rate gyroscope fabricated in conventional CMOS," in *Symposium Gyro Technology*, Stuttgart, Germany, Sep. 16–17, 1997, pp. 3.0–3.8.

153. S. Alper and T. Akin, "A single-crystal silicon symmetrical and decoupled MEMS gyroscope on an insulating substrate," *J. Microelectromech. Syst.*, vol. 14, no. 4, pp. 707–717, Aug. 2005.

154. J. G. L. Woon-Tahk Sung, Sangkyung Sung and T. Kang, "Design and performance test of a MEMS vibratory gyroscope with a novel AGC force rebalance control," *J. Micromech. Microeng.*, vol. 17, no. 10, pp. 1939–1948, Oct. 2007.

155. S. Alper, K. Azgin, and T. Akin, "A high-performance silicon-on-insulator MEMS gyroscope operating at atmospheric pressure," *Sensors and Actuators A: Physical*, vol. 135, no. 1, pp. 34 – 42, Mar. 2006.

Index

CPSIA information can be obtained at www.ICGtesting.com
Printed in the USA
LVOW100542180512

282274LV00002B/32/P